International Conference "Structural and Phase Transformations in Materials: Theory, Computer Modelling and Experiment" (SPTM 2017)

IOP Conference Series: Materials Science and Engineering
Volume 192

Ekaterinburg, Russia
23 – 27 March 2017

ISBN: 978-1-5108-3961-8
ISSN: 1757-8981

Printed from e-media with permission by:

Curran Associates, Inc.
57 Morehouse Lane
Red Hook, NY 12571

Some format issues inherent in the e-media version may also appear in this print version.

Printed by Curran Associates, Inc. (2017)

For permission requests, please contact the Institute of Physics
at the address below.

Institute of Physics
Dirac House, Temple Back
Bristol BS1 6BE UK

Phone: 44 1 17 929 7481
Fax: 44 1 17 920 0979

techtracking@iop.org

Additional copies of this publication are available from:

Curran Associates, Inc.
57 Morehouse Lane
Red Hook, NY 12571 USA
Phone: 845-758-0400
Fax: 845-758-2633
Email: curran@proceedings.com
Web: www.proceedings.com

Table of contents

Volume 192

International Conference "Structural and Phase Transformations in Materials: Theory, Computer Modelling and Experiment"

23–27 March 2017, Ekaterinburg, Russian Federation

Accepted papers received: 7 April 2017
Published online: 19 April 2017

Preface

Invited Talks

Papers

IOP Conf. Series: Materials Science and Engineering **192** (2017) 011001 doi:10.1088/1757-899X/192/1/011001

FOREWORD

The international conference ***"Structural and phase transformations in materials: theory, computer modelling and experiment"*** (SPTM 2017) was held on the Chusovaya base on March 23-27, 2017 (~ 80 kilometers from the fourth-largest city in Russia, Ekaterinburg). The conference was organized by the Ural Federal University (Laboratory of Multi-Scale Mathematical Modeling).

The conference was devoted to discussions of the current state and new scientific achievements on the subject of structural and phase transformations in materials, including the following topics:

- Theory of the Processes
- Experiments and Their Results
- Software and Hardware for Modeling
- Results of Computer Simulation.

The present event paid its special attention to the phase transformation phenomena such as dendritic growth, nucleation and pattern formation including theory, experiment and software developments.

The organizing Committee was pleased to welcome more than 40 participants from 8 countries attending the conference (Belarus, China, France, Germany, India, Morocco, Russian Federation, and Ukraine). The Organizers have selected 34 peer reviewed research papers in the present issue of ***IOP Conference Series: Materials Science and Engineering***. We hope that the present papers will be of interest for readers and represent a thorough research skeleton for inspiration and future developments.

Yours sincerely,

Dmitri V. Alexandrov, Peter K. Galenko,

Ilya O. Starodumov and Victoria V. Polyakova

April 7, 2017

IOP Conf. Series: Materials Science and Engineering **192** (2017) 011001 doi:10.1088/1757-899X/192/1/011001

CONFERENCE PHOTOS

SPTM 2017 participants near the Ural Federal University on March 23, 2017.

IOP Conf. Series: Materials Science and Engineering **192** (2017) 011001 doi:10.1088/1757-899X/192/1/011001

SPTM 2017 participants near the Ural Federal University on March 23, 2017.

SPTM 2017 participants after the oral talk session on March 25, 2017.

IOP Conf. Series: Materials Science and Engineering **192** (2017) 011001 doi:10.1088/1757-899X/192/1/011001

SPTM 2017 participants on the Chusovaya base on March 23, 2017.

SPTM 2017 participants on the Chusovaya base on March 27, 2017.

IOP Conf. Series: Materials Science and Engineering **192** (2017) 011001 doi:10.1088/1757-899X/192/1/011001

SPTM 2017 Organizing Committee on March 23, 2017.

(from left to right: Dmitri Alexandrov, Victoria Polyakova, Peter Galenko and Ilya Starodumov).

IOP Conf. Series: Materials Science and Engineering **192** (2017) 011001 doi:10.1088/1757-899X/192/1/011001

ORGANIZERS

Ural Federal University

Laboratory of Multi-Scale Mathematical Modeling

ORGANIZING COMMITTEE

Chairman: Professor, Doctor of Science, Dmitri Alexandrov (Ural Federal University, Russia)

Vice-Chairman: Professor, Doctor of Science, Peter Galenko (University of Jena, Germany)

Members (Ural Federal University): Ilya Starodumov and Victoria Polyakova

PROGRAM COMMITTEE

Chairman: Dmitri Alexandrov (Prof. Dr., Ural Federal University, Ekaterinburg, Russia)

Vice-Chairman: Peter Galenko (Prof. Dr., University of Jena, Jena, Germany)

Members:

Jianrong Gao (PhD, Northeastern University, Shenyang, China)

Phanikumar Gandham (PhD, Indian Institute of Technology Madras, Chennai, Tamilnadu, India)

Olga Gusakova (PhD, Belarusian State University, Minsk, Belarus)

Oleg Kazak (PhD, Technische Universitat Ilmenau, Germany)

Valery Golod (PhD, Peter the Great St. Petersburg Polytechnic University, St. Petersburg, Russia)

Leonid Klyuev (PhD, Immers Ltd., Moscow, Russia)

Mikhail Krivilyov (Dr., Udmurt State University, Izhevsk, Russia)

Evgeny Kharanzhevsky (Prof. Dr., Udmurt State University, Izhevsk, Russia)

Irina Nizovtseva (PhD, Ural Federal University, Ekaterinburg, Russia)

Alexander Ivanov (PhD, Ural Federal University, Ekaterinburg, Russia)

Nikolai Kropotin (PhD, MKM Ltd., Izhevsk, Russia)

Alexey Malygin (PhD, Ural Federal University, Ekaterinburg, Russia)

Valentin Kropotin (MKM Ltd., Izhevsk, Russia)

Vladimir Ankudinov (Udmurt State University, Izhevsk, Russia)

Ilya Starodumov (Ural Federal University, Ekaterinburg, Russia)

Irina Alexandrova (Ural Federal University, Ekaterinburg, Russia)

Evgeny Pavlyuk (Ural Federal University, Ekaterinburg, Russia)

Peer review statement

All papers published in this volume of *IOP Conference Series: Materials Science and Engineering* have been peer reviewed through processes administered by the proceedings Editors. Reviews were conducted by expert referees to the professional and scientific standards expected of a proceedings journal published by IOP Publishing.

On the theory of evolution of particulate systems

Yuri A Buyevich[1] **and Dmitri V Alexandrov**[2]

[1] CRSS, University of California, Santa Barbara, CA 93117, USA

[2] Department of Theoretical and Mathematical Physics, Laboratory of Multi-Scale Mathematical Modeling, Ural Federal University, Ekaterinburg, 620000, Russian Federation

E-mail: dmitri.alexandrov@urfu.ru

Abstract. An analytical method for the description of particulate systems at sufficiently long times is developed. This method allows us to obtain very simple analytical expressions for the particle distribution function. The method under consideration can be applied to a number of practically important problems including evaporation of a polydisperse mist, dissolution of dispersed solids, combustion of dispersed propellants, physical and chemical transformation of powders and phase transitions in metastable materials.

1. Introduction

The most simple evolution problems for a particulate assemblage arise in numerous processes where 1) there is no exchange by particles between the assemblage and external forces, 2) new particles do not originate, and 3) fluctuations in the transformation rate of a single particle are negligible. Such processes are wide-spread in multifarious fields of modern technology, and they range from basically unsophisticated processes, in which particles either increase or decrease their size due to heat and mass exchange with their environment, to much more complicated conversion processes, such as combustion of pulverized propellants or chemical transformation of powders.

A significant common feature of such processes consists in the fact that all of them can be described with the help of the simplest possible version of the Fokker-Planck evolutionary equation which does not include 1) terms involving second derivatives of the population density distribution function, and 2) any source terms. Even though, dealing with such an equation presents essential difficulties, and evaluation of the distribution function and various properties of the particulate assemblage and its environment at different moments of time usually entrails a great deal of cumbersome and tedious numerical calculations.

An alternative method is advanced in this paper which allows us to get very simple analytical results for the particle distribution function at sufficiently long times having elapsed after the beginning of an evolution process. This method was successfully applied to a number of practically important problems (see, among others, [1–3]). It is based on introducing a set of self-similar functions that exponentially decrease in amplitude, but do not change their shape, during the process. The distribution function can be represented as a sum of such self-similar functions. At long times, only the term with a minimal rate of change remains, whereas other terms eventually degenerate. From the physical point of view, this means that the particulate assemblage forgets particulars of its initial distribution as time goes by, and only its relatively rough features endure.

2. The essence of the method

Let us consider an assemblage of particles that are continuously distributed over some physical character designated by the parameter ξ. Most commonly, parameter ξ characterizes particle size; but it can also to have other physical meaning, such as a degree of physical transformation or chemical conversion of particles. At any moment of time, the current distribution of the assemblage over this parameter can be described with the help of the particle population density distribution function, $f(t, \mathbf{r}; \xi)$, which is regarded as normalized to the number concentration, $n(t, \mathbf{r})$, of particles.

2.1. General formulation

Under the assumptions enumerated in the beginning of this paper, the evolution of the distribution function over time and in space is governed by a first-order kinetic equation

$$\frac{\partial f}{\partial t} + \nabla \cdot (\mathbf{v} f) + \frac{\partial}{\partial \xi}(wf) = 0, \; w = \frac{d\xi}{dt}. \tag{1}$$

This equation represents perhaps the simplest version of the general Fokker-Planck equation. Quantity w is understood as a deterministic rate of transformation of a single particle that induces a change in parameter ξ. It is presumed known as a function of ξ, and it is also dependent on time and coordinates (either explicitly or, more often implicitly, though some other functions of t and $\mathbf{r}$). In addition, we assume that this transformation rate is expressible as a product of two separate functions separately depending on ξ and on t, $\mathbf{r}$, that is,

$$w \equiv w(t, \mathbf{x}; \xi) = -M(t, \mathbf{x}) N(\xi). \tag{2}$$

This equation actually covers a great variety of possible applications (see, among others, [4]). Exact forms of the functions in this equation are dictated by the very nature of the transformation process under study and by the kinetic peculiarities of this process. Function M usually depends implicitly on its argument, through certain variables that are employed to describe the current instantaneous state of the ambient medium into which the particles are submerged. The role of such variables is commonly played by the medium temperature, $T(t, \mathbf{r})$, and by concentrations $c_j(t, \mathbf{r})$ of various admixtures and reagents that affect particle transformation. It is significant that all such variables depend on moments of the particle distribution function, and this mere fact gives rise to a strongly nonlinear feedback between the current states of the ambient medium and the particulate assemblage itself. It is fairly well evident that equation (1) must be supplemented with equations for these variables which ultimately stem from the mass and energy conservation laws. As a symbolic representative of such equations, we formulate an equation

$$\frac{\partial c_j}{\partial t} + \nabla \cdot (\mathbf{v} c_j) = \eta_j \int w(t, \mathbf{x}; \xi) f(t, \mathbf{x}; \xi) \chi_j(\xi) d\xi. \tag{3}$$

The integration is performed here over the whole range of allowable values of ξ. In this equation, η_j is a coefficient, and χ_j is a weight function, whose exact forms are determined by the physical essence of the transformation process. The term on the right-hand side of equation (3) describes an effective source of quantity c_j per unit volume that is caused by the transformation. It is significant that more than one equation of the type of (3) may be relevant for the transformation problem under study. For instance, there are equations for temperature and vapour concentration that influence the evaporation kinetics for a polydisperse mist.

Imposing initial conditions

$$f(0, \mathbf{x}; \xi) = f_0(\mathbf{x}; \xi), \; \int f_0(\mathbf{x}; \xi) d\xi = n_0(\mathbf{x}), \; c_j(0, \mathbf{x}) = c_{0j}(\mathbf{x}) \tag{4}$$

brings to an end the formation of an initial value problem for equations (1) and (3). It is worth noting that the wanted solution to these equations can also be subjected to conditions at a certain surface $\mathbf{r} = \mathbf{r}_0(t)$, instead of initial conditions at $t = 0$.

2.2. *Self-similar distribution functions*

When taking into account the fact that the rate of transformation, w, is expressed in accordance with equation (2), and also using the Fourier method of separating variables, we look for elementary solutions to kinetic equation (1), so that the full solution takes the form

$$f(t, \mathbf{x}; \xi) = \mathbf{S}\left\{F(t, \mathbf{x}; s)R(\xi; s)\right\}, \tag{5}$$

where s is proportional to a separation variable, and symbol $\mathbf{S}\{...\}$ denotes either summation or integration over s. Equation (1) then splits into two equations

$$\frac{dF}{dt} + \nabla \cdot (\mathbf{v}F) + 2sMF = 0, \quad \frac{d}{d\xi}(NR) + 2sR = 0, \tag{6}$$

where $2s$ being understood as a separation variable, while equation (4) reduces to

$$\frac{\partial c_j}{\partial t} + \nabla \cdot (\mathbf{v}c_j) = \eta_j \mathbf{S}\left\{F(t, \mathbf{x}; s)M(t, \mathbf{x}) \int N(\xi)R(\xi; s)\chi_j(\xi)d\xi\right\}. \tag{7}$$

An initial condition for c_j preserves its original form as written down in equation (4). To formulate an initial condition for unknown functions $F(t, \mathbf{r}; s)$, we represent the initial population density in the same manner as that used in equation (5), that is,

$$f_0(\mathbf{x}; \xi) = \mathbf{S}\left\{F_0(\mathbf{x}; s)R(\xi; s)\right\}. \tag{8}$$

Hence it becomes obvious that we are free to take, without loss in generality,

$$F_0(0, \mathbf{x}; s) = F_0(\mathbf{x}; s) \tag{9}$$

provided that functions $R(\xi; s)$are normalized in an appropriate manner (say, to unity).

Thus the main benefit of using the Fourier method consists in the fact that the problem of finding elementary distribution functions $R(\xi; s)$ becomes completely separated from the problem of determining variations of the particle population density in space and time. In its turn, the latter problem involves only integrals of the elementary distribution functions over ξ that appear in equation (7), but it by no means includes these functions themselves. Neither the aforementioned elementary functions, nor their moments, depend on time, and consequently, these functions can be properly termed as "self-similar distribution functions".

2.3. *Systems of ideal mixing*

Other statements of initial conditions are possible as well. For example, in order to provide for high transformation rates, a fluid medium containing suspended particles is often subjected to vigorous mixing, so that different parts of the fluid-particle system are more or less at the same conditions. In systems with ideal mixing in which the spatial dependence of all variables disappears by definition, it may be more convenient to require functions $F_0(\mathbf{x}; s) \equiv F_0(s)$ to equal unity for any s. Then, functions $R(\xi, s)$ must be uniquely determined by the initial population density, $f_0(\mathbf{x}, \xi) \equiv f_0(\xi)$. In what follows, we shall use this alternative statement of initial conditions, instead of that in equation (9),

$$F(0) = F_0(s) = 1, \text{ and } f_0(\xi) = \mathbf{S}\left\{R(\xi; s)\right\}. \tag{10}$$

A general solution to the equation (6) reads

$$R(\xi; s) = \frac{C(s)}{N(\xi)} \exp\left(-2s \int_{\xi_0}^{\xi} \frac{d\xi}{N(\xi)} \right), \tag{11}$$

where $C(s)$ being looked upon as a given function of s. The relevant solution to the first equation (6) that satisfies condition (10) is

$$F(t) = \exp\left(-2s \int_{0}^{t} M(t)dt \right). \tag{12}$$

Equations (11) aand (12) completely determine the wanted population density. Indeed, using them in equations (8), or (10), and (5) results in two final formulae

$$N(\xi)f_0(\xi) = \mathbf{S}\left\{ C(s) \exp\left(-2s \int_{\xi_0}^{\xi} \frac{d\xi}{N(\xi)} \right) \right\}, \tag{13}$$

which determines $C(s)$, if a factor of $N(\xi)$ in equation (2) for the transformation rate and initial population density $f_0(\xi)$ is known, and

$$N(\xi)f(t;\xi) = \mathbf{S}\left\{ C(s) \exp\left(-2s \int_{0}^{t} M(t)dt \right) \exp\left(-2s \int_{\xi_0}^{\xi} \frac{d\xi}{N(\xi)} \right) \right\}, \tag{14}$$

which describes the particle population density at subsequent moments of time. Thus, equation (14) gives the generalized solution of the problem under study for systems of ideal mixing, if a factor of $M(t)$ that is involved in equation (2) is also known.

It follows from equation (13) that $C(s)$ may be regarded as known quantity, as far as initial population density and function $N(\xi)$ that appear in the left-hand side of this equation are definite. There are two major possibilities. The first possibility matches the case where the initial population density is approximated as a sum of suitable functions of the type of (11), each of which is characterized by its own value of s. In this case, corresponding values of $C(s)$ represent approximation coefficients that must be presumed known, by their very definition.

The other major possibility fits the situation where the left-hand side of equation (13) is expressed as an integral of the function that appears within the curly brackets over a certain continuous range of variable s. In such a situation, equation (13) actually represents an integral transform that relates the said function with known product $N(\xi)f_0(\xi)$. In this case, the task of finding $C(s)$ is in fact reduced to finding an integral transform that is inverse with respect to the transform mentioned. Some examples of application of this technique are presented in the following sections of this paper.

Meanwhile, it is worth noting an important feature of the population distribution densities. For definiteness, let us assume that $N(\xi)$ is always positive. Then $F(t)$ may be either positive or negative, depending on the physical nature of an evolution process under study. The integral of positive $M(t)$ over time that is involved in equation (14) is commonly a monotonously increasing function of time. If the initial particle population density has been approximated in accordance with equation (13), as a sum of a few terms corresponding to different values of s, only the term that is related to the smallest value of s is likely to prevail at sufficiently large times in the analogous sum that appears in equation (14). This means that the particle population density can be more or less accurately approximated at large times by this term taken alone, whereas the influence of other terms eventually vanishes. If $M(t)$ is negative, only the term involving the large value of s is essential at long times, and the behaviour of the population density is in a sense opposite to that described above for the case of positive $M(t)$.

3. Evaporation of a polydisperse mist

Evolution of an assemblage of suspended droplets is governed by heat and mass exchange of the droplets with the suspending gaseous mixture. Two main limiting cases are to be singled out. The first one concerns slow evaporation under roughly isothermal conditions in a relatively cool gas. In this case, the evaporation kinetics is dictated by the rate with which vapour is removed away from the surfaces of evaporating droplets, whereas heat transfer does not play a significant role. This case is characteristic of most natural evaporation processes occurring in fogs and clouds. In the opposite limiting case, the evaporation kinetics is limited by the rate with which heat is supplied to the droplet surfaces to compensate for heat absorption at evaporation, whereas mass transfer is inessential. This case is specific to fast evaporation in a hot gaseous mixture whose temperature is sufficiently close or even exceeds the liquid boiling point. It pertains to a number of industrial situations frequently encountered in power engineering, such as evaporative cooling of gases.

3.1. Slow diffusion-controlled evaporation

If the mean temperature of the gaseous mixture that contains evaporating droplets is low enough, the evaporation is slow, and its thermal effect can be ignored. Consequently, the kinetics of evaporation of individual droplets is determined by the transport of vapour away from the droplet surfaces. Furthermore, if the droplet radii considerably exceed the mean free path length of molecules in the ambient gas, this transport can be described as a diffusion process. Under conditions of ideal mixing, there are then only one parameter that determines the state of a droplet - its radius $\xi = r$, and the only relevant variable characteristic of the current state of the gaseous mixture - the mean vapour concentration, $c(t)$, in the bulk.

We make use of this comparatively unsophisticated evaporation process that proceeds under ideal mixing conditions in order to demonstrate, in rather a meticulous way, different possible approaches to treating the evolution of a particulate system.

Equation (2) can now be specified as follows

$$w = \frac{dr}{dt} = -M(t)R(t), \; M = \frac{D}{\rho}\left[c^*(t) - c(t)\right], \; N = \frac{1}{r}, \tag{15}$$

where $c^*(t)$ is the equilibrium vapour concentration, D is the vapour diffusivity, and ρ is the liquid density. For the case of evaporation, we have $c < c^*$, and $w < 0$, while the inverted inequalities conform to condensation. Equation (3) has to be formulated as

$$\frac{dc}{dt} = -\rho \int_0^\infty f(t,r)\left(4\pi r^2 \frac{dr}{dt}\right) dr \tag{16}$$

and it describes the material balance as is maintained in the system under study in a differential form. It can be replaced right away by an equation that reflects the integral material balance

$$c(t) + \frac{4\pi}{3}\rho \int_0^\infty f(t,r) r^3 dr = c_0 + \frac{4\pi}{3}\rho n_0 \langle r^3 \rangle_0, \tag{17}$$

where subscript 0 referring to the initial state, n being the number concentration of droplets, and angular brackets signifying integration over r with a weight function of $f(t,r)$.

Equation (17) is supposed to be a result of integration of differential material balance equation (16). However, the relation between equations (16) and (17) is not strictly evident, and it needs some comments. Differentiation of the last equation with respect to time formally yields an equation

$$\frac{dc}{dt} = -\frac{4\pi}{3}\rho \int_0^\infty \frac{\partial f}{\partial t} r^3 dr$$

because the variable of integration, r, must not of course be differentiated. With the aid of kinetic equation (1), this equation can be transformed to give, in succession,

$$-\int_0^\infty \frac{\partial f}{\partial t} r^3 dr = \int_0^\infty \frac{\partial}{\partial r}\left(\frac{dr}{dt} f\right) r^3 dr = \frac{dr}{dt} f(t,r) r^3 \Big|_{r=0}^{\infty} - 3\int_0^\infty f(t,r)\left(r^2 \frac{dr}{dt}\right) dr.$$

Thus, we recover equation (16) when the first term in the right-hand side turns to zero, which happens if the particle population density 1) tends to zero faster than $1/r^2$ as r goes to infinity, and 2) diverges at $r \to 0$ slower than $1/r^2$. Apart from other things, this simple consideration shows that one should be rather careful while dealing with distribution functions.

Equation (11) reduces in the case under study to the following form

$$R(r;s) = 2Csr \exp\left(-sr^2\right), \tag{18}$$

where a factor of s having been added to the constant coefficient in this equation. We shall start with addressing the simplest case where the initial population density can be approximated with sufficient accuracy by function (18) with a certain fixed s.

Instead of using formal expression (12) for $F(t;s)$, we come back to the first equation (6) and to condition (10) that read, with allowance made for equation (15),

$$\frac{dF}{dt} + \frac{2D}{\rho} s\left[c^*(t) - c(t)\right] F = 0, \; F(0) = 1. \tag{19}$$

Using representation (5) of the wanted solution, we have from equation (17)

$$c(t) = c_0 + \frac{4\pi}{3}\rho n_0 \langle r^3\rangle_0 \left[1 - F(t;s)\right] \tag{20}$$

whence, after inserting this in equation (19), we obtain

$$\frac{dF}{dt} + a(t)F + bF^2 = 0, \; a(t) = \frac{2D}{\rho}\left[c^*(t) - c_0 - \frac{4\pi}{3}\rho n_0 \langle r^3\rangle_0\right] s,$$
$$b = \frac{8\pi}{3} D n_0 \langle r^3\rangle_0 s, \; n_0 = \int_0^\infty f(0;r)dr = \int_0^\infty R(r;s)dr = C. \tag{21}$$

When deriving this equation, the self-similarity property of the population density $f = FR$ with R defined by equation (18) has been accounted of, according to which moments of $f(t,r)$ remain invariable during the evolution process. In particular, the third moment equals

$$\langle r^3\rangle = \frac{1}{n(t)}\int_0^\infty f(t,r) r^3 dr = \langle r^3\rangle_0 = \frac{1}{n_0}\int_0^\infty f(0;r) r^3 dr = \frac{3\sqrt{\pi}}{4s^{3/2}} \tag{22}$$

as can be easily seen from the above relations.

It is not difficult to solve equation (21) at an arbitrary dependence of c^* on t which might be dictated by the thermal regime of evaporation. If true isothermal conditions are maintained in the sense that $c^* = 0$, a is also independent of time, and we readily obtain from equation (21)

$$F(t;s) = \left[\left(1+\frac{b}{a}\right)\exp\left(at\right) - \frac{b}{a}\right]^{-1} \approx \left(1+\frac{b}{a}\right)\exp\left(-at\right) \tag{23}$$

an approximate equality holding true at sufficiently large times, that is, at $at \gg 1$. The particle population distribution density and the number concentration of droplets can be presented in an exceedingly simple form, approximate equalities being again valid at large times,

$$\begin{aligned} f(t,r) &= 2sn_0F(t;s)\exp\left(-sr^2\right) \approx 2\left(1+\frac{b}{a}\right)sn_0r\exp\left(-at-sr^2\right), \\ n(t) &= n_0F(t) \approx \left(1+\frac{b}{a}\right)n_0e\exp\left(-at\right), \; s \equiv \left(\frac{4}{3\sqrt{\pi}}\langle r^3\rangle_0\right)^{2/3}. \end{aligned} \tag{24}$$

It is worthwhile to stress once again that population distribution density (24) is self-similar in the sense that the droplet assemblage evolution affects merely the total number of droplets which eventually disappear during the process, but not the distribution moments of any order.

In a more general case, the initial population density cannot be represented by function (18). Notwithstanding, it can be sometimes described with a good accuracy as a sum of a few elementary self-similar functions of the type of that in equation (18), that is,

$$f(0,r) = 2\sum_{j=1}^{J} C_js_jr\exp\left(-s_jr^2\right), \; \sum_{j=1}^{J} C_j \equiv n_0. \tag{25}$$

In order to obtain $f(t,r)$, each elementary function $R(t;s_j) = 2C_js_jr\exp(-s_jr^2)$ that appears in this equation must be multiplied by a proper $F(t;s_j)$, and then the products have to be summed up, according definition (5) and quite similarly to equation (25). The problem of calculating $F(t;s_j)$ is essentially the same as before, and it is formulated in equation (19) where s_j substitutes for s. The resultant equations for different $F(t;s_j)$ lead at once to the following relations

$$s_j \ln F_j = s_j \ln F_i, \text{ and } F_j^{s_i} = F_i^{s_j}, \; F_j \equiv F(t;s_j) \tag{26}$$

valid for any couple of integers i and j that may vary from unity to J.

Equation (26) proves that only a single function among $F(t;s_j)$ must be found to determine all of them. For the sake of definiteness, we arrange the set of s_j in such a manner that $s_j > s_i$ at $j > i$, and we also choose $F_1(t) \equiv F(t;s_1)$ to be determined first. After a simple manipulation, which is in essence precisely the same as that has led to equation (21), we arrive at a new problem

$$\begin{aligned} &\frac{dF_1}{dt} + \left(a_1(t) + \sum_{j=1}^{J} b_jF_1^{s_j/s_1}\right)F_1 = 0, \; F_1(0) = 1, \\ &a_1(t) = \frac{2D}{\rho}\left[c^*(t) - c_0 - \pi^{3/2}\rho\sum_{j=1}^{J}\frac{C_j}{s_j^{3/2}}\right]s_1, \; b_j = \frac{2\pi^{3/2}DC_j}{s_j^{3/2}}s_1. \end{aligned} \tag{27}$$

This equation can always be integrated numerically. Besides, if c^* is constant, and $s_j = k_js_1$, where k_j are integral numbers, the right-hand side of this equation represents a polynomial. After expanding this polynomial into elementary multipliers, equation (27) can be integrated analytically to yield an algebraic transcendental equation that relates F_1 to t in an implicit form.

The population size distribution density is expressible in the case under consideration as

$$f(t;r) = 2\sum_{j=1}^{J} C_j s_j F_1^{s_j/s_1}(t) r \exp\left(-s_j r^2\right). \tag{28}$$

Equation (28) permits a straightforward evaluation of different characteristics of the droplet assemblage. For instance, we have for the droplet number concentration and for the third distribution moment that determines the mean droplet volume as functions of time

$$\begin{aligned} n(t) &= \sum_{j=1}^{J} C_j F_1^{s_j/s_1}(t) \to C_1 F_1(t), \ t \to \infty, \\ \langle r^3 \rangle(t) &= \frac{3\sqrt{\pi}}{4n(t)} \sum_{j=1}^{J} \frac{C_j}{s_j^{3/2}} F_1^{s_j/s_1}(t) \to \frac{3\sqrt{\pi}}{4s_1^{3/2}}, \ t \to \infty, \end{aligned} \tag{29}$$

which agrees asymptotically with equation (22). The bulk vapour concentration can now be found from equations (17) and (29)

$$c(t) = c_0 + \pi^{3/2}\rho \sum_{j=1}^{J} \frac{C_j}{s_j^{3/2}} \left[1 - F_1^{s_j/s_1}(t)\right]. \tag{30}$$

Because of the obvious inequality $F_1 < 0$ at $t > 0$, it can be clearly seen that the first term of sum (28) becomes to dominate as time grows, no matter how insignificant it might be at $t = 0$, and how large coefficients C_j have been assigned to describe the other terms of this sum at the initial moment. More generally, only the influence of that part of the initial distribution that is described by function (18) with $s = s_1$, that is, with the smallest s_j, persists to cause a noticeable effect on droplet assemblage properties after sufficiently long time intervals, as it is shown in equation (30). In other words, after a period of time, the system forgets particulars of its initial state, and consequently, a final self-similar evolution stage virtually establishes itself. It is not difficult to see that the original maxima of the population size distribution density attained at smaller values of r degenerate faster than those corresponding to larger r. This also means that any moment $\langle r^m \rangle$ of this distribution must increase with time due to the assemblage evaporation, until the self-similar stage is attained where any moment becomes approximately invariable.

At last, we turn to the case where the initial size distribution density is described by an arbitrary continuous function, $f(0,r) = f_0(r)$. In this case, the very form of self-similar distributions (18) suggests looking for the particle population distribution density, $f(t,r)$, as an integral transformation with a kernel belonging to the same class as function (18). Therefore, we take

$$f(t,r) = r\int_0^\infty A(s) F(t;s) \exp\left(-sr^2\right) ds \tag{31}$$

and, after accepting that $F(0;s) = 1$ in compliance with equations (19) and (27), we also obtain

$$f_0(r) = f(0,r) = r\int_0^\infty A(s) \exp\left(-sr^2\right) ds. \tag{32}$$

Our immediate task consists in determining function $A(s)$ that is involved in these equations, in terms of the initial size distribution that is presumed given. With the introduction of a new

IOP Conf. Series: Materials Science and Engineering **192** (2017) 012001 doi:10.1088/1757-899X/192/1/012001

independent variable, $z = r^2$, we become able to rewrite equation (32) in a familiar form of Laplace transformation, that is,

$$g(z) \equiv \left.\frac{f_0(r)}{r}\right|_{r=\sqrt{z}} = \int_0^\infty A(s)\exp(-sz)\,ds. \tag{33}$$

This equation allows us to regard function $A(s)$ as the inverse Laplace transform of the known function, $g(z)$, that is defined in terms of the initial distribution. Couples of functions connected by the Laplace transform are well tabulated, so that $A(s)$ can be found for a great many $f_0(t)$ without need to undertake any calculation.

It seems natural, when using an analogy with equation (26), to take $F(t;s) = F^s(t)$ where $F(t)$ is a new unknown variable. In this case, after simple transformations that are similar to those employed above, we come out with a problem

$$\frac{dF}{dt} + \left(\frac{a(t)}{s} + \pi^{3/2} D \int_0^\infty A(s) F^s \exp\left(-sr^2\right)\frac{ds}{s^{5/2}}\right) F = 0, \; F(0) = 1, \tag{34}$$

where ratio $a(t)/s$ is independent of s, with $a(t)$ being defined in accordance with equation (21). In contrast to equations (21) and (27), this equation is an integro-differential. Function $F(t)$ found as a solution to problem (34) finally determines the particle size distribution density in accordance with equation (31). However, the handling of this integro-differential equation seems to be somewhat easier than either analytical or numerical analysis of conventional integro-differential equation that can be immediately obtained from equations (1) and (17), when supplemented with equation (15). Below, we briefly demonstrate the derivation of this equation for reference sake.

Applying the well-known method of characteristics to equations (1) and (15), one obtains

$$f(t,r) = \frac{r}{\left[r^2 + \theta(t)\right]^{1/2}} f_0\left(\left[r^2 + \theta(t)\right]^{1/2}\right), \; \theta(t) = \frac{2D}{\rho}\int_0^t \left[c^*(t) - c(t)\right] dt.$$

Substituting this into equation (17), one arrives further at a rather complicated integro-differential equation for function $\theta(t)$ which we do not attempt to write down here. The bulk vapour concentration is related to the unknown variable of this equation, $\theta(t)$,

$$c(t) = c^*(t) - \frac{\rho}{2D}\frac{d\theta}{dt}.$$

In the generalized case, solving the last integro-differential equation is hardly an easier task than applying conventional methods directly to the kinetic equation. These time-consuming conventional methods are usually based on either dividing the assemblage into separate fractions, or analyzing moments of the population size distribution density. Being much simpler, the calculation scheme developed above possesses certain evident advantages over schemes resulting from these methods. In particular, it can readily be used for the express evaluation of various significant characteristics of an evolving particulate system, and it can be simplified even more.

The further simplification becomes possible due to the fact that the time necessary for the final self-similar evolution stage to establish itself often happens to be much shorter than the duration of the evolution process of a particulate system being considered. Consequently, it is this self-similar stage that frequently governs the process at large, and approximately determines the time dependence of different system's characteristics. This is a rather fortunate occurrence, since the population size distribution density is then most advantageously described by exceedingly simple equation of the type of (24), whereas mean properties of the ambient medium can be found from equations of the type of equation (20) for vapour concentration.

Among important characteristics of particulate systems, the "degree of transformation"

$$y(t) = \frac{n(t)\langle r^3\rangle(t)}{n_0\langle r^3\rangle_0} \tag{35}$$

seems to be of special practical concern. At the self-similar stage, the third distribution moment does not depend on time, and the degree of transformation equals $F(t)$ as defined by equation (21) for the particular case of slow diffusion-controlled evaporation just considered. In this particular case, $y(t)$ is the fraction of liquid that remains in the form of droplets by moment t, and at the same time, it coincides with the fraction of the total number of droplets that have not evaporated by the moment indicated. In other situations, the degree of transformation can of course have a different physical meaning.

3.2. Fast thermally-controlled evaporation

Absorption of heat by evaporating droplets influences the kinetics of evaporation as well. Allowance for this influence implies 1) an appropriate correction to be introduced to formula (15) for the evaporation rate of a single droplet that would account for the effect caused by thermal factors, and 2) a heat balance equation to be considered alongside material balance equation (16) or (17). However, the said influence is practically insignificant, if the mean system temperature, T, remains well below the liquid saturation temperature, T_s. However, the situation drastically changes as T approaches, or even becomes larger than T_s. If this is the case, it is the heat transfer to the droplet surfaces that limit the evaporation rate of individual droplets, and the resistance due to diffusion of vapour from these surfaces may be regarded as negligible. We again consider the evaporation process at conditions of ideal mixing.

There are theoretical and empirical formulae for the rate of evaporation of a single droplet in the thermally-controlled regime. For definiteness, we make use of one of available empirical expressions for this rate that leads to the formula

$$w = \frac{dr}{dt} = -M(t)R(r), \;\; M = \frac{k}{\rho L}\langle r\rangle_0^m \left(T - T_s\right), \;\; N = \frac{1}{r^m}, \tag{36}$$

where k is a constant that plays the role of an effective heat exchange coefficient, and m is a constant exponent. Formula (36) is valid in cases where the actual system temperature is larger than the saturation temperature, or the liquid boiling point, T_s. Equation (16) has now be replaced by the heat balance equation

$$c_p\frac{dT}{dt} = \mu L\frac{dy}{dt}, \;\; c_p = c_g + \mu(1-y)c_v. \tag{37}$$

Here, μ is the ratio of an initial mass of the droplets to the mass of gas, and c_p is the isobaric heat capacity of the gas-vapour mixture expressed in terms of the specific heat capacities of the gas and vapour, c_g and c_v, respectively. Function $y = y(t)$ stands for the degree of transformation as defined by equation (35). Equation (37) establishes the fact that heat delivered to the droplets from the mixture is spent entirely on evaporation. For simplicity, the mixture pressure is assumed invariable, so that the saturation temperature is constant.

The self-similar population size distribution density is once again looked for in the form

$$f(t,r) = F(t)R(r), \;\; F(t) \equiv y(t). \tag{38}$$

The problem for $R(r)$ reads

$$\frac{d}{dr}\left(\frac{R}{r^m}\right) + s(m+1)R = 0, \ \int_0^\infty R(r)dr = 0, \tag{39}$$

$s(m+1)$ being a separation constant, and its solution can be written out as

$$R(r) = n_0 s(m+1) r^m \exp\left(-sr^{m+1}\right), \ s = \Gamma^{m+1}\left(\frac{m+2}{m+1}\right)\langle r\rangle_0^{-(m+1)}, \tag{40}$$

$\Gamma(x)$ being the Eulerian gamma-function. The problem for function $F(t)$ is

$$\frac{dF}{dt} + \lambda\vartheta F = 0, \ F(0) = 1, \ \vartheta \equiv T - T_s, \ \lambda = \frac{(m+1)k}{\rho L}\langle r\rangle_0^m s. \tag{41}$$

A trouble with this problem consists in the fact that the current temperature difference depends on time implicitly, through the degree of transformation $y(t) = F(t)$, as it is evidenced by equation (37). However, the said equation can be readily integrated to yield

$$\vartheta = \vartheta_0 - \frac{L}{c_v}\ln\frac{c_g + \mu(1-y)c_v}{c_g}, \ \vartheta \equiv T_0 - T, \tag{42}$$

T_0 being an initial system temperature. Equations (41) and (42) lead to a problem for $y(t)$

$$\frac{dy}{dt} + \lambda\left[\vartheta_0 - \frac{L}{c_v}\ln\frac{c_g + \mu(1-y)c_v}{c_g}\right]y = 0, \ y(0) = 1. \tag{43}$$

It is not difficult to integrate this equation numerically. However, a sufficiently accurate solution to problem (43) can be obtained by expanding the logarithmic term into a Taylor series and by retaining only the terms up to the second order inclusive. As a result, we obtain

$$\frac{dy}{dt} + ay + by^2 = 0, \ y(0) = 1, \ a = \lambda\left(\vartheta_0 - \vartheta^*\right), \ b = \lambda\vartheta^*, \ \vartheta^* = \frac{\mu L}{c_g}. \tag{44}$$

The wanted solution to this problem coincides with the presented in equation (23). Allowing for the new definition of a and b in equation (44), we rewrite it as follows

$$y(t) = \frac{\vartheta_0 - \vartheta^*}{\vartheta_0 \exp\left[\lambda\left(\vartheta_0 - \vartheta^*\right)t\right] - \vartheta^*}. \tag{45}$$

Hence we derive an approximate solution that is valid at long times

$$y(t) \approx \left(1 - \frac{\vartheta^*}{\vartheta_0}\right)\exp\left[-\lambda\left(\vartheta_0 - \vartheta^*\right)t\right], \ t \gg \frac{1}{\lambda\left(\vartheta_0 - \vartheta^*\right)}, \ \vartheta_0 > \vartheta^*. \tag{46}$$

Equation (46) is formulated for the case when initial mixture temperature T_0 is high enough to ensure complete evaporation of all the droplets that where present in the system at the initial moment. In the opposite case, the mixture attains a thermodynamic equilibrium state at $t \to \infty$, and equilibrium state is characterized by a relation that follows directly from equation (45)

$$y_\infty = 1 - \frac{\vartheta_0}{\vartheta^*}, \ \vartheta_0 < \vartheta^*. \tag{47}$$

In a special case where the initial superheat or the two-phase mixture is just sufficient to evaporate all suspended droplets, equation (46) must be replaced by

$$y(t) = \frac{1}{1 + \lambda \vartheta_0 t}, \ \vartheta_0 = \vartheta^*. \tag{48}$$

Inserting the degree of transformation from equation (45) into equation (37) leads to a new equation that describes the temperature variation in the evaporation process

$$\vartheta \frac{d^2\vartheta}{dt^2} - \left(\frac{d\vartheta}{dt}\right)^2 + \lambda \vartheta^2 \frac{d\vartheta}{dt} = 0, \ \vartheta(0) = \vartheta_0 \tag{49}$$

the solution of which reads as follows

$$\vartheta(t) = \vartheta_0 \left(1 - \frac{\vartheta^*}{\vartheta_0}\right) \left\{1 - \frac{\vartheta^*}{\vartheta_0} \exp\left[-\lambda \left(\vartheta_0 - \vartheta^*\right) t\right]\right\}^{-1}. \tag{50}$$

Analytical results derived in this section have been checked by comparing them with experimental data reported by different authors, and agreement was always satisfactory.

4. Dissolution of dispersed solids

Another example of particulate processes to which the method developed can be most advantageously applied concerns various technological schemes which involve dissolution of a polydisperse assemblage of particles in a liquid mixture. The dissolution processes are even somewhat simpler to treat than evaporation processes that have been addressed in the preceding section, since the dependence of the transformation rate for a single particle can be usually described with the help of less sophisticated relations. In particular, if particles dissolve in a well-stirred tank, the ideal mixing approximation applies again. If mixing is vigorous enough, the dissolution rate of a single particle often happens to be independent of particle size [5–9], and it can be expressed by the following formula

$$w = \frac{dr}{dt} = -M(t)N(r), \ M = \frac{k}{\rho}\left[c^*(t) - c(t)\right], \ N = 1, \tag{51}$$

where k now designates a constant mass transfer coefficient, ρ is now the particle material density, and c and c^* are the mean concentration of the solute and its equilibrium value at the saturation conditions, respectively. In the case under consideration, this formula substitutes for the previously employed kinetics expressed by equations (15) and (36). It holds true under conditions of co-current two-phase mixture flow, as well as in mixer-type apparatuses in which there is a periodic external supply of solids and withdrawal of solution. In both the cases mentioned, the mass balance equation that replaces equation (17) can be formulated as

$$c(t) = c_0 + \beta_0 \left[1 - y(t)\right], \tag{52}$$

where c_0 is an initial solute concentration, β_0 is an initial mass of solids within a unit mixture volume, and $y(t)$ is the transformation rate as identified in equation (35).

When dropping out particulars of the calculation, which strongly resembles those in the preceding section, but are actually even simpler, we present an expression for the self-similar population density that substitutes for equations (18) or (40)

$$R(r) = n_0 s \exp\left(-sr\right), \ s = \langle r \rangle_0^{-1} \tag{53}$$

and states that unknown variable $F(t) = y(t)$ again satisfies an equation of the type of (43), with coefficients

$$a(t) = \frac{k}{\rho}\left[c^*(t) - c_0 - \beta_0\right] s,\ b = \frac{k\beta_0}{\rho} s. \tag{54}$$

In a particular case (say, under isothermal and isobaric conditions), where the equilibrium concentration does not depend on time, the wanted solution is given by equation (23), and it can be conveniently rewritten in the form

$$y(t) = \left[\frac{\exp(\lambda t)}{1-\alpha} + \frac{1}{\alpha}\right]^{-1} \approx (1-\alpha)\exp(-\lambda t),\ t \gg \frac{1}{\lambda};\ \alpha = \frac{c^* - c_0}{\beta_0},\ \lambda = \frac{k}{\rho}(1-\alpha)\langle r\rangle_0^{-1}. \tag{55}$$

Under conditions of counter-current flow, equations (51) and (52) give way to

$$M = -\frac{k}{\rho}\left[c^*(t) - c_0 - \beta_0 y(t)\right],\ N = 1,\ c(t) = c_0 + \beta_0 y(t). \tag{56}$$

A similar analysis results now in the same equation (55) for the degree of transformation, but with coefficient λ being now specified as $\lambda = (k/\rho)\alpha\langle r\rangle_0^{-1}$, and this is the only substantial difference between dissolution in well mixed co- and counter-current turbulent flows.

The analytical formulae listed that pertain to self-similar processes were used to correlate different experimental data, and the resultant correlations were surprisingly good.

5. Conclusions

The theory developed here can be advantageously applied in studies involving different physical processes. So, for instance, combustion of dispersed fuels and phase transitions in supercooled (supersaturated) liquids may be mentioned.

It is well-known that there exists a very wide variety of combustion processes of polydisperse, either liquid or solid, fuels which are extremely popular in everyday life and different fields of industry. These processes are effected in multifarious furnaces and other devices of different origin and design, ranging from relatively primitive ovens and boilers, through various sophisticated furnaces of modern power plants, to combustion chambers of rocket propellants [10,11]. Due to a great diversity of fuel physical state and chemical composition, as well as to an exceedingly broad range of possible operating conditions, the kinetics of these processes can be governed by quite different physical and chemical transformations. As a result, there are great many regime-dependent relations that describe the rate of burning of a single particle under different thermal and flow conditions [12,13].

Another example consists in investigating the bulk formation of the new phase in the supercooled (supersaturated) layer upstream of the phase transition front at the stage of simultaneous nucleation and growth of the solid phase elements. Still now, no systematic study of this phenomenon was performed because no generalized methods were detailed for investigating the initial, intermediate and concluding stages of this phase transition process as a whole. A number of recent successful attempts in this way devoting to a theoretical description of the aforementioned stages (nucleation [14–19], instability of the solid-liquid phase interface [20–24], dendritic growth [25–29] and mushy layer evolution [30–38]) may be mentioned.

The present paper shows that analytic investigation of a number of major processes and factors in the theory of evolution of particulate systems involves fundamental difficulties and requires the use of qualitatively new approaches and concepts. These aspects also require detailed work in future studies.

6. References

[1] Alyab'eva A V, Buyevich Yu A and Mansurov V V 1994 *J. Phys. II France* **4** 951–957
[2] Buyevich Yu A, Alexandrov D V and Mansurov V V 2001 *Macrokinetics of Crystallization* (New York: Begell House)
[3] Alexandrov D V 2016 *J. Phys.: Condens. Matter* **28** 035102
[4] Alexandrov D V 2016 *J. Phil. Mag. Lett.* **96** 132–141
[5] Mullin J W 1972 *Crystallization* (London: Butterworth)
[6] Vollmer U and Raisch J 2001 *Control Eng. Practice* **9** 837–845
[7] Alexandrov D V and Malygin A P 2013 *J. Phys. A: Math. Theor.* **46** 455101
[8] Alexandrov D V and Malygin A P 2014 *Modelling Simul. Mater. Sci. Eng.* **22** 015003
[9] Rachah A, Noll D, Espitalier F and Baillon F 2016 *Math. Meth. Appl. Sci.* **39** 1101–1120
[10] Hernandez-Atonal F D, Ryu C, Sharifi V N and Swithenbank J 2007 *Chem. Eng. Sci.* **62** 627–635
[11] Luyben W L 2013 *Energy Fuels* **27** 6316–6321
[12] Buyevich Yu A, Korolyova N A and Natalukha I A 1993 *Int. J. Heat Mass Trans.* **36** 2223–2231
[13] Buyevich Yu A, Korolyova N A and Natalukha I A 1993 *Int. J. Heat Mass Trans.* **36** 2233–2238
[14] Alexandrov D V and Nizovtseva I G 2014 *Proc. R. Soc. A* **470** 20130647
[15] Alexandrov D V 2014 *J. Phys. A: Math. Theor.* **47** 125102
[16] Shneidman V A 2013 *Phys. Rev. E* **88** 010401
[17] Alexandrov D V 2015 *J. Phys. A: Math. Theor.* **48** 035103
[18] Sagui C and Grant M 1999 *Phys. Rev. E* **59** 4175–4187
[19] Rubinstein I and Zaltzman B 2000 *Phys. Rev. E* **61** 709–717
[20] Alexandrov D V and Ivanov A O 2000 *J. Crystal Growth* **210** 797–810
[21] Alexandrov D V 2004 *Int. J. Heat Mass Trans.* **47** 1383–1389
[22] Worster M G 1992 *J. Fluid Mech.* **237** 649–669
[23] Alexandrov D V and Malygin A P 2011 *Int. J. Heat Mass Trans.* **54** 1144–1149
[24] Guba P and Worster M G 2006 *J. Fluid Mech.* **553** 419–443
[25] Alexandrov D V, Galenko P K and Herlach D M 2010 *J. Crystal Growth* **312** 2122–2127
[26] Alexandrov D V and Galenko P K 2013 *Phys. Rev. E* **87** 062403
[27] Pelcé P 1988 *Dynamics of Curved Fronts* (Boston: Academic Press)
[28] Alexandrov D V and Galenko P K 2014 *Physics-Uspekhi* **57** 771–786
[29] Bouissou Ph and Pelcé P 1989 *Phys. Rev A* **40** 6673–6680
[30] Alexandrov D V aand Malygin A P 2006 *Doklady Earth Sciences* **411** 1407–1411.
[31] Alexandrov D V and Aseev D L 2006 *Comput. Mater. Sci.* **37** 1–6
[32] Huppert H 1990 *J. Fluid Mech.* **212** 209-240
[33] Aseev D L and Alexandrov D V 2006 *Doklady Physics* **51** 291–295
[34] Alexandrov D V, Nizovtseva I G, Malygin A P, Huang,H-N and Lee D 2008 *J. Phys.: Condens. Matter* **20** 114105
[35] Huguet L, Alboussière T., Bergman M I, Deguen R, Labrosse S and Lesoeur G 2016 *Geophys. J. Int.* **204** 1729–1755
[36] Alexandrova I V, Alexandrov D V, Aseev D L and Bulitcheva S V 2009 *Acta Physica Polonica A* **115** 791–794
[37] Alexandrov D V and Malygin A P 2012 *Int. J. Heat Mass Trans.* **55** 3196–3204
[38] Alexandrov D V and Nizovtseva I G 2008 *Int. J. Heat Mass Trans.* **51** 5204–5208

Acknowledgments

This work was supported by the Russian Foundation for Basic Research (grant no. 16-08-00932) and the Ministry of Education and Science of the Russian Federation (project no. 1.9527.2017).

IOP Conf. Series: Materials Science and Engineering **192** (2017) 012002 doi:10.1088/1757-899X/192/1/012002

A relaxation time of secondary dendritic branches to their steady-state growth

E A Titova[1], D V Alexandrov[1] and P K Galenko[2]

[1] Department of Theoretical and Mathematical Physics, Laboratory of Multi-Scale Mathematical Modeling, Ural Federal University, Ekaterinburg 620000, Russian Federation
[2] Friedrich-Schiller-Universität-Jena, Physikalisch-Astronomische Fakultät, 07743 Jena, Germany

E-mail: ekatitova@mail.ru

Abstract. In the present article, we use the selection theory to estimate the non-stationarity time of evolving dendrites to their steady-state growth.

1. Introduction

The microstructure predictions in solidification processes have the deep scientific and practical roots (see [1–5] and references therein). A development of experimental methods in solidification processing made possible to reach a much broader range of measurable crystal growth velocities, temperature gradients and cooling rates [6]. For instance, splats and films can be quenched with the cooling rate of the order of 10^7 K/s, the temperature gradients can have the order of 10^8 K/s in laser annealing of sample surfaces, and containerless methods of droplets processing provide the deep undercoolings having the values of 200-400 K prior to the primary crystallization. A large driving force for transformation, arising in such methods, leads to fast solidification from a liquid metastable state as well as the high-speed solid-state transformations of metastable crystalline phases [7]. For instance, the experimentally measured solidification velocity has the order of $10^{-1} - 10^2$ m/s in droplets processed by the electromagnetic levitation facility [7,8]. As such, the total duration of primary solidification in small droplets is rather short and estimated as [8]: $10^{-5} - 10^{-3}$ s.

For the analytical calculations of rapid solidification regimes and theoretical estimations of microstructural parameters of dendritic and eutectic crystals, different models of non-equilibrium crystallization are used, as a rule, formally developed for the steady state scenario [9–11]. However, the steady state approximation for conditions of rapid solidification of small samples (films, splats, droplets) is questionable and such approach is often criticized. A very common view is that a steady state scenario may be expected to exist but not in rapidly solidifying small samples [12]: the stationary regime of solidification can be reached at long times that is possible in usual experimental circumstances (directional solidification) or well-known classic technologies (continuous casting and low intensive brazing). In other words, during the short periods of time, the steady state is not achieved and, consequently, the quasi-stationary approximation in modeling of rapid solidification does fail. The detailed calculations of solidification regimes lead, however, to the remarkable behavior of the transient time between the non-stationary and stationary regimes of solidification: the non-stationary time sharply depends on the interface

Table 1. Material parameters for succinonitrile (SCN) and nickel (Ni)

Parameter		SCN	Ni
Melting temperature	T_A (K)	331.233 [27]	1728 [28]
Latent heat	L (J/(m^3)	$4.781 \cdot 10^7$ [27]	$2.67 \cdot 10^6$ [30]
Heat capacity	c_p (J/(m^{3}K))	$2.08 \cdot 10^6$ [27]	$6.39 \cdot 10^3$ [30]
Solid-liquid interface energy	σ (J/m^2)	$8.9 \cdot 10^{-3}$ [27]	0.275 [30]
Capillary length	d_0 (m)	$2.821 \cdot 10^{-9}$ [27]	$4.92 \cdot 10^{-10}$ [29]
Surface energy stiffness	α_d (-)	0.0825 [27]	0.27 [30]
Thermal diffusivity	D_T (m^2/s)	$1.13 \cdot 10^{-7}$ [27]	$1.2 \cdot 10^{-5}$ [30]
Selection constant	σ_0 (-)	0.1 [22]	0.191 [29]
Relaxation time for $\partial\phi/\partial t$ $^{(*)}$	τ_ϕ (s)	$2.5 \cdot 10^{-7}$	$5.5 \cdot 10^{-7}$
Maximum speed of ϕ-field$^{(*)}$	$V_\phi{}^B$ (m/s)	11.5	25.25
Diffusion coefficient of $\phi^{(*)}$	ν_ϕ (m^2/s)	$5.5 \cdot 10^{-5}$	$3.5 \cdot 10^{-4}$

$^{(*)}$The numerical value of this parameter is chosen in the present work.

velocity and undercooling in solidifying samples [13]. In particular, for the planar solid-liquid interface, it has been shown that the non-stationary time rapidly decreases as the crystal growth velocity increases taking the values of the order of milli- or even micro-seconds with the velocity of the order of centimeters per second [13].

Moreover, using the theoretical model developed in [13], it is straightforward to show that the non-stationary time sharply increases with the decrease of crystal growth velocity taking the values of the order of seconds or even minutes for the interface velocity of the order of micrometer per second. Such remarkable dependence should, obviously, exist not only for the planar interface but also for many other crystal growth shapes. As one of the main crystal growth forms is the dendritic crystal shape (for slow and rapid regimes of rapid solidification [1–3]), we provide analytical treatments for obtaining the non-stationary time which could be quantitatively estimated in comparison with the time for primary solidification of experimentally investigated samples. Our studies are limited to two dimensional dendritic growth. Many of properties of metallic samples (for instance, the time for homogenization) strictly depend on the interdendritic spaces between secondary branches of dendrite. Applicability of the steady-state models to the growth of secondary branches is an actual task. Therefore, the present analysis gives some estimations of a non-stationary period of growing secondary branches in the slow and rapid solidification conditions.

2. Relaxation time to a steady-state regime

The relaxation time τ_T characterizes how fast the dendrite tip velocity,

$$v(t) \propto A_v(t)\exp(-A_m t/\tau_T) \tag{1}$$

relaxes to the steady-state regime of dendritic growth [14]. Here $A_v(t)$ is the time-dependent amplitude of relaxation, A_m is the factor which depends on the model parameters, t is the time and the relaxation time τ_T represents a relaxation parameter to the steady state regime of thermal dendrite. Let us consider the symmetric model of dendritic growth in the positive z direction with a constant velocity v. Let $\zeta = \zeta(x,t)$ and ρ designate the instantaneous coordinate of the crystallization front and the radius of curvature of the dendrite tip, respectively. It is convenient to measure lengths in units of ρ and times in units of ρ/v. In the case of zero surface tension, the dendrite shape represents the Ivantsov parabola [14–17]

$$\zeta_{iv} = -\frac{x^2}{2}.$$

In the case of nonzero surface tension, this solution should be modified by a shape correction $\zeta_0(x)$. The steady-state solution in the form of $\zeta_{iv} + \zeta_0(x)$ exists only if the so-called selection parameter

$$\sigma* = \frac{d_0}{\rho P_g} = \frac{2d_0 D_T}{\rho^2 v} = \frac{d_0 v}{2D_T P_g^2} \tag{2}$$

takes a special value determined from the microscopic solvability theory [18–25]. Here

$$P_g = \frac{\rho v}{2D_T} \tag{3}$$

is the growth Péclet number, d_0 is the capillary length, and D_T is the thermal diffusivity.

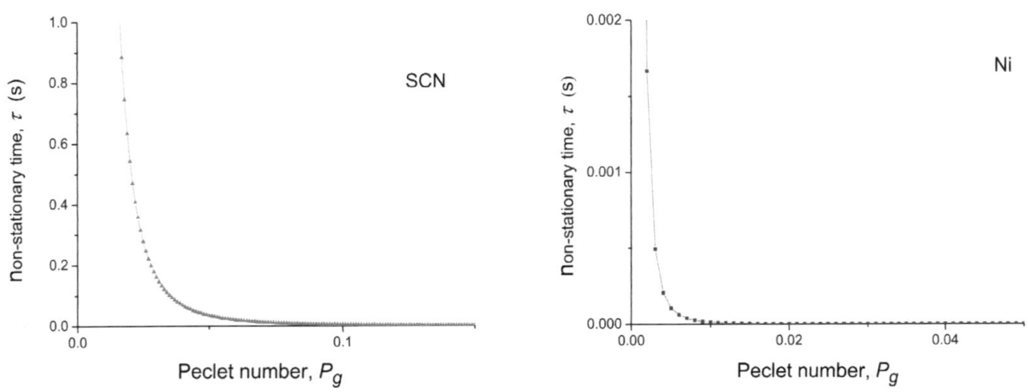

Figure 1. The thermal dendrite growth from the undercooled nickel and SCN melts. The relaxation time τ_T is a function of the growth Péclet number $P_g = \rho v/2D_T$. Calculations are made using Eqs. (9) and (10).

It is important to know what is the real relaxation time when the parabolic dendrite shape can be considered as steady-state. In order to answer this question let us introduce a small time-dependent correction $\zeta_1(x,t)$ to the steady-state solution and present the secondary branches $\zeta(x,t)$ as:

$$\zeta(x,t) = \zeta_{iv} + \zeta_0(x) + \zeta_1(x,t). \tag{4}$$

The time-dependent addition $\zeta_1(x,t)$ has been found in [26] on the basis of WKB theory. The result is

$$\zeta_1(x,t) = f(2z)^{1/8}\sigma^{*1/4}\mathrm{Im}\left\{\exp\left[\frac{(2z)^{1/4}}{c^3\sqrt{\sigma^*}}\left(1 - i\frac{z-t}{\sqrt{2z}}\right)^{3/2}\right]\right\}, \tag{5}$$

where $c = \sqrt{3/2}$, $z = x^2/2$, and f is a constant of unity order. Rising the complex number $1 - i(z-t)/\sqrt{2z}$ to the power $3/2$, separating the real and imaginary parts of expression (5) and introducing the dimensional variables

$$\tau = \frac{\rho t}{v}, \quad \xi = x\rho, \tag{6}$$

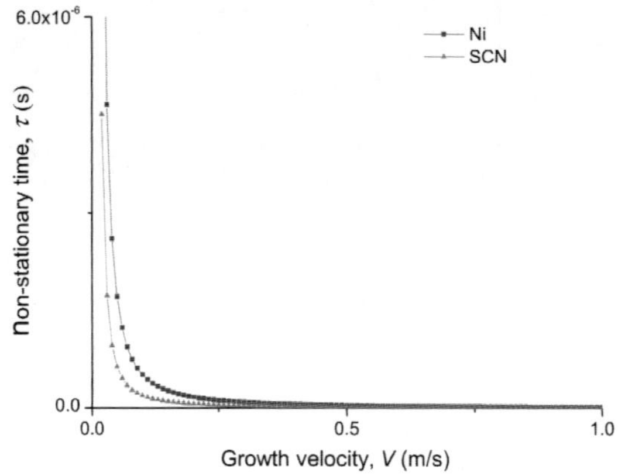

Figure 2. Growth of the thermal dendrite from the undercooled nickel and SCN melts. The relaxation time τ_T is a function of the growth velocity. Calculations are made by Eqs. (3) and (10).

one can represent the final result as

$$\zeta_1(x,t) = A\exp\left\{-\left[(\beta\delta)^{4/3} + \left(\frac{(\beta\delta)^{2/3}}{2} - \frac{\tau}{\tau_T}\right)^2\right]^{3/4}\right\}, \tag{7}$$

where the following designations are introduced

$$\tau_T = \frac{\rho}{v(\beta\delta)^{2/3}}, \quad \beta = \frac{\sqrt{\xi}}{c^3\sqrt{\rho\sigma^*}}, \quad \alpha = \frac{\xi}{2\rho} - \frac{v\tau}{\xi}, \quad \delta = |\cos(\varphi/2)|,$$

$$A = f\left(\frac{\xi\sigma^*}{\rho}\right)^{1/4}\sin\left[\beta\sin(\varphi/2)\right], \quad \varphi = \arctan\left(\frac{\alpha^2-3}{1-3\alpha^2}\right). \tag{8}$$

To obtain a dependence of the thermal relaxation time τ_T of secondary dendritic branches on the growth Péclet number P_g let us express the dendrite tip velocity v from the solvability condition derived in Refs. [21–23] for arbitrary Péclet numbers

$$v(P_g) = \frac{2D_T\alpha_d^{7/4}\sigma_0 P_g^2}{d_0(1+a_1\sqrt{\alpha_d}P_g)^2}, \tag{9}$$

where α_d is the surface energy stiffness, σ_0 is the selection constant and $a_1 = \sqrt{8\sigma_0/7}(3/56)^{3/8}$. Taking into account expression (8) and considering a vicinity of dendritic tip ($\xi \sim \rho$) one can get

$$\tau_T \approx \frac{2D_T P_g c^2}{\delta^{2/3}v^2(P_g)}\left(\frac{d_0 v(P_g)}{2D_T P_g^2}\right)^{1/3}, \tag{10}$$

with the following estimated constants: $\alpha \approx 1/2$ and $\delta \approx 0.738$. These estimations were made for the thermal dendrite with the following values of the parameters: the relaxation time $\tau_T \approx 10^{-10}...10^{-7}$(s), the scaled coordinate $\xi \approx \rho \approx 10^{-6}$ (m) and the growth velocity $v \leq 10^{-2}$

(m/s). Then, using equation (8), one gets $\alpha \approx 0.5 - v\tau/\rho$ with $v\tau/\rho \leq 10^{-3} \ll 1/2$ and obvious estimation for the parameter δ. Expression (10) determines the thermal relaxation time as a function of the growth Péclet number P_g. Together with the dendrite tip velocity v given by the selection criterion (9) this makes possible to quantitatively estimate the relaxation time to the steady-state regime of dendrite growth.

3. Conclusions

In summary, the growth of secondary dendritic branches represents a complex phenomenon with many effects and parameters for a wide range of undercoolings. Therefore, we plan to advance the present formalism for estimation of non-stationary time to include the kinetic effects [23], non-equilibrium effects (such as solute trapping, solute drag, trapping of disorder [16]) and three dimensional dendritic structure.

4. References

[1] Trivedi R and Kurz W 1994 *Int. Mater. Rev.* **39** 49–74
[2] Herlach D M 1994 *Mater. Sci. Eng. R* **12** 177–272
[3] Glicksman M E, Koss M B and Winsa E A 1994 *Phys. Rev. Lett.* **73** 573–576
[4] Alexandrova I V, Alexandrov D V, Aseev D L and Bulitcheva S V 2009 *Acta Physica Polonica A* **115** 791–794
[5] Herlach D 2015 *Crystals* **5** 355–375
[6] Herlach D M, Galenko P and Holland-Moritz D 2007 *Metastable solids from undercooled melts* (Amsterdam: Elsevier)
[7] Hartmann H, Galenko P K, Holland-Moritz D, Kolbe M, Herlach D M and Shuleshova O 2008 *Journal of Applied Physics* **103** 073509
[8] Eckler K 1992 (Ph.D. thesis), (Ruhr-Universität)
[9] Galenko P K and Danilov D A 1997 *Phys. Lett. A* **235** 271–280
[10] Galenko P K and Herlach D M 2006 *Phys. Rev. Lett.* **96** 150602
[11] Gao J, Han M, Kao A, Pericleous K, Alexandrov D V and Galenko P K 2016 *Acta Mater.* **103** 184–191
[12] Das A K 1998 *Defect and Diffusion Forum* **162** 97–118
[13] Galenko P K and Danilov D A 2000 *J. Cryst. Growth* **216** 512–526
[14] Langer J S 1980 *Rev. Mod. Phys.* **52** 1–28
[15] Pecle P and Pomeau Y 1986 *Stud. Appl. Math.* **74** 245–258
[16] Galenko P K, Danilov D A and Alexandrov D V 2015 *Int. J. Heat Mass Trans.* **89** 1054–1060
[17] Alexandrov D V and Galenko P K 2017 *Physica A* **469** 420–428
[18] Pecle P and Bensimon D 1987 *Nucl. Phys. B* **2** 259–270
[19] Alexandrov D V and Pinigin D A 2013 *Technical Physics* **58** 309–315
[20] Alexandrov D V and Galenko P K 2013 *J. Phys. A: Math. Theor.* **46** 195101
[21] Alexandrov D V and Galenko P K 2013 *Phys. Rev. E* **87** 062403
[22] Alexandrov D V and Galenko P K 2014 *Phys. Usp.* **57** 771–786
[23] Alexandrov D V and Galenko P K 2015 *Phys. Chem. Chem. Phys.* **17** 19149–19161
[24] Alexandrov D V, Danilov D A and Galenko P K 2016 *Int. J. Heat Mass Trans.* **101** 789–799
[25] Galenko P K, Danilov D A, Reuther K, Alexandrov D V, Rettenmayr M and Herlach D M 2017 *J. Cryst. Growth* **457** 349–355
[26] Barber M N, Barbieri A and Langer J S 1987 *Phys. Rev. A* **36** 3340–3349
[27] Koss M B, LaCombe J C, Tennenhouse L A, Glicksman M E and Winsa E A 1999 *Metall. Mater. Trans.* A **30** 3177–3190
[28] Hultgren R, Desai P D, Hawkins D T, Gleiser M, Kelley K K and Wagman D D 1973 *Selected values of the thermodynamic properties of the elements* (Ohio: ASM International, Metals Park)
[29] Gao J, Kao A, Bojarevics V, Pericleous K, Galenko P K and Alexandrov D V 2016 *J. Cryst. Growth* http://dx.doi.org/10.1016/j.jcrysgro.2016.11.069
[30] Galenko P K, Phanikumar G, Funke O, Chernova L, Reutzel S, Kolbe M and Herlach D M 2007 *Materials Science and Engineering A* **448–451** 649–653

Acknowledgments

This work was supported by the Russian Science Foundation (grant no. 16-11-10095).

On the theory of unsteady-state crystallization with a mushy layer

D V Alexandrov, I V Alexandrova and A A Ivanov

Department of Theoretical and Mathematical Physics, Laboratory of Multi-Scale Mathematical Modeling, Ural Federal University, Ekaterinburg, 620000, Russian Federation

E-mail: dmitri.alexandrov@urfu.ru

Abstract. The self-similar and unsteady-state crystallization processes of binary melts with a mushy layer are considered. Two analytical methods of solution of the mushy layer nonlinear model with the moving phase transition boundaries are detailed. The obtained analytical solutions are in good agreement with experimental data.

1. Introduction

It is well-known that crystallization processes play a key role in materials science and determine different properties of solidified liquids and melts [1–3]. A standard description of crystallization by means of the classical Stefan thermodiffusion model with a planar solid-liquid interface works well far from always. This is explained by the fact that, in the certain circumstances, a supercooled two-phase (mushy) layer arises ahead of a moving crystallization front [4–6]. As a result of this phenomenon, after a lapse of time, a crystallization domain becomes divided into three regions: solid phase, mushy layer and liquid phase. In other words, a mushy layer represents a region of morphological instability [7–10]. Due to the effect of constitutional supercooling, different elements of the solid phase in the form of nuclei or dendrite-like structures become capable to evolve in the mushy layer. Their evolution accompanied by the released latent heat of solidification will compensate the constitutional supercooling and the mushy layer as a whole will evolve in a non-equilibrium manner [11–13]. If the latent heat of crystallization completely compensates the constitutional supercooling, the mushy layer is termed as quasiequilibrium. A theoretical description of this crystallization scenario was developed in [14–16]. A mathematical model describing these solidification conditions represents a set of non-linear differential equations supplemented by the corresponding boundary conditions imposed at the moving interfaces. In the case of steady-state solidification regime, this model admits an exact solution [17–19] in the presence and absence of different non-linear effects such as a weak convection, thermodiffusion and temperature-dependent diffusivity. A time-dependent solidification scenario frequently met in practice is very important too. Two original methods of analytical solutions of the non-linear transient mushy layer equations were developed in a series of papers [20–24]. The present study is concerned with generalization of these methods for the description of unsteady-state crystallization regimes.

The outline of this paper is as follows: Section 2 is devoted to a theoretical description of self-similar solidification process with a planar front and as well as with a mushy layer; an analytical

IOP Conf. Series: Materials Science and Engineering **192** (2017) 012003 doi:10.1088/1757-899X/192/1/012003

approach for the transient solidification conditions is described in Section 3; a summary of results and conclusions are presented in Section 4.

2. Self-similar crystallization scenario

Let us consider a unidirectional crystallization process of a binary mixture from a cooled wall shown in Figure 1 (z being the solidification direction). The semi-infinite domain $z > 0$ filled with the liquid phase is initially maintained at the constant temperature T_∞ and has the uniform impurity concentration $C = c_0$. Let us also neglect the diffusion transport in the solid phase as well as the gravitational force. The temperature of the cooled wall $z = 0$ equals $T = T_B$, which is lower than the liquidus temperature.

2.1. Self-similar regime with a planar front

The temperature (T) and concentration (C) distributions in the solid ($0 < z < h(t)$) and liquid ($z > h(t)$) phases are governed by the following transfer equations

$$\rho_s C_{ps} \frac{\partial T}{\partial t} = k_s \frac{\partial^2 T}{\partial z^2}, \ 0 < z < h(t), \tag{1}$$

$$\rho_l C_{pl} \frac{\partial T}{\partial t} = k_l \frac{\partial^2 T}{\partial z^2}, \ \frac{\partial C}{\partial t} = D \frac{\partial^2 C}{\partial z^2}, \ z > h(t), \tag{2}$$

where ρ_s and ρ_l are the densities of the solid and liquid phases, C_{ps} and C_{pl} are the specific heats in these phases, D is the diffusion coefficient, and $h(t)$ is the solid-liquid interface coordinate, which moves into the liquid region and depends on time t.

The wall temperature is fixed

$$T = T_B, \ z = 0, \tag{3}$$

and the far-field boundary conditions take the form

$$T \to T_\infty, \ C \to c_0, \ z \to \infty. \tag{4}$$

At the moving phase transition interface, we have the following liquidus and balance conditions

$$T = -mC, \ z = h(t), \tag{5}$$

$$\rho_s L \frac{dh}{dt} = k_s \left(\frac{\partial T}{\partial z} \right)_{z=h^-} - k_l \left(\frac{\partial T}{\partial z} \right)_{z=h^+}, \ z = h(t), \tag{6}$$

$$C \frac{dh}{dt} + D \left(\frac{\partial C}{\partial z} \right)_{z=h^+} = 0, \ z = h(t). \tag{7}$$

Here m is the liquidus slope, L is the latent heat of solidification, and subscripts h^- and h^+ designate the solid and liquid sides of the moving front $h(t)$.

The moving-boundary problem (1)-(7) can be solved by means of introducing the self-similar variable η and the parabolic growth rate constant λ

$$\eta = \frac{z}{\sqrt{4Dt + s}}, \ \lambda = \frac{h(t)}{\sqrt{4Dt + s}}, \tag{8}$$

where $s = h^2(0)/\lambda^2$, and λ represents the self-similar solid-liquid interface.

The self-similar solutions of equations (1)-(7) can be written out as [20, 25]

$$T(\eta) = T_B + \frac{(T_h - T_B)\mathrm{erf}(\varepsilon_s \eta)}{\mathrm{erf}(\varepsilon_s \lambda)}, \ \eta < \lambda, \tag{9}$$

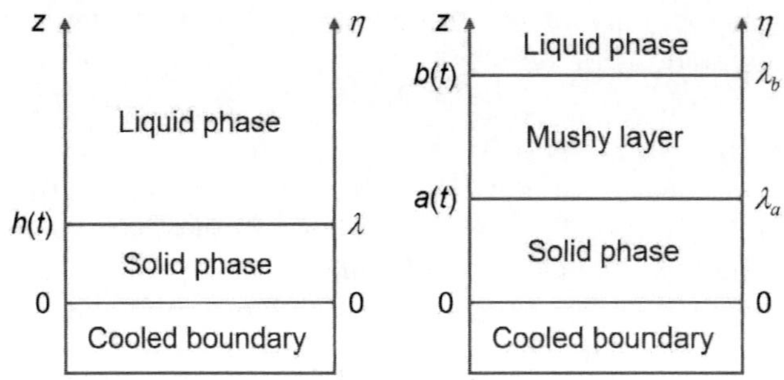

Figure 1. A scheme of solidification process with a planar front and mushy layer.

$$T(\eta) = T_\infty + \frac{(T_h - T_\infty)\mathrm{erfc}(\varepsilon_l \eta)}{\mathrm{erfc}(\varepsilon_l \lambda)}, \ \eta > \lambda, \tag{10}$$

$$C(\eta) = c_0 + \frac{(C_h - c_0)\mathrm{erfc}(\eta)}{\mathrm{erfc}(\lambda)}, \ \eta > \lambda, \tag{11}$$

where T_h and C_h designate the temperature and concentration at the front $\eta = \lambda$, $\varepsilon_s = \sqrt{DC_{ps}\rho_s/k_s}$, and $\varepsilon_l = \sqrt{DC_{pl}\rho_l/k_l}$.

Three unknown parameters T_h, C_h and λ can be found from the boundary conditions (5)-(7) in the form

$$T_h = -mC_h, \ C_h - c_0 = C_{fi}(\lambda) \equiv \frac{c_0 F(\lambda)}{1 - F(\lambda)}, \tag{12}$$

$$mC_{fi}(\lambda)\left[\frac{\beta}{F(\varepsilon_l\lambda)} + \frac{1}{G(\varepsilon_s\lambda)}\right] = \frac{T_1}{G(\varepsilon_s\lambda)} - \frac{\beta T_0}{F(\varepsilon_l\lambda)} - \frac{L}{C_{ps}}. \tag{13}$$

Here we used the following designations

$$\beta = \frac{\rho_l C_{pl}}{\rho_s C_{ps}}, \ F(x) = \sqrt{\pi}x\exp(x^2)\mathrm{erfc}(x), \ G(x) = \sqrt{\pi}x\exp(x^2)\mathrm{erf}(x).$$

The self-similar front position is determined by equation (13). Note that a behavior of λ as well as the temperature differences $T_1 = -mc_0 - T_B$ and $T_0 = T_\infty + mc_0$ has been demonstrated in [25].

The constitutional supercooling appears ahead of the planar solid-liquid interface when the concentration gradient becomes greater than the temperature one at $z = h^+$, that is

$$\frac{\partial T}{\partial z} < -m\frac{\partial C}{\partial z}, \ z = h^+. \tag{14}$$

Substitution of analytical solutions (10) and (11) into inequality (14) gives

$$C_{fi}(\lambda) > C_i(\lambda) \equiv \frac{(T_0/m)\,\varepsilon_l^2 F(\lambda)}{F(\varepsilon_l\lambda) - \varepsilon_l^2 F(\lambda)}. \tag{15}$$

If this inequality holds true, the constitutional supercooling exists ahead of the self-similar solidification front $\eta = \lambda$ and, as a consequence, we must use a mushy layer model instead of the frontal model above.

2.2. Self-similar regime with a mushy layer

In the case of a mushy layer solidification scenario (Figure 1), the heat and mass transfer process in this layer, $a(t) < z < b(t)$, depends on the solid fraction φ and is described by the following transfer equations

$$(\rho C_p)_m \frac{\partial T}{\partial t} = \frac{\partial}{\partial z}\left(k_m \frac{\partial T}{\partial z}\right) + \rho_s L \frac{\partial \varphi}{\partial t},\ T = -mC,\ a(t) < z < b(t), \tag{16}$$

$$\chi \frac{\partial C}{\partial t} = \frac{\partial}{\partial z}\left(D\chi \frac{\partial C}{\partial z}\right) + C\frac{\partial \varphi}{\partial t},\ \chi = 1-\varphi,\ a(t) < z < b(t), \tag{17}$$

where χ represents the liquid fraction of a mushy layer and the thermal properties of a mush are assumed to be volume-fraction weighted averages of the properties of the solid and liquid phases [26], i.e.

$$(\rho C_p)_m = \chi\rho_l C_{pl} + (1-\chi)\rho_s C_{ps},\ k_m = \chi k_l + (1-\chi)k_s.$$

The heat and mass balance boundary conditions imposed at the solid - mushy layer ($z = a(t)$) and mushy layer - liquid ($z = b(t)$) interfaces can be written out as follows

$$\rho_s L \chi_a \frac{da}{dt} = k_s\left(\frac{\partial T}{\partial z}\right)_{z=a^-} - k_m\left(\frac{\partial T}{\partial z}\right)_{z=a^+},\ C_a\chi_a\frac{da}{dt} = -D\chi_a\left(\frac{\partial C}{\partial z}\right)_{z=a^+},$$
$$\rho_s L(1-\chi_b)\frac{db}{dt} = k_m\left(\frac{\partial T}{\partial z}\right)_{z=b^-} - k_l\left(\frac{\partial T}{\partial z}\right)_{z=b^+}, \tag{18}$$
$$C_b(1-\chi_b)\frac{db}{dt} = D\chi_b\left(\frac{\partial C}{\partial z}\right)_{z=b^-} - D\left(\frac{\partial C}{\partial z}\right)_{z=b^+},$$

where subscripts a^- and a^+ designate the solid and mushy layer sides of the phase transition boundary $z = a(t)$, whereas b^- and b^+ designate the mushy layer and liquid sides of the second boundary $z = b(t)$.

A condition of marginal equilibrium reads as

$$\frac{\partial T}{\partial z} = -m\frac{\partial C}{\partial z},\ z = b^+. \tag{19}$$

The last condition demonstrates that none of the liquid layer, $z > b(t)$, is supercooled.

Let us introduce the following self-similar variables by analogy with (8)

$$\lambda_a = \frac{a(t)}{\sqrt{4Dt+s}},\ \lambda_b = \frac{b(t)}{\sqrt{4Dt+s}}, \tag{20}$$

where the parabolic growth rate constants λ_a and λ_b determine the self-similar coordinates of a mushy layer.

A nonlinear set of equations and boundary conditions (16)-(19) can be rewritten in the self-similar variables (8) and (20) as

$$-2\eta(1-\varphi)\frac{dC}{d\eta} = -\frac{d\varphi}{d\eta}\frac{dC}{d\eta} + (1-\varphi)\frac{d^2C}{d\eta^2} - 2\eta C\frac{d\varphi}{d\eta},\ \alpha = \frac{mC_{ps}}{L},\ \lambda_a < \eta < \lambda_b, \tag{21}$$

$$-2\alpha\varepsilon_s^2\left[\beta + (1-\beta)\varphi\right]\eta\frac{dC}{d\eta} = \left[\Lambda + \Lambda_1\varphi\right]\alpha\frac{d^2C}{d\eta^2} + \alpha\Lambda_1\frac{d\varphi}{d\eta}\frac{dC}{d\eta} + 2\varepsilon_s^2\eta\frac{d\varphi}{d\eta},\ \lambda_a < \eta < \lambda_b, \tag{22}$$

$$C = c_0 + C_i(\lambda_b),\ \eta = \lambda_b;\ \Lambda = \frac{k_l}{k_s},\ \Lambda_1 = 1 - \Lambda, \tag{23}$$

$$\left[\frac{\varepsilon_s^2 \chi_b}{\alpha} + \frac{k_m(\chi_b)}{k_s} C_b - \frac{C_i(\lambda_b)}{F(\lambda_b)}\right](1 - \chi_b) = 0,\ \eta = \lambda_b, \tag{24}$$

$$\frac{dC}{d\eta} = -2\lambda_b \left[\frac{k_l}{k_m(\chi_b)} \frac{C_i(\lambda_b)}{F(\lambda_b)} + \frac{\varepsilon_s^2}{\alpha} \frac{k_s}{k_m(\lambda_b)}(1 - \chi_b)\right],\ \eta = \lambda_b, \tag{25}$$

$$\left[2\lambda_a \frac{\varepsilon_s^2 \chi_a}{\alpha} - \frac{k_m(\chi_a)}{k_s} \frac{dC}{d\eta}\right] G(\varepsilon_s \lambda_a) + 2\lambda_a \varepsilon_s^2 \left[C + \frac{T_B}{m}\right] = 0,\ \eta = \lambda_a, \tag{26}$$

$$\left[\frac{dC}{d\eta} + 2\lambda_a C\right] \chi_a = 0,\ \eta = \lambda_a. \tag{27}$$

The boundary condition (24) implies that either $\chi_b = 1$ or

$$\chi_b \equiv \chi_i = \frac{c_0\left[C_i(\lambda_b)/C_{fi}(\lambda_b) - 1\right]}{\varepsilon_s^2/\alpha + C_b\left(k_l/k_s - 1\right)}. \tag{28}$$

This makes possible to choose the boundary condition in the form $\chi_b = \chi_i$ if $0 \leq \chi_i < 1$ and $\chi_b = 1$ otherwise [25]. The boundary condition (26) must be considered in the same manner.

Let us seek for the solution of the aforementioned problem in the form of series in η, i.e.

$$\begin{aligned} \varphi(\eta) &= \varphi_0 + \eta\varphi_1 + \eta^2\varphi_2 + \eta^3\varphi_3 ..., \\ C(\eta) &= C_0 + \eta C_1 + \eta^2 C_2 + \eta^3 C_3 + ... \end{aligned} \tag{29}$$

Now substituting expansions (29) into equations (21) and (22), we arrive at two equations in the zero-order approximation in η

$$\varphi_1 C_1 = 2(1 - \varphi_0)C_2,\ 2C_2\left[\Lambda + (1 - \Lambda)\varphi_0\right] + (1 - \Lambda)\varphi_1 C_1 = 0.$$

These equations show that the solution is trivial, that is $\varphi_1 = 0$ and $C_2 = 0$.

The first-order approximation leads to equations for φ_2 and C_3. Their solution reads as follows

$$\begin{aligned} \varphi_2 &= (1 - \varphi_0)\left[\Lambda + (1 - \Lambda)\varphi_0 - \varepsilon_s^2\left(\beta + (1 - \beta)\varphi_0\right)\right], \\ C_3 &= L_0(\lambda_b)C_1,\ L_0(\lambda_b) = -\frac{(1 - \Lambda)(1 - \varphi_0) + \varepsilon_s^2\left[\beta + (1 - \beta)\varphi_0\right]}{3}. \end{aligned} \tag{30}$$

Substitution of solutions (30) into the boundary condition $\varphi(\lambda_b) = \varphi_b = \varphi_0 + \lambda_b^2\varphi_2$ leads to a quadratic equation for $\varphi_0(\lambda_b)$. Its solution reads

$$\varphi_0(\lambda_b) = \varphi_0^{\pm} = \frac{\pm\sqrt{b_1^2 - 4a_1c_1} - b_1}{2a_1}, \tag{31}$$

$$a_1(\lambda_b) = \lambda_b^2\left[1 - \Lambda - \varepsilon_s^2(1 - \beta)\right],\ b_1(\lambda_b) = \left[2\Lambda\lambda_b^2 - \lambda_b^2 - 1 + \varepsilon_s^2\lambda_b^2(1 - 2\beta)\right],$$

$$c_1(\lambda_b) = \varphi_b(\lambda_b) - \lambda_b^2\left(\Lambda - \varepsilon_s^2\beta\right),\ \varphi_b(\lambda_b) = 1 - \chi_b(\lambda_b).$$

The boundary condition (23) implies that

$$C_0(\lambda_b) = c_0 + C_i(\lambda_b) - \lambda_b C_1\left[1 + \lambda_b^2 L_0(\lambda_b)\right]. \tag{32}$$

Let us consider the case when $\chi_a \neq 0$. Then we have from condition (27)

$$C_1 = -\frac{2\lambda_a\left[c_0 + C_i(\lambda_b)\right]}{1 + 2\lambda_a^2 - 2\lambda_a\lambda_b + L_0(\lambda_b)\left[3\lambda_a^2 + 2\lambda_a^4 - 2\lambda_a\lambda_b^3\right]}. \tag{33}$$

The boundary condition (25) leads to

$$C_1(\lambda_b) = H_1(\lambda_b),\ H_1(\lambda_b) = -\frac{H(\lambda_b)}{k_m(\lambda_b)\left[1 + 3\lambda_b^2 L_0(\lambda_b)\right]}, \tag{34}$$

$$H(\lambda_b) = 2\lambda_b\left[k_l\frac{C_i(\lambda_b)}{F(\lambda_b)} + \frac{\varepsilon_s^2 k_s}{\alpha}(1 - \chi_b)\right].$$

Now the boundary condition (26) gives

$$\left[2\lambda_a\frac{\varepsilon_s^2\chi_a}{\alpha} - \frac{k_m(\chi_a)}{k_s}\left(C_1 + 3\lambda_a^2 C_3\right)\right]G(\varepsilon_s\lambda_a) + 2\lambda_a\varepsilon_s^2\left[C_0 + \lambda_a C_1 + \lambda_a^3 C_3 + \frac{T_B}{m}\right] = 0. \tag{35}$$

Equations (33)-(35) enable us to find λ_a and λ_b.

When $\lambda_a \ll \lambda_b$, one can obtain an explicit solution. In this case, we get from expressions (33) and (34)

$$\lambda_a(\lambda_b) = \frac{\pm\sqrt{r_2^2 - 4r_1 H_1} - r_2}{2r_1}, \tag{36}$$

$$r_1(\lambda_b) = H_1(\lambda_b)\left(2 + 3L_0(\lambda_b)\right),\ r_2(\lambda_b) = 2\left[c_0 + C_i(\lambda_b)\right] - 2H_1(\lambda_b)\lambda_b\left(1 + \lambda_b^2 L_0(\lambda_b)\right).$$

Note that our numerical solutions show that equation (36) is valid in the case of its positive root. In this case, equation (35) determines the growth constant λ_b.

Now let us consider the case $\varphi_a = 1$ ($\chi_a = 0$). The coefficints φ_0, φ_2, C_0, C_1 and C_3 can be found as before by means of expressions (30)-(32) and (34), while λ_a is determined by equation $\varphi_0(\lambda_b) + \lambda_a^2\varphi_2(\lambda_b) = 1$. Its solution gives

$$\lambda_a(\lambda_b) = \sqrt{\frac{1 - \varphi_0(\lambda_b)}{\varphi_2(\lambda_b)}}. \tag{37}$$

Note that the analytical solutions of self-similar solidification scenario are described by expressions (24) and (29)-(37).

The obtained analytical solution is demonstrated in Figure 2. A self-similar mushy layer appears a result of constitutional supercooling at a critical point $T_1 > 1.776^{\circ}$C and $\lambda > 0.154$. The method of approximate analytical solutions under consideration is in good agreement with experiments [27]. More detailed analysis of a behavior of analytical solutions in the vicinity of a critical point can be found in [20, 25].

An important point is that we can construct an explicit analytical solution in the limiting case $\varphi \to 1$. In this case, equations (21), (24), (26) and (27) will be satisfied automatically, and equation (22) leads to

$$\frac{d^2C}{d\eta^2} = -2\varepsilon_s^2\eta\frac{dC}{d\eta}.$$

Its integration gives

$$C(\eta) = A\int_{\lambda_b}^{\eta}\exp\left(-\varepsilon_s^2 z_1^2\right)dz_1 + B, \tag{38}$$

where constants A and B are determined from the boundary conditions (23) and (25) so that

$$A = -2\lambda_b\exp\left(\varepsilon_s^2\lambda_b\right)\left[\frac{k_l}{k_s}\frac{C_i(\lambda_b)}{F(\lambda_b)} + \frac{\varepsilon_s^2}{\alpha}\right],\quad B = c_0 + C_i(\lambda_b).$$

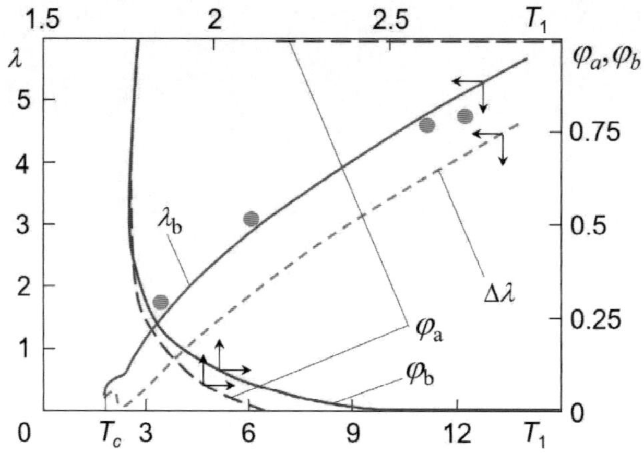

Figure 2. The parabolic growth rate constants λ, λ_b, the mushy layer width $\Delta\lambda = \lambda_b - \lambda_a$ and the volume fractions φ_a and φ_b as functions of the temperature difference T_1. The supercooling occurs at $T_1 > T_c \approx 1.776$°C and $\lambda > 0.154$. The model parameters are [25]: $T_\infty = 15$°C, $c_0 = 14$, $m = 0.4$°C, $L = 3.35 \cdot 10^5$ J kg^{-1}, $D = 10^{-9}$ m^2 s^{-1}, $k_s = 2.219$ J m^{-1} s^{-1} °C^{-1}, $k_l = 0.544$ J m^{-1} s^{-1} °C^{-1}, $\rho_s = 920$ kg m^{-3}, $\rho_l = 1000$ kg m^{-3}, $C_{ps} = 2.01 \cdot 10^3$ J kg^{-1} °C^{-1}, $C_{pl} = 4.187 \cdot 10^3$ J kg^{-1} °C^{-1}. The circles show experimental data [27].

3. Unsteady-state crystallization scenario

Let us now consider a unidirectional process of binary system crystallization (e.g. sea water and ice) in the case of nonstationary boundary conditions at $z = 0$ (see Figure 1). Let the temperature at the cooled boundary $z = 0$ is an arbitrary function of time, i.e. $T = T_{at}(t)$ (e.g. atmospheric temperature in the case of sea water freezing).

3.1. Unsteady-state crystallization regime with a planar front

First we consider the thermally controlled heat transfer process, which is described by the temperature conductivity equation (1) in the solid phase (ice). If we treat the liquid phase (ocean) as isothermal, its temperature may be regarded as constant, i.e. $T = T_w$ at $z \geq h(t)$. Analyzing experimental data [23,28,29] and taking into account numerical solutions of the Stefan problem [30,31], we conclude that the temperature distribution in the solid phase, in some cases, is a nearly linear function of the spatial coordinate z. In this case, the temperature conductivity equation in the solid phase becomes $\partial^2 T/\partial z^2 = 0$. Physically it means that a relaxation time of the temperature field is many times smaller than a characteristic time of the phase interface motion. Thus, the temperature distribution obeys

$$T(z,t) = T_{at}(t) + \frac{T_w - T_{at}(t)}{h(t)} z, \ 0 \leq z \leq h(t). \tag{39}$$

The coordinate $h(t)$ of front motion is determined from the thermal balance condition, which takes the form

$$k_s \frac{\partial T}{\partial z} = \rho_s L \frac{dh}{dt}, \ z = h(t). \tag{40}$$

Combining equations (39) and (40), we get

$$h(t) = \sqrt{\frac{2k_s}{\rho_s L}\left(T_w t - \int_0^t T_{at}(\alpha) d\alpha\right)}, \tag{41}$$

where we consider the initial condition $h(0) = 0$.

The ice width and transient temperature profiles plotted accordingly to expressions (39) and (41) are demonstrated in Figure 3. It is easily seen that the frontal approximation differs from experiments. This discrepancy can be explained by the fact that, in real situations, the freezing process occurs in the presence of a phase transition (mushy) layer. The physical model of this process and its analytical solution are given in the next subsection.

3.2. Unsteady-state crystallization regime with a mushy layer

Let us now analyze the transient solidification scenario with a mushy layer shown in Figure 1. As before, we consider the linear temperature distribution in the solid phase, i.e.

$$T(z,t) = T_{at}(t) + C_1(t)z, \ 0 < z < a(t), \tag{42}$$

where C_1 represents an arbitrary function of time t.

The temperature distribution in the mushy layer will be regarded as linear too

$$T(z,t) = T_1(t) + T_2(t)z, \ a(t) < z < b(t), \tag{43}$$

where the time-dependent coefficients T_1 and T_2 are determined below. Note that the linear temperature profiles (42) and (43) follow from experimental data and numerical calculations [28–31].

We use here the Scheil equation [32,33] to describe the mass transfer in a mushy region

$$\frac{\partial}{\partial t}\left[(1-\varphi)C\right] = 0, \ a(t) < z < b(t), \tag{44}$$

where C stands for the solute concentration (brine salinity) of a mushy layer.

The temperature and concentration fields in the phase transition layer are connected by means of the liquidus equation of the form

$$T = -mC, \ a(t) < z < b(t), \tag{45}$$

where m, as before, is the liquidus slope.

The boundary conditions imposed at the solid phase (ice) - mushy layer interface can be written as [25]

$$\varphi = \varphi_a, \ (T)_{z=a^-} = (T)_{z=a^+}, \ z = a(t), \tag{46}$$

$$L_V(1-\varphi_a)\frac{da}{dt} = k_s\left(\frac{\partial T}{\partial z}\right)_{z=a^-} - \left[k_s\varphi_a + k_l(1-\varphi_a)\right]\left(\frac{\partial T}{\partial z}\right)_{z=a^+}, \ z = a(t), \tag{47}$$

$$C_a(1-\varphi_a)\frac{da}{dt} = -D(1-\varphi_a)\left(\frac{\partial C}{\partial z}\right)_{z=a^+}, \ z = a(t), \tag{48}$$

where $L_V = \rho_s L$ and subscript a designates the solid phase - mushy region interface.

At the mushy layer - liquid phase (ocean) interface, the boundary conditions take the form

$$\varphi = \varphi_b, \ (T)_{z=b^-} = (T)_{z=b^+} = T_w, \ z = b(t), \tag{49}$$

$$L_V\varphi_b\frac{db}{dt} = \left[k_s\varphi_b + k_l(1-\varphi_b)\right]\left(\frac{\partial T}{\partial z}\right)_{z=b^-}, \ z = b(t), \tag{50}$$

where T_w is the constant temperature at $z \geq b(t)$ (constant sea water temperature) and subscript b denotes the mushy layer - liquid boundary.

IOP Conf. Series: Materials Science and Engineering **192** (2017) 012003 doi:10.1088/1757-899X/192/1/012003

Integration of equation (44) gives the solid fraction

$$\varphi(z,t) = 1 + \frac{(\varphi_b - 1)T_w}{T_1(t) + zT_2(t)}, \tag{51}$$

where expressions (43), (45) and (49) are taken into account.

Substituting now expressions (42), (43), (45) and (51) into the boundary conditions (46)-(50), we obtain

$$\varphi_a(t) = 1 + \frac{(\varphi_b - 1)T_w}{T_{at}(t) + C_1(t)a(t)}, \tag{52}$$

$$C_1(t) = \frac{(1-\varphi_a)L_V}{k_s}\frac{da}{dt} + \left[\varphi_a + K(1-\varphi_a)\right]T_2(t),\ K = \frac{k_l}{k_s}, \tag{53}$$

$$T_{at}(t) + C_1(t)a(t) = T_w + T_2(t)\left[a(t) - b(t)\right], \tag{54}$$

$$\left[T_2(t)\left(b(t) - a(t)\right) - T_w\right](1-\varphi_a)\frac{da}{dt} = D_w(1-\varphi_a)T_2(t), \tag{55}$$

$$T_2(t) = \frac{L_V\varphi_b}{\Phi}\frac{db}{dt},\ \Phi(\varphi_b) = k_s\varphi_b + k_l(1-\varphi_b), \tag{56}$$

$$T_1(t) = T_w - b(t)T_2(t). \tag{57}$$

The solid fraction at the boundary solid phase - mushy layer can be found from expressions (52), (54) and (56) as

$$\varphi_a(t) = 1 + \frac{(\varphi_b - 1)T_w}{T_w + (L_V\varphi_b/\Phi)(db/dt)\left[a(t) - b(t)\right]}. \tag{58}$$

Now combining expressions (55) and (56), we get

$$\left[(a(t) - b(t))\frac{L_V\varphi_b}{\Phi}\frac{db}{dt} + T_w\right]\frac{da}{dt} = -\frac{DL_V\varphi_b}{\Phi}\frac{db}{dt}. \tag{59}$$

Substitution of functions $C_1(t)$ and $T_2(t)$ from equations (53) and (55) into condition (54) leads to the nonlinear differential equation of the form

$$L_VT_w(\varphi_b - 1)\left[(1-K)\frac{\varphi_b}{\Phi}\frac{db}{dt} - \frac{1}{k_s}\frac{da}{dt}\right]a(t)$$

$$= \left(T_w - T_{at}(t) - \frac{L_V\varphi_b}{\Phi}b(t)\frac{db}{dt}\right)\left[(a(t) - b(t))\frac{L_V\varphi_b}{\Phi}\frac{db}{dt} + T_w\right]. \tag{60}$$

Here we used expression (58).

Equations (59) and (60) represent the nonlinear system for the determination of the moving boundaries $a(t)$ and $b(t)$. Below we consider three possible analytical solutions of these equations.

First of all, we consider the case when both boundaries evolve slowly [28, 29]. If this is really the case, neglecting the term proportional to $(da/dt)(db/dt)$ in equation (59), we come to

$$a(t) = -\frac{DL_V\varphi_b}{T_w\Phi}b(t), \tag{61}$$

where $a(0) = b(0) = 0$. Combining now equations (59) and (60) and substituting $a(t)$ from expression (61), we arrive at

$$b(t) = \sqrt{\frac{2}{I}\left(T_wt - \int_0^t T_{at}(\alpha)d\alpha\right)},\ I = \frac{L_V\varphi_b}{\Phi}\left[1 - \frac{DL_V\varphi_b}{T_w\Phi}\left(\frac{L_VD}{k_sT_w} + 1 - K\right)(\varphi_b - 1)\right]. \tag{62}$$

Secondly, if we consider the case of small temperature fluctuations, i.e.

$$|T_{at}(t)| \sim |T_w| \ll \frac{L_V \varphi_b}{\Phi} b(t) \frac{db}{dt},$$

we have from equations (59) and (60)

$$b(t) = Aa(t), \; a(t) = \sqrt{\frac{2D}{A-1} t}, \; A = \sqrt{\frac{T_w(\varphi_b - 1)(1-K)\Phi}{DL_V \varphi_b}}. \tag{63}$$

Analytical solutions (63) describe the case when the phase transition interfaces $a(t)$ and $b(t)$ evolve as the square root of time and the solidification process is far from its initial stage [25,34,35].

Thirdly, let us consider the case when the solid phase - mushy layer boundary moves slower than the mushy layer - liquid phase boundary, i.e. when $da/dt \ll db/dt$. In this case, substitution of expression in square brackets from (59) into (60) gives

$$\frac{T_w}{2D}(\varphi_b - 1)(1-K)a^2(t) - \frac{L_V \varphi_b}{2\Phi} b^2(t) + T_w t - \int_0^t T_{at}(\alpha) d\alpha = 0.$$

Again, taking into consideration $a(t) \ll b(t)$, we finally obtain

$$a(t) = -D \int_0^t \frac{(T_w - T_{at}(\alpha))\, d\alpha}{T_{at}(\alpha) b(\alpha)}, \; b(t) = \sqrt{\frac{2\Phi}{L_V \varphi_b} \left(T_w t - \int_0^t T_{at}(\alpha) d\alpha \right)}. \tag{64}$$

Thus, expressions (61)-(64) completely determine the evolution of phase transition boundaries $a(t)$ and $b(t)$ while the temperature distribution is governed by expressions (42), (43), (53), (56) and (57). The solute concentration and solid phase distributions therewith can be found from expressions (45) and (51).

The theory under consideration is compared with experimental data in Figure 3. The phase interfaces $a(t)$ and $b(t)$ are plotted in accordance with analytical solutions (61) and (62). All functions describing the mushy layer crystallization scenario essentially differ from the frontal solutions. Note that analytical solutions are in good agreement with experimental data.

4. Concluding remarks

Let us emphasize in conclusion that the unsteady-state interface dynamics determined by expressions (61), (62) and (64) becomes self-similar (i.e. becomes proportional to the square root of time) when the external (atmospheric) temperature T_{at} is constant. This conclusion connects the self-similar theory developed in Section 2 and the unsteady-state theory presented in Section 3. What is more, the phase transition boundaries $a(t)$ and $b(t)$ lie between their values corresponding to the maximum (T_{max}) and minimum (T_{min}) temperatures, that is

$$\sqrt{\frac{2}{I}(T_w - T_{max})\, t} \le b(t) \le \sqrt{\frac{2}{I}(T_w - T_{min})\, t}$$

in the case of interface dynamics (61) and (62).

The theory of interface dynamics in the presence of a phase transition (mushy) layer under consideration can be generalized to take into account the processes of nucleation and growth of particles in a metastable region. To do this, one can use the recently developed analytical theories of transient nucleation for single-component [36–38] and binary [39, 40] systems and analytical approaches under consideration.

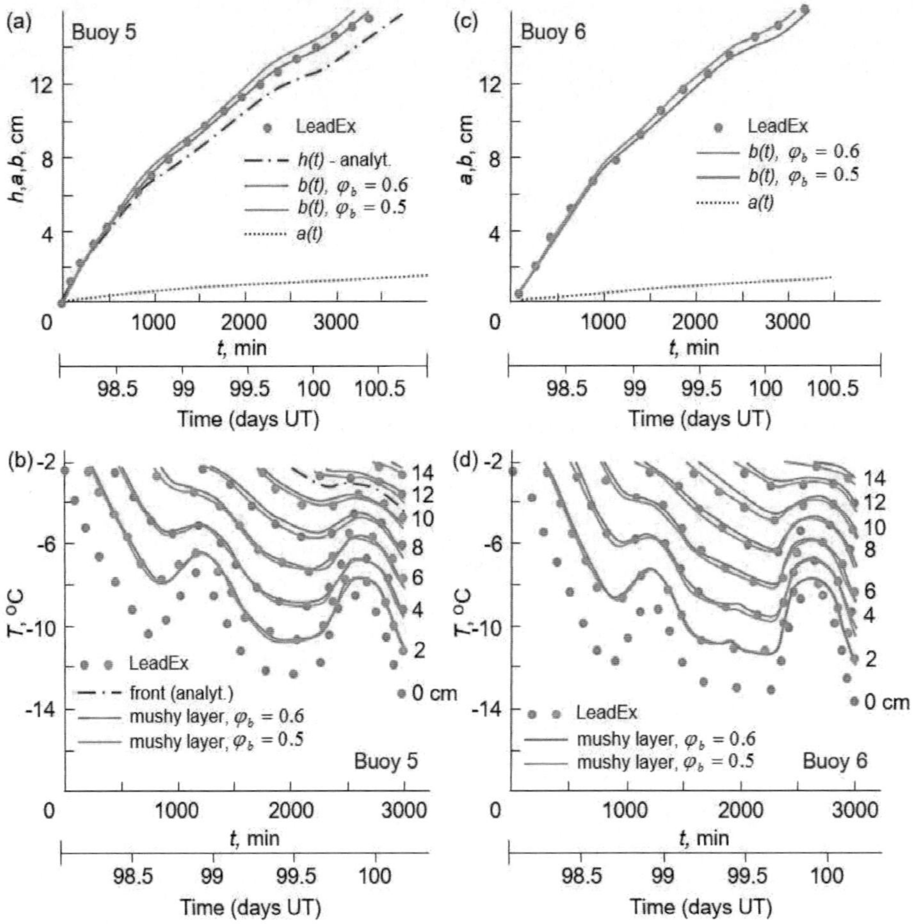

Figure 3. Ice thickness and temperature distributions as functions of time for (a), (b) buoy 5 and (c), (d) buoy 6 at lead 3 in accordance with the LeadEx experiment and the theory under consideration [28, 29]. The solid phase (ice) - liquid (ocean) interface is plotted by the dot-dashed line. Numbers at the curves corresponding to each line show the depths (in centimeters) measured from the ice/atmosphere interface $z = 0$. The temperature at $z = 0$ cm is the atmospheric temperature ($T_{at}(t)$). The time scale used by the LeadEx group is expressed in decimal days of 1992, abbreviated as UT and designated on the figure. The time origin in minutes corresponds to 0221, day 98 UT. Physical parameters used in calculations: $T_w = -2^{\mathrm{O}}$C, $L_V = 3072{\cdot}10^5$ W·s·m^{-3}, $k_s = 2.03$ W·m$^{-1}{\cdot}^{\mathrm{O}}$C^{-1}, $k_l = 0.56$ W·m$^{-1}{\cdot}^{\mathrm{O}}$C^{-1}, $D = 1.2{\cdot}10^{-9}$ m^2·s^{-1}.

5. References

[1] Chalmers B 1961 *Physical Metallurgy* (New York: John Wiley)
[2] Flemings M 1974 *Solidification Processing* (New York: McGraw Hill)
[3] Kurz W and Fisher D J 1989 *Fundamentals of Solidification* (Aedermannsdorf: Trans. Tech. Publ.)
[4] Ivantsov G P 1951 *Dokl. Akad. Nauk SSSR* **81** 179–182
[5] Alexandrova I V, Alexandrov D V, Aseev D L and Bulitcheva S V 2009 *Acta Physica Polonica A* **115** 791–794
[6] Hoglund D E, Thompson M O and Aziz M J *Phys. Rev. B* **58** 189–199
[7] Mullins W W and Sekerka R F 1964 *J. Appl. Phys.* **35** 444–451
[8] Alexandrov D V and Ivanov A O 2000 *J. Crystal Growth* **210** 797–810
[9] Worster M G 1992 *J. Fluid Mech.* **237** 649–669

[10] Alexandrov D V 2004 *Int. J. Heat Mass Trans.* **47** 1383–1389
[11] Mansurov V 1990 *Mathl. Comput. Modelling* **14** 819–821
[12] Aseev D L and Alexandrov D V 2006 *Doklady Physics* **51** 291–295
[13] Alexandrov D V and Malygin A P 2012 *Int. J. Heat Mass Trans.* **55** 3196–3204
[14] Hills R N, Loper D E and Roberts P H 1983 *Q. J. Appl. Math.* **36** 505–539
[15] Fowler A C 1985 *IMA J. Appl. Math.* **35** 159–174
[16] Borisov V T 1987 *Theory of Two-Phase Zone of Metallic Ingot* (Moscow: Metallurgia)
[17] Alexandrov D V 2001 *J. Crystal Growth* **222** 816–821
[18] Alexandrov D V and Aseev D L 2005 *J. Fluid Mech.* **527** 57–66
[19] Aseev D L and Alexandrov D V 2006 *Acta Mater.* **54** 2401–2406
[20] Alexandrov D V and Malygin A P 2006 *Int. J. Heat Mass Trans.* **49** 763–769
[21] Alexandrov D V and Malygin A P 2006 *Dokl. Earth Sciences* **411** 1407–1411
[22] Alexandrov D V and Nizovtseva I G 2008 *Int. J. Heat Mass Trans.* **51** 5204–5208
[23] Alexandrov D V, Aseev D L, Nizovtseva I G, Huang H-N and Lee D 2007 *Int. J. Heat Mass Trans.* **50** 3616–3623
[24] Alexandrov D V, Nizovtseva I G, Malygin A P, Huang H-N and Lee D 2008 *J. Phys.: Condens. Matter* **20** 114105
[25] Worster M G 1986 *J. Fluid Mech.* **167** 481–501
[26] Batchelor G K 1974 *Ann. Rev. Fluid Mech.* **6** 227–255
[27] Huppert H E and Worsster M G 1985 *Nature* **314** 703–707
[28] Morison J, McPhee M, Muench R, Burke S, Coon M, D'Asaro E, Davidson K, Dillon T, Fairall C, Fernando J, Fett R, Francis J, Grenfell T, Jessup A, Lau P, Levine M, Onstott R, Overland J, Paulson C, Perovich D, Pinkel R, Richter-Menge J, Shuchman R, Smith D, Stanton T, Waggoner A, Weaver R and Wettlaufer J 1993 *Eos. Trans. AGU* **74** 393–397
[29] Wettlaufer J S, Worster M G and Huppert H E 2000 *J. Geophys. Res.* **105** 1123–1134
[30] Lee D and Alexandrov D V 2010 *Int. J. Pure Appl. Math.* **58** 381–416
[31] Lee D, Alexandrov D V and Huang H-N 2012 *Numer. Math. Theor. Meth. Appl.* **5** 157–185
[32] Scheil E 1942 *Z. Metall.* **34** 70–72.
[33] Kerr R C, Woods A W and Worster M G 1990 *J. Fluid Mech.* **216** 323–342
[34] Alexandrov D V and Ivanov A A 2009 *Int. J. Heat Mass Trans.* **52** 4807–4811
[35] Alexandrov D V, Ivanov A A and Malygin A P 2009 *Acta Physica Polonica A* **115** 795–799
[36] Alexandrov D V and Malygin A P 2013 *J. Phys. A: Math. Theor.* **46** 455101
[37] Alexandrov D V and Nizovtseva I G 2014 *Proc. R. Soc. A* **470** 20130647
[38] Alexandrov D V and Malygin A P 2014 *Modelling Simul. Mater. Sci. Eng.* **22** 015003
[39] Alexandrov D V 2014 *J. Phys. A: Math. Theor.* **47** 125102
[40] Alexandrov D V 2014 *Phil. Mag. Lett.* **94** 786–793

Acknowledgments

This work was supported by the Russian Science Foundation (grant no. 16-11-10095).

IOP Conf. Series: Materials Science and Engineering **192** (2017) 012004 doi:10.1088/1757-899X/192/1/012004

General set of traveling-wave solutions for amplitude equations in the phase field crystal model

I G Nizovtseva[1,2], P K Galenko[1,2]

[1]Friedrich-Schiller-Universität Jena, Physikalisch-Astronomische Fakultät, D-07743 Jena, Germany

[2]Ural Federal University, Department of Mathematical Physics, Laboratory of Multi-Scale Mathematical Modeling, 620000 Ekaterinburg, Russian Federation

nizovtseva.irina@gmail.com

Abstract. Fronts dynamics of periodic crystalline state, which invades the homogeneous state (liquid phase), are analysed. These fronts are considered as traveling waves of atomic density amplitudes. The propagation of amplitudes is described by the hyperbolic equation of an extended Allen-Cahn type for which the complete set of analytical traveling-wave solutions are obtained by tanh-method. The set of solutions includes previously known traveling waves for the parabolic Allen-Cahn equation of both extended and standard form.

1. Introduction

The phase field crystal model (PFC-model) has been used to examine the dynamics of liquid-solid transformations, grain boundary migration and dislocation motion [1,2]. The PFC model is a continuum model that describes processes on atomic length scales and pattern on the nano- and micro-length scales [3]. This model is characterized by a free energy which is represented by a functional of a conserved atomic density field that is periodic in the solid phase and uniform in a liquid state.

One of the simplest ways to analyze the PFC-model is to use the amplitude equations [4,5,6] which represent smooth profiles over picks of the density field. Taking into account slow and fast degrees of freedom for the crystal-liquid interface propagation, the amplitude equation of the PFC-model is described by the following partial differential equation (PDE) [7]:

$$\frac{\partial u}{\partial t} + \tau\frac{\partial^2 u}{\partial t^2} = \nabla^2 u - \mathrm{K}_0 u + bu^2 - u^3. \quad (1)$$

The following notations are introduced in Eq. (1): $u(\vec{r}, t)$ is the amplitude of atomic density (non-conserved order parameter), $\vec{r}$ is the radius-vector, t is the time, the parameter b can be written as

$$b = \frac{2a}{\sqrt{15v|\Delta B_0|}}, \quad (2)$$

in which the driving force ΔB_0 describes

$$\Delta B_0 = \begin{cases} \Delta B_0 > 0, \; the\ transition\ from\ a\ metastable\ state\ with\ K_0 = +1, \\ \Delta B_0 < 0, \; the\ transition\ from\ an\ unstable\ state\ with\ K_0 = -1, \end{cases} \tag{3}$$

a and v are the coefficients in the free energy density which has the form of Landau-de Gennes potential:

$$f(u) = \frac{K_0 \Delta B_0}{2} u^2 - \frac{2a}{3} u^3 + \frac{15v}{4} u^4. \tag{4}$$

In the equilibrium, the two states (liquid and solid) have the equal energy with the parameter $b = 8a^2 / 135\, v$ and the crystalline front has zero velocity. Finally, it should be noted that Eq. (1) can be considered as an extended Cahn-Allen equation which transforms to its standard form at $b = 0$ and $\tau = 0$ [8] that was suggested for the anti-phase boundary motion and then used in a wide spectrum of mathematical and physical applications [9], for instance, in the description of free-boundary problems by phase-field method [10]. In its complete form, Eq. (1) has been applied in the field of fast phase transitions [11,12], whose validity has been verified by comparison with experimental data [13], in molecular dynamics simulations [14] and by coarse graining derivations of the phase field equations [15]. For the parabolic type of the extended Allen-Cahn equation, i.e., for $\tau = 0$, traveling wave has already been obtained by Wazwaz [16].

Generally, PDE can be analyzed using an important class of traveling wave solutions which, in their particular form, include tanh-functions [16]. Particular solutions of Eq. (1) have also been found in the form of tanh-function [7]. However, so far, by the moment there exists no general set of exact traveling waves for the hyperbolic equation of Allen-Cahn type (1). Therefore, the main purpose of the present work is to find a complete set of traveling waves as the set of exact analytical solutions of Eq. (1). This complete set will be found using the tanh-method [16,17,18], which nowadays represents one of the simplest and the most convenient ways in searching for solutions of traveling waves. As a final result, the complete set of solutions will be checked on the existence of tanh-functions in the obtained traveling waves.

2. The tanh-method and traveling-wave solutions

One of the important solutions for the analysis of phase transformations is related to traveling waves [7,9,16-19]. To treat the non-linear PDEs, the traveling waves are obtained by the first integral method [19,20,21,22] (which can be considered as one of particular cases of the direct method [23], generalizing the use of equivalent methods in finding the exact solutions of PDE, which were reduced to ODE [24]), and also using G'/G-expansion method [25], the rank analytical technique [26,27], and the phase-plane analysis [28,29].

In the present work, we use the tanh-method as a useful tool for the computation of the exact travelling waves by introducing a power series in tanh-function (function of hyperbolic tangent). The efficiency of the tanh-method has been illustrated in Refs. [16,17,18] by applying it for a variety of selected equations, such as nonlinear equations of the Fischer's type and generalized Korteveg-de-Vries-equation (KdV-equation). Moreover, its modification, the tanh-coth-method [18,30], is used to derive the solitons and kink solutions for some of the well-known nonlinear parabolic partial differential equations (the Newell-Whitehead-, Fitz-Hugh-Nagumo-, and Burgers-Fisher-equation). The tanh-coth-method extends a set of the possible solutions and provides abundant solitons and kink solutions in addition to the existing ones. As a result, the power of the tanh-method is confirmed as the most direct and effective algebraic methods [30,31] for finding the exact solutions of nonlinear differential equations.

Let's consider spatially one-dimensional equation (1) for the atomic density amplitude $u(x,t)$, which is evolving in time t along spatial coordinate x. Following Wazwaz [18], we introduce a new independent variable

$$\xi = \frac{x-ct}{\delta}, \tag{5}$$

which describes propagation of the amplitude with the velocity c and transforms the amplitude $u(x,t) \rightarrow U(\xi)$. This transformation re-writes the derivatives as follows:

$$\frac{\partial}{\partial t} = -\frac{c}{\delta}\frac{d}{d\xi}, \quad \frac{\partial^2}{\partial t^2} = \frac{\partial}{\partial t}\left(-\frac{c}{\delta}\frac{d}{d\xi}\right) = \frac{c^2}{\delta^2}\frac{d^2}{d\xi^2}, \tag{6}$$

$$\frac{\partial}{\partial x} = \frac{1}{\delta}\frac{d}{d\xi}, \quad \frac{\partial^2}{\partial x^2} = \frac{1}{\delta^2}\frac{d^2}{d\xi^2}, \quad \frac{\partial^3}{\partial x^3} = \frac{1}{\delta^3}\frac{d^3}{d\xi^3}. \tag{7}$$

Using (6) and (7), spatially one-dimensional equation (1) in the new variable looks like:

$$\frac{c}{\delta}\frac{dU(\xi)}{d\xi} - \tau\frac{c^2}{\delta^2}\frac{d^2U(\xi)}{d\xi^2} + \frac{1}{\delta^2}\frac{d^2U(\xi)}{d\xi^2} - U^3(\xi) - K_o U(\xi) + bU^2(\xi) = 0. \tag{8}$$

To solve (8), we shall apply now the tanh-method [16], introducing the finite expansion:

$$U(\xi) = S(Y) = \sum_{k=0}^{M} a_k Y^k + \sum_{k=1}^{M} b_k Y^{-k}, \tag{9}$$

$$M \in \mathbb{Z}, Y = \tanh(\xi). \tag{10}$$

Using new variable (5), the solution (10) and derivatives,

$$\frac{d}{d\xi} = (1-Y^2)\frac{d}{dY},$$

$$\frac{d^2}{d\xi^2} = (1-Y^2)\left(-2Y\frac{d}{dY} + (1-Y^2)\frac{d^2}{dY^2}\right),$$

one may find that Eq. (8) takes the following form

$$\frac{c}{\delta}(1-Y^2)\frac{dS}{dY} + \frac{1}{\delta^2}(1-\tau c^2)\left[(1-Y^2)(-2Y\frac{dS}{dY} + (1-Y^2)\frac{d^2S}{dY^2})\right] - S^3 - K_o S + bS^2 = 0. \tag{11}$$

The parameter M from Eq. (8) can be determined using the analysis of Wazwaz [16]. Indeed, balancing the linear terms of the highest order with the highest order nonlinear terms in Eq. (11) one can get $3M = 4 + M - 2$, therefore, $M = 1$, so the expansion (9) becomes

$$S(Y) = a_0 + a_1 Y, \tag{12}$$

with the following derivatives

$$S'(Y) = \sum_{k=0}^{M}(a_k Y^k)' = a_1, S''(Y) = 0. \tag{13}$$

Now, opening the braces in Eq. (11), using Eqs. (12) and (13), we collect the coefficients of powers of Y^n in the resulting equations as follows

$$\left(\frac{2}{\delta^2}(1-\tau c^2)a_1 - a_1{}^3\right)Y^3 - \left(\frac{c}{\delta}a_1 + 3a_0a_1{}^2\right)Y^2 + \left(\frac{2}{\delta^2}(1-\tau c^2)a_1 - 3a_0{}^2a_1 - K_o a_1 + 2ba_0a_1\right)Y$$
$$+\left(a_1\frac{c}{\delta} - a_0{}^3 - a_0K_o + ba_0{}^2\right)Y^0 = 0. \quad (14)$$

Equation (14) has a solution if the braces ahead of Y^k are placed to zero. Thus, the following system of equations for the parameters a_k, k=0..*M*, c and δ is obtained:

$$Y^3: \frac{2}{\delta^2}(1-\tau c^2)a_1 - a_1{}^3 = 0, \quad (15)$$

$$Y^2: -\frac{c}{\delta}a_1 - 3a_0a_1{}^2 + ba_1{}^2 = 0, \quad (16)$$

$$Y^1: -\frac{2}{\delta^2}(1-\tau c^2)a_1 - 3a_0{}^2a_1 - K_o a_1 + 2a_0a_1b = 0, \quad (17)$$

$$Y^0: a_1\frac{c}{\delta} - a_0{}^3 - a_0K_o + ba_0{}^2 = 0. \quad (18)$$

The system of equations (15)-(18) has a trivial solution a_1=0 and $a_0 = b/2\,(1 \pm \sqrt{1-4K_0/b^2})$ with the arbitrary values of c and δ. In this case, the amplitude has constant profile $u(x,t) = a_0$. This homogeneous solution has no interest for us because we are looking for amplitude's profiles of atomic density, which are moving through metastable/unstable homogeneous state (liquid phase). In the case $a_1 \neq 0$, equations (15)-(18) look like:

$$a_1{}^2 = \frac{2}{\delta^2}(1-\tau c^2), \quad (19)$$

$$a_1 = \frac{\frac{c}{\delta}}{b-3a_0}, \quad (20)$$

$$3a_0{}^2 - 2a_0b + \frac{2}{\delta^2}(1-\tau c^2) + K_o = 0, \quad (21)$$

$$a_0{}^3 - ba_0{}^2 + a_0K_o - a_1\frac{c}{\delta} = 0. \quad (22)$$

Determination of the parameters a_0, a_1, c and δ from Eqs. (19)-(22) leads us to the amplitude profiles of the form:

$$u(x,t) = a_0 + a_1 \tanh\left[\frac{x-ct}{\delta}\right]. \quad (23)$$

Concrete values for a_0, a_1, c and δ represent different types of solutions, that is shown in the next two sections.

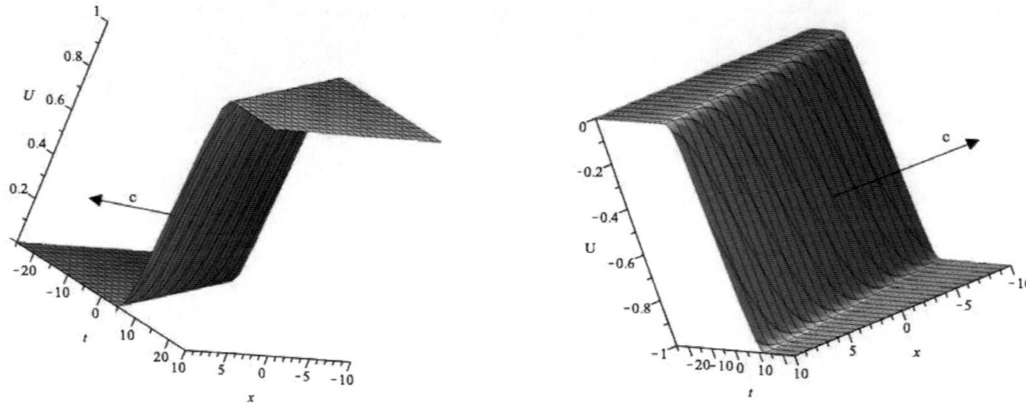

(a) (b)

Figure 1. Smooth and continuous $\varphi -$ profiles calculated with the usage of *Solutions 3* and *4* from Table 1. Calculation were developed for fixed $\tau = 0.5$ and for the values of a_0, a_1 and δ given by $K_o = 1$ and under the condition of $b^2 \geq 4K_0$ as from Eq. (2). (a) $\varphi -$ profile moves in direction of the x –axis with the constant negative velocity *c*. (b) $\varphi -$ profile moves in direction of the x –axis with the constant positive velocity *c*.

3. Set of solutions

A complete set of solutions consists of 12 decisions for the parameters K_o and b, which are defined by Eqs. (2) and (3). This number of decisions follows from the degrees of Eqs. (19)-(22), where each (19) and (21) assume the existence of two roots (4 roots in total), multiplied by 3 decisions from the cubic equation (22). These 12 decisions can be divided into three sets, every one of which contains 4 similar by notation type of solution. All the coefficients from Eq. (23) are obtained for the following far field boundary conditions, $\xi \to \pm\infty: u \equiv const$, namely, for $u = 0$ or $u = \pm 1$.

3.1. Solutions 1-4

The first set of solutions can be recognized by the signs of the parameters a_1, c. As a result, solutions 1-4 are presented in Table 1.

Table 1. First set of solutions: solutions 1-4.

Values	Solutions 1 and 2	Solutions 3 and 4
a_0	$\frac{1}{2}b$	$\frac{1}{2}b$
a_1	$\pm\frac{1}{2}\sqrt{b^2 - 4K_0}$	$\frac{1}{2}\sqrt{b^2 - 4K_0}$
δ	$\frac{4}{\sqrt{2b^2 - 8K_0 + b^4\tau - 4b^2\tau K_0}}$	$-\frac{4}{\sqrt{2b^2 - 8K_0 + b^4\tau - 4b^2\tau K_0}}$
c	$\frac{b\sqrt{b^2 - 4K_0}}{\sqrt{2b^2 - 8K_0 + b^4\tau - 4b^2\tau K_0}}$	$\pm\frac{b\sqrt{b^2 - 4K_0}}{\sqrt{2b^2 - 8K_0 + b^4\tau - 4b^2\tau K_0}}$

With $b = 0$ and $K_0 = -1$, table 1 shows that $a_0 = 0, a_1 = \pm 1, \delta = \pm\sqrt{2}$ and $c = 0$. In this particular case, Eq. (23) predicts stationary profiles:

$$u(x,t) = \pm\tanh\left(\mp\frac{x}{\sqrt{2}}\right). \tag{24}$$

This profile is consistent with the steady solution of hyperbolic Allen-Cahn equation ($\tau \neq 0$) and parabolic Allen-Cahn equation ($\tau = 0$), which are obtained from (1) for the above accepted parameters $b = 0$ and $K_o = -1$.

3.2. *Solutions 5-8*

The second group of solution consist of solutions 5-8. It has the similar structure. In order to simplify the representation, we shall introduce the following notations for this set of solutions:

$$A_{5-8} = \Big[-\left(10b^2 + 6b\sqrt{b^2-4K_0} + 162\tau K_0{}^2 - 36K_0 - 72b^2\tau K_0 + 8b^4\tau\right)\cdot \left(8b^2K_0 - b^4 - 6bK_0\sqrt{b^2-4K_0} + b^3\sqrt{b^2-4K_0} - 18K_0{}^2\right)\Big]^{1/2}, \tag{25}$$

$$B_{5-8} = 5b^2 + 3b\sqrt{b^2-4K_0} + 81\tau K_0{}^2 - 18K_0 - 36b^2\tau K_0 + 4b^4\tau, \tag{26}$$

$$G_{5-8} = \sqrt{2b^2 + 2b\sqrt{b^2-4K_0} - 4K_0}. \tag{27}$$

Now using the parameters (25) - (27), we rewrite Solutions 5-8 in new designations:

$$a_0 = \frac{1}{4}b + \frac{1}{4}\sqrt{b^2-4K_0}, \quad a_1 = \frac{1}{4}G_{5-8}, \quad \delta = \frac{2A_{5-8}}{B_{5-8}K_0}, \quad c = -\frac{1}{4}\left(-\frac{1}{2}b + \frac{3}{2}\sqrt{b^2-4K_0}\right)G_{5-8}\frac{A_{5-8}}{B_{5-8}K_0}, \tag{28}$$

$$a_0 = \frac{1}{4}b + \frac{1}{4}\sqrt{b^2-4K_0}, \quad a_1 = \frac{1}{4}G_{5-8}, \quad \delta = -\frac{2A_{5-8}}{B_{5-8}K_0}, \quad c = \frac{1}{4}\left(-\frac{1}{2}b + \frac{3}{2}\sqrt{b^2-4K_0}\right)G_{5-8}\frac{A_{5-8}}{B_{5-8}K_0}, \tag{29}$$

$$a_0 = \frac{1}{4}b + \frac{1}{4}\sqrt{b^2-4K_0}, \quad a_1 = -\frac{1}{4}G_{5-8}, \quad \delta = \frac{2A_{5-8}}{B_{5-8}K_0}, \quad c = \frac{1}{4}\left(-\frac{1}{2}b + \frac{3}{2}\sqrt{b^2-4K_0}\right)G_{5-8}\frac{A_{5-8}}{B_{5-8}K_0}, \tag{30}$$

$$a_0 = \frac{1}{4}b + \frac{1}{4}\sqrt{b^2-4K_0}, \quad a_1 = -\frac{1}{4}G_{5-8}, \quad \delta = -\frac{2A_{5-8}}{B_{5-8}K_0}, \quad c = -\frac{1}{4}\left(-\frac{1}{2}b + \frac{3}{2}\sqrt{b^2-4K_0}\right)G_{5-8}\frac{A_{5-8}}{B_{5-8}K_0}. \tag{31}$$

3.3. Solutions 9-12

The third group consists of Solutions 9-12. It has the similar structure to the first group once again and is also similar
to the previous group of Solutions 5-8. Introducing the parameters,

$$A_{9-12} = \left[-\left(-10b^2 + 6b\sqrt{b^2-4K_0} - 162\tau{K_0}^2 + 36K_0 + 72b^2\tau K_0 - 8b^4\tau\right)\cdot\left(-8b^2K_0 + b^4 - 6bK_0\sqrt{b^2-4K_0} + b^3\sqrt{b^2-4K_0} + 18{K_0}^2\right)\right]^{1/2}, \tag{32}$$

$$B_{9-12} = -5b^2 + 3b\sqrt{b^2-4K_0} - 81\tau{K_0}^2 + 18K_0 + 36b^2\tau K_0 - 4b^4\tau, \tag{33}$$

$$G_{9-12} = \sqrt{2b^2 - 2b\sqrt{b^2-4K_0} - 4K_0}, \tag{34}$$

we, finally, rewrite parameters for a_0, a_1, c and δ for Solution 9-12 in the following form

$$a_0 = \frac{1}{4}b - \frac{1}{4}\sqrt{b^2-4K_0}, \quad a_1 = \frac{1}{4}G_{9-12}, \quad \delta = -\frac{2A_{9-12}}{BK_0}, \quad c = \frac{1}{4}\left(-\frac{1}{2}b - \frac{3}{2}\sqrt{b^2-4K_0}\right)G_{9-12}\frac{A_{9-12}}{BK_0}, \tag{35}$$

$$a_0 = \frac{1}{4}b - \frac{1}{4}\sqrt{b^2-4K_0}, \quad a_1 = \frac{1}{4}G_{9-12}, \quad \delta = \frac{2A_{9-12}}{BK_0}, \quad c = -\frac{1}{4}\left(-\frac{1}{2}b - \frac{3}{2}\sqrt{b^2-4K_0}\right)G_{9-12}\frac{A_{9-12}}{BK_0}, \tag{36}$$

$$a_0 = \frac{1}{4}b - \frac{1}{4}\sqrt{b^2-4K_0}, \quad a_1 = -\frac{1}{4}G_{9-12}, \quad \delta = -\frac{2A_{9-12}}{BK_0}, \quad c = -\frac{1}{4}\left(-\frac{1}{2}b - \frac{3}{2}\sqrt{b^2-4K_0}\right)G_{9-12}\frac{A_{9-12}}{BK_0}, \tag{37}$$

$$a_0 = \frac{1}{4}b - \frac{1}{4}\sqrt{b^2-4K_0}, \quad a_1 = -\frac{1}{4}G_{9-12}, \quad \delta = \frac{2A_{9-12}}{BK_0}, \quad c = \frac{1}{4}\left(-\frac{1}{2}b - \frac{3}{2}\sqrt{b^2-4K_0}\right)G_{9-12}\frac{A_{9-12}}{BK_0}. \tag{38}$$

4. Particular solutions

For the amplitude's equation (8) the whole set of 12 solutions of the form of Eq. (23) is presented by the coefficients summarized in Table 1, Eqs. (28) -(31) and Eqs. (35) -(38). Now, we compare special and particular cases of these solutions with corresponding results, which were obtained earlier.

Using the first integral method [19], traveling wave solutions have been obtained for the hyperbolic Allen-Cahn equation [20]. This equation is consistent with Eq. (1), if $b = 0$ and $K_0 = -1$. Indeed, if we substitute these values for b and K_0 into solutions (28)-(31), then solutions of the form (23) corresponds to those ones obtained in [20]. The graphical representation for this particular case is shown in Fig. 1, which gives a view for the atomic density profile, that invades the homogeneous phase with positive and negative values of c. Another pair of solutions could be obtained for $a_0 = a_1 = \pm 0.5$. Thus, in general, we have obtained 4 bounded solutions, which correspond to 4 bounded solutions of Ref. [20]. It should be noticed, that in [20] another 4 unbounded solutions were obtained. These solutions were extracted by us in [20] from the general set of solutions due to its physical and mathematical insolvency, namely, due to the absence of the physical sence and the violation of the mathematical problem statement. In the current work we do not obtain the unbounded solutions since tanh-method build (9)-(10) on the bounded set *a priori*. Therefore, solution (23) is always bounded.

With the zero relaxation time, namely $\tau = 0$, hyperbolic equation (1) transforms into parabolic partial differential type whose traveling wave solution has been previously found by Wazwaz [18]. Indeed, as it follows from our solutions (35)-(38), if we use $\tau = 0$ and take into account (32)-(34), solutions of [18] are covered for the extended Allen-Cahn equation.

In general, we have obtained traveling wave solutions represented by hyperbolic tanh-functions (23) that confirms the correctness of the particular solutions for the dynamical problem of fast diffuse interfaces [32-33].

5. Conclusions

We considered the atomic density amplitudes which represented by an extended hyperbolic Allen-Cahn equation (1). Using the tanh-method [16-18,30,31], we obtained traveling wave solutions for Eq. (1) as kink-profiles (step-profiles), which invade metastable or unstable homogeneous states (liquid phase). The kink-profiles are described by tanh-functions (23), which in its order confirms the correctness of the particular solutions [32,33], which were chosen for equations from the fields of rapid solidification and fast transformations. We have shown, that the presently obtained solutions indeed include as particular case the previously known sets of traveling waves, which have been found for (*i*) the extended parabolic Allen-Cahn equation [16] and (*ii*) the hyperbolic Allen-Cahn equation with a standard free energy density describing only the transitions from the unstable state [20].

6. References

[1] Elder K R and Grant M 2004 Modeling elastic and plastic deformations in nonequilibrium processing using phase field crystals *Phys. Rev. E* **70** 051605-1-18

[2] Elder K R, Provatas N, Berry J, Stefanovich P and Grant M 2007 Phase-field crystal modeling and classical density functional theory of freezing *Phys. Rev. B* **75** 064107-1-14

[3] Elder K R, Rossi G, Kanerva P, Sanches F, Ying S.-C, Granato E, Achim C V and Ala-Nissila T 2012 Patterning of heteroepitaxial overlayers from nano to micron scales *Phys. Rev. Lett.* **108** 226102-1-5

[4] Goldenfeld N, Athreya B P and Danzig J A 2005 Renormalization group approach to multiscale simulation of polycrystalline materials using the phase field crystal model *Phys. Rev. E* **72** 020601(R)-1-4

[5] Athreya B P, Goldenfeld N and Danzig J A 2006 Renormalization-group theory for the phase-field crystal equation 2006 *Phys. Rev. E* **74** 011601-1-13

[6] Elder K R, Huang Zh F and Provatas N 2010 Amplitude expansion of the binary phase-field-crystal model *Phys. Rev. E* **81** 011602-1-11

[7] Galenko P K, Sanches F I and Elder K R 2015 Traveling wave profiles for a crystaline front invading liquid states *Physica D* **308** 1-10

[8] Allen S M and Cahn J W 1979 A microscopic theory for antiphase boundary motion and its application to antiphase domain coarsening *Acta Metall.* **27** 1085-1095

[9] van Saarlos W 2003 Front propagation into unstable states *Phys. Rep.* **386** 29-222

[10] Provatas N and Elder K 2010 Phase-Field Methods in Materials Science and Engineering (Weinheim: Wiley-VCH)

[11] Galenko P and Jou D 2005 Diffuse-interface model for rapid phase transformations in nonequilibrium systems *Phys. Rev. E* **71** 046125-1-13

[12] Galenko P K, Danilov D A, and Lebedev V G 2009 Phase-field-crystal and Swift-Hohenberg equations with fast dynamics *Phys. Rev. E* **75** 051-110

[13] Galenko P K and Herlach D M 2006 *Phys. Rev. Lett.* **96** 150602

[14] Yang Y, Humadi H, Buta D, Laird B B, Sun D, Hoyt J J and Asta M 2011 Atomistic simulations of nonequilibrium crystal-growth kinetics from alloy melts *Phys. Rev. Lett.* **107** 025505

[15] Jou D and Galenko P 2013 Coarse graining for the phase-field model of fast phase transitions

Phys. Rev. E **88** 042151
[16] Wazwaz A M 2004 The tanh method for travelling wave solutions of nonlinear equations *App. Math. Comp.* **154** 713-23
[17] Fan E, Hon Y C 2002 Generalized tanh method extended to special types of nonlinear equations *Z. Naturforsch* **57a** 692-700
[18] Wazwaz A M 2007 The tanh-coth method for solitons and kink solutions for nonlinear parabolic equations *App. Math. Comp.* **188** 1467-75
[19] Feng Zh 2002 The first-integral method to study the Burgers-Korteweg-de Vries equation *J. Phys. A: Math. Gen.* **35** 343–349
[20] Nizovtseva I G, Galenko P K and Alexandrov D V 2016 The hyperbolic Allen-Cahn equation: exact solutions *J. Phys. A: Math. Theor.* **49** 435201-1-14
[21] Lu B, Zhang HQ and Xie F D 2010 Traveling wave solutions of nonlinear partial differential equations by using the first integral method *App. Math. Comp.* **216** 1329-36
[22] Feng Zh and Wang X 2003 The first integral method to the two-dimensional Burgers--Korteweg-de Vries equation *Physics Letters A* **308** 173-78
[23] Fan E. 2002 Multiple travelling wave solutions of nonlinear evolution equations using a unified algebraic method *J. Phys. A: Math. Gen.* **35** 6853-72
[24] Kudryashov N A 2009 Seven common errors in finding exact solutions of nonlinear differential equations *Commun. Nonlinear Sci. Numer. Simulat.* **14** 3507-29
[25] Kim H and Sakthivel R 2010 Travelling wave solutions for time-delayed nonlinear evolution equations *App. Math. Lett.* **23** 527-32
[26] Feng X 2000 Exploratory approach to explicit solution of nonlinear evolution equations *Int. J. Theor. Phys.* 2000 **39(1)** 207-22
[27] Hu J and Zhang H 2001 A new method for finding exact traveling wave solutions to nonlinear partial differential equations *Physics Letters A* **286(2-3)** 175-79
[28] Rehman T, Gambino G and Choudhury S R 2013 Smooth and non-smooth traveling wave solutions of some generalized Camassa-Holm equations *Commun. Nonlinear Sci. Numer. Simulat.* **19(6)** 1746-69
[29] Malfliet W 1992 Solitary wave solutions of nonlinear wave equations *Am. J. Phys.* **60(7)** 650-654
[30] Wazwaz A M 2008 The extended tanh method for new compact and noncompact solutions for the KP-BBM and the ZK-BBM equations *Chaos, Solitons and Fractals* **38(5)** 1505-16
[31] Khalter A H, Malfliet W, Callebaut D K and Kamel E S 2002 The tanh method, a simple transformation and exact analytical solutions for nonlinear reaction-diffusion equations *Chaos Soliton. Fract.* **14** 513-522
[32] Galenko P K, Abramova E V, Jou D, Danilov D A, Lebedev V G and Herlach D M 2011 Solute trapping in rapid solidification of a binary dilute system: a phase-field study *Phys. Rev. E.* **84** 041143-1-17
[33] Salhoumi A and Galenko P K 2016 Gibbs-Thomson condition for the rapidly moving interface in a binary system, *Physica A* **447** 161-171

Acknowledgements
The authors acknowledge the support from the Russian Science Foundation (project no. 16-11-10095). I.G. Nizovtseva specially acknowledges the support of Alexander von Humboldt Foundation (ID 1160779) and Council on grants and scholarships of the President of the Russian Federation.

IOP Conf. Series: Materials Science and Engineering **192** (2017) 012005 doi:10.1088/1757-899X/192/1/012005

Stabilization of stochastic operating modes in the flow reactor

I Bashkirtseva

Ural Federal University, Lenina, 51, Ekaterinburg, 620000, Russia

E-mail: irina.bashkirtseva@urfu.ru

Abstract. A conceptual multistable model of the flow reactor proposed by Volter and Salnikov is studied in the presence of random disturbances. A stochastic excitability of this system, even in a zone of stable equilibria, is explained by the high stochastic sensitivity. Such high stochastic sensitivity of the equilibrium can be a basic reason of the noise-induced destruction of the operating mode. For the stabilization of the stochastic flow reactor, a theory of the synthesis of the assigned low stochastic sensitivity is suggested. It is shown how to use this control approach and construct an appropriate regulator which suppresses unwanted large-amplitude stochastic oscillations and provide a proper operation of the flow reactor.

1. Introduction

Many manufacturing processes are described by nonlinear dynamic systems with equilibrium operating modes. Traditionally, a deterministic stability of the equilibrium is considered as a condition of the proper operation. However, such stability can be insufficient, especially in nonlinear systems. In excitable systems, even weak noise can destroy a stable operating mode, and cause unexpected consequences [1, 2]. In the study of the probabilistic reasons of noise-induced phenomena, the stochastic sensitivity analysis [3, 4] can be a useful tool.

Along with the analysis of nonlinear stochastic systems, control problems attract attention of many researchers [5-8]. A theory of the control based on the stochastic sensitivity synthesis was developed in [8-11].

Mathematical models of flow chemical reactors demonstrate a wide variety of the nonlinear dynamic regimes with complex oscillations [12-14]. In the present paper, we consider a conceptual dynamic model of the flow reactor proposed by Volter and Salnikov [15] in a presence of random disturbances.

In Section 2, dynamic regimes of the deterministic model are shortly discussed. In Section 3, an influence of the random noise on this model is studied. A stochastic excitability in a zone of stable equilibria is shown. For the stabilization of the stochastic flow reactor, a general mathematical approach based on the stochastic sensitivity synthesis is briefly presented in Section 4. In Section 5, it is shown how to apply this approach to the synthesis of the appropriate regulator which reduces the stochastic sensitivity and suppresses unwanted large-amplitude stochastic oscillations in the randomly forced flow reactor.

2. Deterministic model of the flow chemical reactor

Consider the following dynamic model of the thermochemical processes in the flow reactor [15] of ideal mixing:

$$\begin{aligned} \dot{x} &= -x \exp\left(-\frac{1}{y}\right) + l(a - x), \\ \dot{y} &= x \exp\left(-\frac{1}{y}\right) + m(b - y). \end{aligned} \tag{1}$$

Here, x is a concentration of the reagent, and y is a temperature in the reactor. Parameters a and b characterize the concentration and temperature at the reactor inlet; l, m are positive parameters.

Following [15] we fix $l = 0.5, m = 0.25, b = 0.165$ and study the system dynamics under the variation of the parameter $a \in [1.578, 1.584]$. In this zone, the system (1) exhibits three dynamic regimes: for $a < a_1 = 1.580079$, the system is monostable with an equilibrium as a single attractor; for $a_1 < a < a_2 = 1.582843$, the system is bistable with coexisting equlibrium and limit cycle; for $a > a_2$, the system is monostable with a limit cycle as a single attractor. The critical point a_1 corresponds to the saddle-node bifurcation, and a_2 is a point of the subcritical Hopf bifurcation. In Fig. 1, y-coordinates of attractors and repellers of system (1) are presented.

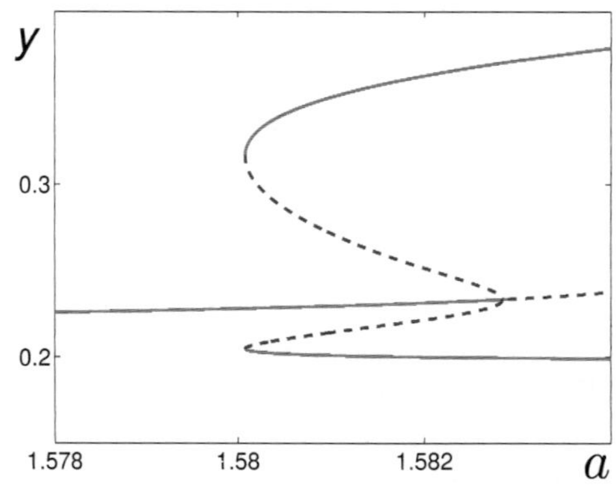

Figure 1. Bifurcation diagram of the deterministic system (1). Stable equilibria and cycles are plotted by solid lines, and unstable equilibria and cycles are shown by dashed lines.

3. Model of the flow reactor with stochastic disturbances

To study an influence of noise on the dynamics of the reactor, we will use the following stochastic model:

$$\begin{aligned} \dot{x} &= -x \exp\left(-\frac{1}{y}\right) + l(a - x) + \varepsilon_1\xi_1(t), \\ \dot{y} &= x \exp\left(-\frac{1}{y}\right) + m(b - y) + \varepsilon_2\xi_2(t). \end{aligned} \tag{2}$$

Here, $\xi_{1,2}(t)$ are the standard Gaussian uncorrelated processes, and $\varepsilon_{1,2}$ are the noise intensities. In what follows, we put $\varepsilon_1 = \varepsilon_2 = \varepsilon$.

Random disturbances deform deterministic dynamics of the system and can destroy a required equilibrium operation mode of the chemical process.

In Fig. 2, a response of the system (2) to random disturbances is presented for two values of the parameter a corresponding to the equilibrium ($a = 1.58$) and self-oscillations ($a = 1.583$) regimes.

As one can see in Fig. 2a, for $\varepsilon = 0.0001$, random trajectories are concentrated near the stable equilibrium, and the system exhibits the small-amplitude stochastic oscillations. Such inevitable small oscillations near the stable equilibrium are allowable. For $\varepsilon = 0.0005$, dynamics of the system (2) essentially changes. Indeed, for this noise intensity, the system begins to demonstrate a new excitable regime with the intermittency of small- and large-amplitude stochastic oscillations. Such breakdowns of the operating mode are impermissible, and require additional stabilization mechanisms.

Note that usually regulators are built for the stabilization of unstable equilibria. However, as we see in this system, a standard deterministic stability of the equilibrium is not enough. In the excitable case presented here, it is required to decrease the stochastic sensitivity of the equilibrium.

More expected reaction of the system in self-oscillation mode is shown in Fig. 2b for $\varepsilon = 0.0005$. Here, the large-amplitude stochastic oscillations are observed. In this case, a regulator has to solve two problems: to stabilize an unstable equilibrium, and provide a low level of its stochastic sensitivity.

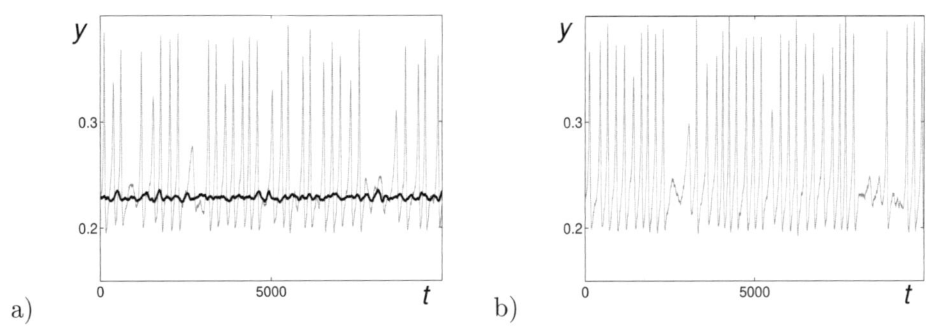

Figure 2. Time series of the uncontrolled stochastic system with a) $a = 1.58$, $\varepsilon = 0.0001$ (black) and $\varepsilon = 0.0005$ (grey); b) $a = 1.583$, $\varepsilon = 0.0005$.

4. Controlling stochastic sensitivity

Consider a general stochastic system with control

$$\dot{x} = f(x, u(x)) + \varepsilon\sigma(x, u(x))\xi(t), \tag{3}$$

where x is an n-dimensional state, u is an l-dimensional control, $f(x, u)$ is an n-vector-function, $\xi(t)$ is an m-dimensional standard Gaussian process, and ε is the noise intensity. The $n \times m$-matrix-function $\sigma(x, u)$ characterizes a dependence of disturbances on state and control.

It is supposed that the unforced and uncontrolled system (3) (with $\varepsilon = 0$, $u = 0$) has an equilibrium $\bar{x}$. The stability of $\bar{x}$ is not required.

Consider a set $\mathcal{U}$ of admissible feedbacks $u = u(x)$ satisfying the following conditions: (a) a function $u(x)$ is continuously differentiable and $u(\bar{x}) = 0$; (b) a feedback $u(x)$ provides an exponential stability of the equilibrium $\bar{x}$ for the closed-loop deterministic system

$$\dot{x} = f(x, u(x)) \tag{4}$$

in some neighbourhood of $\bar{x}$.

For deviations $z(t) = x(t) - \bar{x}$ of states $x(t)$ of system (4) from the equilibrium $\bar{x}$, consider the following first approximation system:

$$\dot{z} = (F + BK)z, \qquad F = \frac{\partial f}{\partial x}(\bar{x}, 0), \quad B = \frac{\partial f}{\partial u}(\bar{x}, 0), \quad K = \frac{\partial u}{\partial x}(\bar{x}). \tag{5}$$

The second condition (b) is equivalent to the exponential stability of the trivial solution of system (5). Here, we consider the following linear feedback regulators:

$$u(x) = K(x - \bar{x}). \tag{6}$$

Denote by $\mathbf{K}$ a set of matrices K which provide an exponential stability of the trivial solution of system (5):

$$\mathbf{K} = \{K|\ Re\lambda_i(F + BK) < 0\}.$$

Here, $\lambda_i(F + BK)$ are the eigenvalues of the matrix $F + BK$. Suppose that the pair (F, B) is stabilizable, so the set $\mathbf{K}$ and class $\mathcal{U}$ are not empty.

Random trajectories $x^\varepsilon(t)$ of the stochastic system (3) deviate from the equilibrium $\bar{x}$. However, using the feedback (6) with a matrix $K \in \mathbf{K}$, one can provide an exponential stability of the equilibrium, localize random states of system (3),(6) near the equilibrium $\bar{x}$, and form a stationary distributed solution $\bar{x}^\varepsilon(t)$.

For small deviations $z(t) = x^\varepsilon(t) - \bar{x}$, the following first approximation stochastic system can be written

$$\dot{z} = (F + BK)z + \varepsilon G\xi(t), \quad G = \sigma(\bar{x}, 0). \tag{7}$$

The variable $y = \frac{z}{\varepsilon}$ characterizes the sensitivity of the solution of system (7) to noise with the intensity ε. For the covariance matrix $V(t) = \mathrm{cov}(y(t), y(t))$, one can write the equation

$$\dot{V} = (F + BK)V + V(F + BK)^\top + S, \ \ S = GG^\top. \tag{8}$$

For any $K \in \mathbf{K}$, this equation has a unique stationary matrix solution W satisfying the following algebraic equation

$$(F + BK)W + W(F + BK)^\top + S = 0. \tag{9}$$

For non-singular noises ($\det S \neq 0$), the solution W of equation (9) is positive definite.

Any solution $V(t)$ of system (8) converges to the corresponding solution W of system (9)

$$\lim_{t\to\infty} V(t) = W.$$

It holds that

$$\mathrm{cov}(\bar{x}^\varepsilon(t), \bar{x}^\varepsilon(t)) \approx \varepsilon^2 W,$$

where $\mathrm{cov}(\bar{x}^\varepsilon(t), \bar{x}^\varepsilon(t))$ is a covariance matrix of solutions $\bar{x}^\varepsilon$ of system (3). So, the matrix W is a simple quantitative characteristic of a response of the nonlinear system (3) to small noises. The matrix W is called a *stochastic sensitivity matrix* of the equilibrium $\bar{x}$. The control of the dispersion of random states around the equilibrium can be realized as a synthesis of the assigned stochastic sensitivity matrix W.

Denote by $\mathbf{M}$ a set of symmetric and positive definite $n \times n$-matrices. For any $K \in \mathbf{K}$, the regulator (6) forms a stable equilibrium of system (3) with the corresponding stochastic sensitivity matrix W_K which is a solution of the equation (9). Consider the following control problem.

Synthesis of stochastic sensitivity

For the assigned matrix $W \in \mathbf{M}$, it is necessary to find a matrix $K \in \mathbf{K}$ guaranteeing the equality $W_K = W$, where W_K is a solution of equation (9).

Consider important notions of an attainability. The element $W \in \mathbf{M}$ is said to be attainable if the equality $W_K = W$ is true for some $K \in \mathbf{K}$. The attainability set

$$\mathbf{W} = \{W \in \mathbf{M} \mid \exists K \in \mathbf{K} \quad W_K = W\}$$

consists of all attainable elements.

An equilibrium $\bar{x}$ is completely stochastic controllable if

$$\forall\, W \in \mathbf{M} \quad \exists\, K \in \mathbf{K} : \quad W_K \equiv W.$$

So, the condition of the complete stochastic controllability of the equilibrium $\bar{x}$ can be written as $\mathbf{W} = \mathbf{M}$.

To describe the attainability set, consider a connection between the assigned matrix W and the feedback coefficient K:

$$BKW + WK^\top B^\top + H(W) = 0, H(W) = FW + WF^\top + S. \tag{10}$$

The following Theorem gives a solution for the problem of the synthesis of the assigned stochastic sensitivity matrix W [9].

Theorem.

Let the noise be non-singular ($\det S \neq 0$).

(a) If the matrix B is quadratic and non-singular then $\mathbf{W} = \mathbf{M}$ and for any matrix $W \in \mathbf{M}$

$$K = \bar{K} + B^{-1}ZW^{-1} \in \mathbf{K}, \qquad \bar{K} = -B^{-1}\left(\frac{1}{2}SW^{-1} + F\right),$$

where Z is an arbitrary skew-symmetric $n \times n$-matrix.

(b) If $\operatorname{rank}(B) < n$ then the element $W \in \mathbf{M}$ is attainable if and only if the matrix W is a solution of the equation

$$P_2 H(W) P_2 = 0. \tag{11}$$

Under these conditions for any matrix $W \in \mathbf{M}$

$$K = \bar{K} + C \in \mathbf{K}, \qquad \bar{K} = B^{+}H(W)\left(\frac{1}{2}P_1 - I\right)W^{-1},$$

where C is an arbitrary $l \times n$-matrix satisfying the condition

$$BCW + WC^\top B^\top = 0.$$

Here, $P_1 = BB^{+}$ and $P_2 = I - P_1$ are projective matrices, an the sign "+" means a pseudo-inversion.

Further, we apply this theory to the stabilization of operating mode for the stochastic flow reactor.

IOP Conf. Series: Materials Science and Engineering **192** (2017) 012005 doi:10.1088/1757-899X/192/1/012005

5. Stabilization of the stochastic flow reactor

As shown above, the system (4) is excitable, and the random noise can generate large-amplitude stochastic oscillations and destroy a normal operation of the system. So, it is important to suppress such oscillations and return the system to normal operating mode with acceptable small-amplitude oscillations around the equilibrium.

Consider now the system (4) with additional control inputs:

$$\begin{aligned} \dot{x} &= -x \exp\left(-\frac{1}{y}\right) + l(a-x) + u_1 + \varepsilon_1\xi_1(t), \\ \dot{y} &= x \exp\left(-\frac{1}{y}\right) + m(b-y) + u_2 + \varepsilon_2\xi_2(t). \end{aligned} \tag{12}$$

A stochastic sensitivity of the equilibrium $(\bar{x}, \bar{y})$ of the uncontrolled system (2) is characterized by the 2×2-matrix $W = \begin{pmatrix} w_{11} & w_{12} \\ w_{21} & w_{22} \end{pmatrix}$. This symmetric positive definite matrix is a unique solution of the matrix equation (9) with $K = 0$, where for system (2) we have

$$F = \begin{pmatrix} f_{11} & f_{12} \\ f_{21} & f_{22} \end{pmatrix}, \quad S = \begin{pmatrix} 1 & 0 \\ 0 & 1 \end{pmatrix},$$

$$f_{11} = -\exp\left(-\frac{1}{\bar{y}}\right) - l, \; f_{12} = -\bar{x}\exp\left(-\frac{1}{\bar{y}}\right)\frac{1}{\bar{y}^2}, \; f_{21} = \exp\left(-\frac{1}{\bar{y}}\right), \; f_{22} = \bar{x}\exp\left(-\frac{1}{\bar{y}}\right)\frac{1}{\bar{y}^2} - m.$$

For $a = 1.58$ considered above, elements of the stochastic sensitivity matrix W are as follows:

$$w_{11} = 1.0947 \cdot 10^4, \quad w_{12} = w_{21} = -2.2545 \cdot 10^3, \quad w_{22} = 5.2001 \cdot 10^2.$$

The large values of elements of the matrix W are the main reason of the stochastic excitability with the generation of large-amplitude oscillations (see Fig. 2a).

To stabilize an operating mode of the stochastically forced reactor, we will use the regulator

$$u_1 = k_{11}(x - \bar{x}) + k_{12}(y - \bar{y}), \qquad u_2 = k_{21}(x - \bar{x}) + k_{22}(y - \bar{y}), \tag{13}$$

which provides much smaller values of the stochastic sensitivity. Here, for example, we restrict ourselves by the diagonal stochastic sensitivity matrices $W = \begin{pmatrix} w & 0 \\ 0 & w \end{pmatrix}$.

As it follows from the Theorem, the coefficients of the regulator (13) which synthesizes such stochastic sensitivity matrix W with any assigned $w > 0$ can be found explicitly. The results of the control based on this stochastic sensitivity synthesis are shown in Fig. 3.

For the value $a = 1.58$, the time series of the controlled stochastic system with $\varepsilon = 0.0005$ are shown in Fig. 3a. Here, the control providing $w = 10$, is switched on at $t = 5000$. As one can see, this regulator returns the system to the operating mode with small-amplitude oscillations. Analogous results are obtained for $a = 1.583$ (see Fig. 3b).

Note that a decrease of the control parameter w allows us to decrease a dispersion of these small-amplitude oscillations (compare time series in Fig. 4 for $w = 0.1$ and $w = 10$).

6. Conclusion

In the present paper, we have demonstrated how to stabilize stochastic operation modes of the flow reactor with the help of the appropriate feedback regulator. We have shown that the high stochastic sensitivity is a main reason for the destruction of the operation even for the weak noise. In order to avoid the unwanted large-amplitude oscillations and return the system to the proper operating mode, one has to use a regulator which reduces the stochastic sensitivity. On the basis of the presented theory of the synthesis of the assigned stochastic sensitivity, we have constructed a stabilizing regulator for the randomly forced flow reactor.

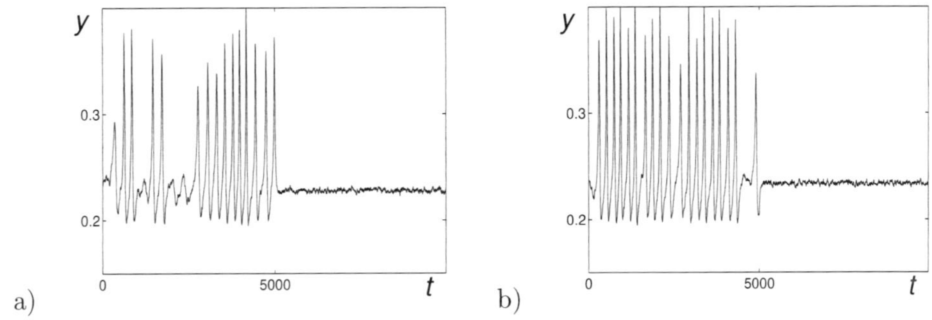

Figure 3. Time series of the controlled stochastic system with $\varepsilon = 0.0005$: a) $a = 1.58$; b) $a = 1.583$. The control is switched on at $t = 5000$, the regulator provides $w = 10$.

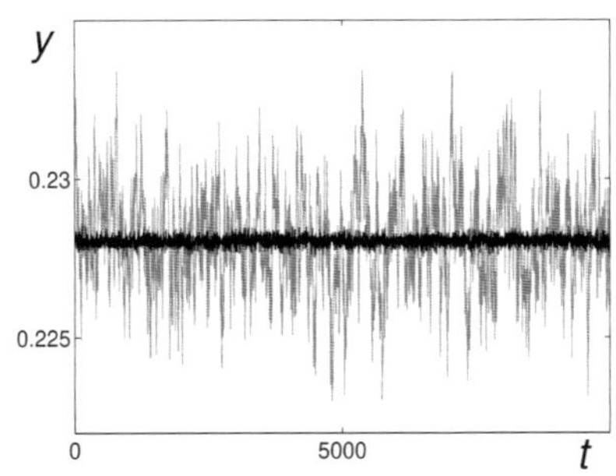

Figure 4. Time series of the controlled stochastic system for $\varepsilon = 0.0005$ with the regulator providing $w = 0.1$ (black) and $w = 10$ (grey). The control is switched on at $t = 0$.

7. References

[1] Lindner B, Garcia-Ojalvo J, Neiman A and Schimansky-Geier L 2004 *Phys. Rep.* **392** 321
[2] Anishchenko V S, Astakhov V V, Neiman A B, Vadivasova T E and Schimansky-Geier L 2007 *Nonlinear Dynamics of Chaotic and Stochastic Systems. Tutorial and Modern Development* (Berlin/Heidelberg: Springer-Verlag)
[3] Bashkirtseva I, Chen G and Ryashko L 2010 *Int. J. Bifurcat. Chaos* **20** 1439
[4] Ryashko L and Bashkirtseva I 2011 *Phys. Rev. E* **83** 061109
[5] Kushner H J 1967 *Stochastic Stability and Control* (New York: Academic Press)
[6] Astrom K J 1970 *Introduction to the Stochastic Control Theory* (New York: Academic Press)
[7] Sun J-Q 2006 *Stochastic Dynamics and Control* (Amsterdam: Elsevier)
[8] Guo L and Wang H 2010 *Stochastic Distribution Control System Design: a Convex Optimization Approach* (New York: Springer-Verlag)
[9] Ryashko L B and Bashkirtseva I A 2008 *Automat. Remote Control* **69** 1171
[10] Bashkirtseva I, Chen G and Ryashko L 2012 *Commun. Nonlin. Sci. Numer. Simulat.* **17** 3381
[11] Bashkirtseva I, Chen G and Ryashko L 2017 *J. Franklin Inst.* **354** 1658
[12] Field R J and Burger M 1985 *Oscillations and Traveling Waves in Chemical Systems* (New York: Wiley)
[13] Kawczynski A L and Nowakowski B 2008 *Phys. Chem. Chem. Phys.* **10** 289
[14] Jakobsen H A 2014 *Chemical Reactor Modeling: Multiphase Reactive Flows* (Berlin: Springer)
[15] Volter B V and Salnikov I Y 1965 *Modeling and Optimization of Catalytic Processes* (Moscow: Nauka) (in Russian)

Acknowledgments
This work was supported by RFBR (16-08-00388).

IOP Conf. Series: Materials Science and Engineering **192** (2017) 012006 doi:10.1088/1757-899X/192/1/012006

Formation of self-assembled micro- and nano-domain structures in uniaxial ferroelectrics

V Ya Shur, A G Shur and A R Akhmatkhanov

School of Natural Sciences and Mathematics, Ural Federal University, Ekaterinburg, 620000, Russian Federation

Abstract. We present the experimental study of appearance of domain wall shape instabilities and self-assembled domain structures in uniaxial ferroelectrics lithium niobate and lithium tantalate covered by artificial dielectric layer. The domain structure evolution has been considered as a manifestation of nucleation processes similar to the first order phase transformation. The necessary conditions for formation of self-assembled domain structures including the highly non-equilibrium switching conditions and stability of concave angles were proposed. The formation of the self-assembled domain shape instabilities under application of the uniform external electric field during switching (domain growth) and backswitching (domain shrinkage) has been studied experimentally with these conditions fulfilled. The formation of the bumps at the vortexes of hexagon domain for diameter above 3 - 5 μm with subsequent oriented growth was obtained in stoichiometric lithium tantalate. The formation of quasi-regular fish-bone domain structure has been revealed during spontaneous backswitching in Mg doped lithium niobate. The resulted structure consisted of the narrow residual domains with width from 200 to 500 nm appeared as a result of finger growth to the center of hexagon domain. The obtained decreasing of the number of the residual domains during growth has been attributed to strong electrostatic interaction of domain walls.

1. Introduction

The study of kinetics of first order phase transitions in highly non-equilibrium conditions became extremely urgent [1]. The formation of self-assembled structures including the dendrite ones was demonstrated experimentally [2] and the phase diagram of the most common structure types was predicted [3–5]. However the first order phase transitions including the crystallization from the melt poses the limited range of control parameters thus prohibiting the realization of all predicted types of structure in one system. This fact stimulates searching of new systems with simple variation of control parameters, *in situ* and high resolution structure visualization. It is known that the evolution of ferroelectric domains during polarization reversal can be considered as an analogy of the first-order phase transformation with electric field as a driving force [6,7]. The highly non-equilibrium switching conditions can be realized in ferroelectrics in wide range of control parameters. The complementary methods of domain visualization with high spatial and temporal resolutions are available [8,9].

The single crystalline lithium niobate ($LiNbO_3$, LN) and lithium tantalate ($LiTaO_3$, LT) represent the most investigated ferroelectric materials. Their great practical importance enabled the growth of the large crystals with extremely high uniformity [8]. The static domain structure can be visualized by scanning probe microscopy and scanning electron microscopy with resolution down to 2 nm and by Raman confocal microscopy in the bulk with resolution about 300 nm. Thus the single crystals of LN and LT are perfect model materials for investigation of self-assembling in highly non-equilibrium switching conditions.

In this paper we review the experimental studies of appearance and growth of domain wall shape instabilities and self-assembled domain structures in uniaxial ferroelectric LN and LT crystals (C_{3v} symmetry) covered by uniform artificial surface dielectric layer.

2. Experiment

The samples represented the 1-mm-thick plates polished to optical grade. The artificial dielectric layer represented the 3-μm-thick photoresist deposited on Z+ polar surface of the samples. The uniform liquid electrodes were used for field application. The single rectangular electric field pulse was applied using the high-voltage amplifier. The domain structure after partial polarization reversal was revealed by shallow selective chemical etching and the relief was visualized using the optical microscope (Olympus BX51, Japan) and scanning probe microscope.

3. The model of formation of dendrite structures in uniaxial ferroelectrics

The domain structure evolution has been considered as a manifestation of various nucleation processes similar to the first order phase transition [10,11]. Within this kinetic approach the neighboring domains are similar to the volumes of different phases divided by interfaces (domain walls). The domain growth is a result of thermally activated generation of one-, two-, and three-dimensional nuclei (*1D*-, *2D*-, and *3D*-nucleation) with preferred orientation of the spontaneous polarization. The shape of growing isolated domain is defined by position of the nucleation sites at the domain walls. The usual hexagon domain shape in LN obtained in slow (equilibrium) switching conditions is caused by determined nucleation at three non-adjacent hexagon vertexes and kink propagation in three Y directions [7]. In equilibrium switching conditions with effective screening the domain shape stability effect is observed which represents the rapid recover of hexagon shape after domain merging [12].

The nucleation probability is governed by local value of the electric field averaged over the nucleus size ("local field" E_{loc}) [6,7]. It is clear that E_{loc} is spatially inhomogeneous and changes essentially during polarization reversal. E_{loc} includes: (1) the external field (E_{ex}) produced by applied voltage, (2) the residual depolarization field (E_{rd}) produced by bound charges and external screening charges, which depends on the domain structure, (3) the bulk screening field (E_b) governed by bulk screening processes [13,14].

$$E_{loc}(r,t) = E_{ex}(r) + E_{rd}(r,t) + E_b(r,t) \qquad (1)$$

The three types of the bulk screening mechanisms have been considered: (1) redistribution of the bulk charges [6,15], (2) reorientation of the defect dipoles [15], (3) injection of the carriers from the electrode through the dielectric gap [16,17]. All bulk screening mechanisms are comparatively slow with time constants ranged from milliseconds to days and months. The retardation of bulk screening results in non-equilibrium switching conditions and qualitative change of the domain structure evolution.

The self-assembled formation of complicated domain shapes is possible only under highly non-equilibrium switching conditions (ineffective screening) and stable concave angles which allows preserving the appeared complex domain shapes.

The ineffective external screening can be realized by the artificial surface dielectric layer [7,18], switching without electrodes by pyroelectric field [19] or spontaneous backswitching after external field switch off by break of external circuit [20]. The stability of the concave angles is obtained for isotropic domain growth due to ionic conductivity which dominated in LN and LT crystals at elevated temperatures (above 150°C), and hampering of domain merging due to electrostatic interaction of the approaching domain walls.

In this paper we have analyzed the formation of the self-assembled domain shape instability under application of the uniform external electric field during switching (domain growth) and backswitching (domain shrinkage).

IOP Conf. Series: Materials Science and Engineering **192** (2017) 012006 doi:10.1088/1757-899X/192/1/012006

4. Experimental results

The domain wall shape instability was studied in stoichiometric LT crystal during the domain growth in the field 2 kV/mm. The process starts from growth of hexagon domain. The formation of six bumps at the domain vortexes was obtained for domain diameter above 3 - 5 μm (Fig. 1a). The formed bumps grow mostly in Y directions without merging (Fig. 1a, b).

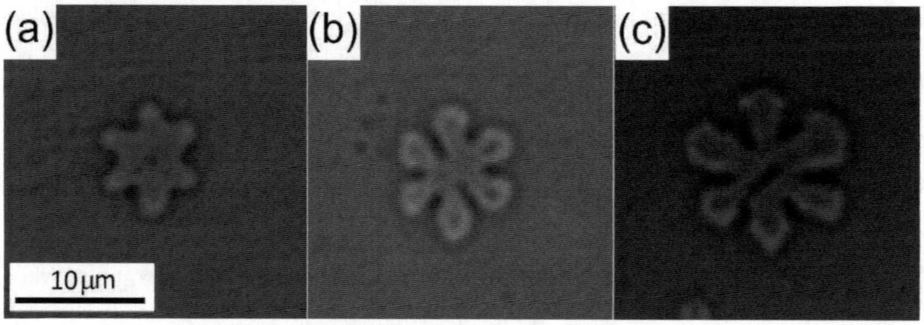

Figure 1. The stages of domain wall shape instability formation. Optical microscopy, transmitted light, phase contrast.

The domain shrinkage was studied in the MgO doped LN crystal. The experiment started by the application of short rectangular pulse of external electric field leading to growth of hexagonal domain. The back switching after external field abrupt switched off with break of external circuit leads to formation of quasi-regular fish-bone structure. The resulting structure consists of residual domains appeared as result of oriented growth of stripe domains ("fingers") from the boundary of hexagon domain to the center (Fig. 2a). The period of stripes is about one micron and the width of residual domains ranged from 200 to 500 nm (Fig. 2b). The nucleation of the short branches is obtained. The obtained increasing of the domain period (decreasing of the number of the fingers) during finger growth to the center can be attributed to strong electrostatic interactions of the neighbouring domain walls.

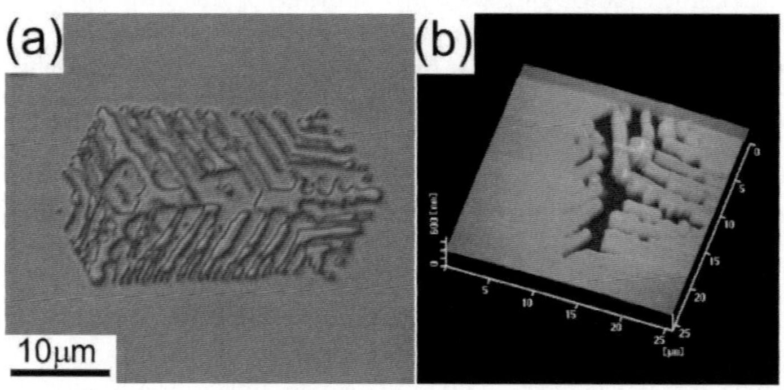

Figure 2. The fish-bone domain structures obtained as a result of spontaneous backswitching in MgO doped LN single crystals, visualized by: (a) optical microscopy, (b) scanning probe microscopy.

5. References

[1] Gibbs J W, Mohan K A, Gulsoy E B, Shahani A J, Xiao X, Bouman C A, De Graef M and Voorhees P W 2015 *Sci. Rep.* **5** 11824
[2] Chen Y, Billia B, Li D Z, Nguyen-Thi H, Xiao N M and Bogno A-A 2014 *Acta Mater.* **66** 219–31
[3] Kobayashi R 1993 *Phys. D Nonlinear Phenom.* **63** 410–23
[4] Brener E, Müller-Krumbhaar H and Temkin D 1996 *Phys. Rev. E* **54** 2714–22
[5] Brener E 2000 *Solid State Ionics* **131** 23–33
[6] Shur V Ya 1996 Fast polarization reversal process: evolution of ferroelectric domain structure in thin films *Ferroelectric Thin Films: Synthesis and Basic Properties. Ferroelectricity and Related Phenomena, vol. 10* ed C A Paz de Araujo, J F Scott and G W Taylor (Amsterdam: Gordon & Breach Science Publ.) pp 153–192
[7] Shur V Ya 2005 Correlated nucleation and self-organized kinetics of ferroelectric domains *Nucleation Theory Applications* ed J W P Schmelzer (WILEY-VCH, Weinheim) pp 178–214
[8] Shur V Ya and Zelenovskiy P S 2014 *J. Appl. Phys.* **116** 66802
[9] Soergel E 2005 *Appl. Phys. B* **81** 729–51
[10] Fatuzzo E and Merz W J 1967 *Ferroelectricity* (Amsterdam: North-Holland)
[11] Miller R and Weinreich G 1960 *Phys. Rev.* **117** 1460–6
[12] Shur V Ya, Akhmatkhanov A R, Chezganov D S, Lobov A I, Baturin I S and Smirnov M M 2013 *Appl. Phys. Lett.* **103** 242903
[13] Shur V Ya, Akhmatkhanov A R, Baturin I S, Nebogatikov M S and Dolbilov M A 2010 *Phys. Solid State* **52** 2147–53
[14] Baturin I S, Akhmatkhanov A R, Shur V Ya, Nebogatikov M S, Dolbilov M A and Rodina E A 2008 *Ferroelectrics* **374** 1–13
[15] Fridkin V M 1980 *Ferroelectric Semiconductors* (New York: Consultants Bureau)
[16] Tagantsev A K, Stolichnov I, Colla E L and Setter N 2001 *J. Appl. Phys.* **90** 1387
[17] Tagantsev A K, Cross L E and Fousek J 2010 *Domains in Ferroic Crystals and Thin Films* (New York, NY: Springer New York)
[18] Akhmatkhanov A R, Shur V Ya, Baturin I S, Zorikhin D V., Lukmanova A M, Zelenovskiy P S and Neradovskiy M M 2012 *Ferroelectrics* **439** 3–12
[19] Shur V Ya, Kosobokov M S, Mingaliev E A and Karpov V R 2015 *AIP Adv.* **5** 107110
[20] Batchko R G, Shur V Ya, Fejer M M and Byer R L 2006 *Appl. Phys. Lett.* **75** 1673-1675

Acknowledgments

The equipment of the Ural Center for Shared Use "Modern nanotechnology" UrFU was used. The research was made possible by Russian Scientific Foundation (Grant 14-12-00826).

On the theory of ternary melt crystallization with a non-linear phase diagram

L V Toropova, G Yu Dubovoi and D V Alexandrov

Department of Theoretical and Mathematical Physics, Laboratory of Multi-Scale Mathematical Modeling, Ural Federal University, Ekaterinburg, 620000, Russian Federation

E-mail: l.v.toropova@urfu.ru

Abstract. The present study is concerned with a theoretical analysis of unidirectional solidification process of ternary melts in the presence of a phase transition (mushy) layer. A new analytical solution of heat and mass transfer equations describing the steady-state crystallization scenario is found with allowance for a non-linear liquidus equation. The model under consideration takes into account the presence of two phase transition layers, namely, the primary and cotectic mushy regions. We demonstrate that the phase diagram nonlinearity leads to substantial changes of analytical solutions.

1. Introduction

Solidification processes of binary and multicomponent melts frequently occur in the presence of a phase transition layer, which divides purely solid and liquid phases. This layer arising as a result of thermal or constitutional supercooling ahead of the growing solid/liquid interface is called a mushy layer [1–4]. This supercooled layer is filled with the evolving solid phase elements in the form of dendrite-like structures and/or nucleating and growing solid particles [5–18]. If the latent heat of solidification releasing as a result of solid phase evolution completely compensates the mushy layer supercooling, such a two-phase layer is called quasiequilibrium. If the latent heat compensates the system supercooling in part, crystallization process evolves in non-equilibrium manner. Mathematical models of these processes and their analytical solutions in binary melts and solutions were considered in some detail in previous studies for a linear form of the phase diagram (quasiequilibrium [19–27] and non-equilibrium [28–30] theories).

If the solidifying system represents a multicomponent melt, the situation changes rather drastically. So, in the case of a ternary system, two mushy layers appear ahead of the solid phase (primary and cotectic mushy layers). A mathematical model of unidirectional crystallization process of a ternary system with a linear phase diagram was developed by Anderson [31] on the basis of experimental data [32]. Exact analytical solutions of this model were determined for the steady-state and self-similar crystallization conditions in [33–35]. Experimental and theoretical investigations demonstrate that the phase diagram nonlinearity substantially changes the process parameters and characteristics [36–38]. So, for example, nonlinearity in the phase diagram of binary systems increases the mushy layer thikness more than twice [37]. The present study is devoted to the question of how the nonlinear phase diagram influences the physical and process parameters of ternary melt solidification with the primary and cotectic mushy layers.

2. The model of directional solidification of ternary systems

Let us consider the process of directional solidification of a ternary system along the spatial axis z (see Figure 1) and designate the concertrations of two substances dissolved in a solvent A as B and C ($A+B+C=1$). As the main material undergoes a phase transition in a layer, which does not coincide with a phase transition region of the second component, two phase transition zones, primary and cotectic, are capable to appear in the solidification process. We denote their lengths as σ_P and σ_C (here subscripts P and C correspond to the primary and cotectic layers). Taking into account that the system phase diagram was discussed in details in refs. [34,35] we will not dwell on this point here and refer the reader to these original publications instead. An important point is that the temperature relaxation time $\tau_T \sim l^2/\kappa$ is much smaller than a characteristic relaxation time of the concentration fields $\tau_B \sim l^2/D_B$ and $\tau_C \sim l^2/D_C$, i.e. $\tau_T \ll \tau_C$ and $\tau_T \ll \tau_B$ (l - is a characteristic length scale, κ is the thermal diffusivity, D_B and D_C are the diffusion coefficients of B and C components). These estimates for the relaxation times show that the temperature derivatives with respect to time τ are much smaller than the other terms of corresponding model equations. With this in mind, we write down a mathematical model of the process based on the previous papers [33–35].

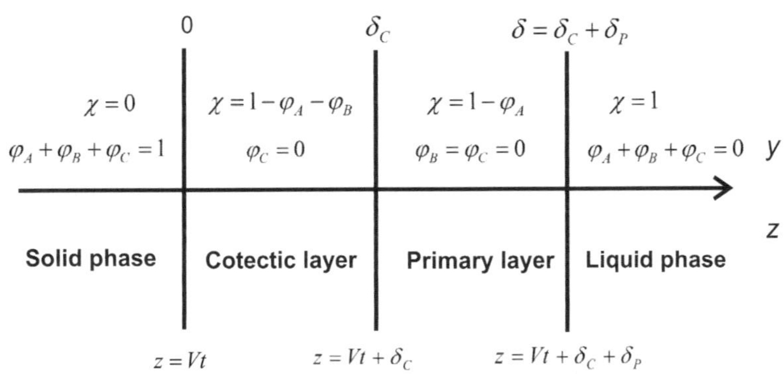

Figure 1. A scheme of the unidirectional crystallization process of a three-component system with two phase transition layers.

The impurity concentrations B_∞ and C_∞ in the liquid phase (melt) and the temperature gradient G_L will be regarded as known

$$B \to B_\infty, \; C \to C_\infty, \; z \to \infty, \tag{1}$$

$$\frac{\partial T}{\partial z} = G_L, \; z > Vt + \delta = Vt + \delta_C + \delta_P, \tag{2}$$

where T is the temperature, and V is the constant solidification rate. In addition, the following diffusion equations in liquid hold true

$$\frac{\partial B}{\partial t} = D_B \frac{\partial^2 B}{\partial z^2}, \; \frac{\partial C}{\partial t} = D_C \frac{\partial^2 C}{\partial z^2}, \; z > Vt + \delta. \tag{3}$$

The boundary conditions at the primary two phase layer - melt interface represent the heat and mass balance and the continuity conditions written out in the form of

$$L_V V \left[\varphi_A\right]_-^+ = \left[\bar{k}\frac{\partial T}{\partial z}\right]_-^+, \; [T]_-^+ = [B]_-^+ = [C]_-^+ = 0, \; z = Vt + \delta, \tag{4}$$

$$BV\left[\varphi_A\right]_-^+ = D_B\left[\chi\frac{\partial B}{\partial z}\right]_-^+,\ CV\left[\varphi_A\right]_-^+ = D_C\left[\chi\frac{\partial C}{\partial z}\right]_-^+,\ z = Vt + \delta. \quad (5)$$

Here L_V is the latent heat of solidification, $\bar{k} = k_L\chi + k_S(1-\chi)$, k_L and k_S are the thermal conductivity coefficients in the melt and solid phases, χ is the liquid phase fraction, φ_A is the solid phase fraction of component A. The symbol $[.]_-^+$ denotes a jump of the corresponding value at the boundary.

The heat and mass transfer equations in the primary two-phase layer, where the phase transition undergoes the component A ($\chi = 1 - \varphi_A$), can be expressed as

$$\frac{\partial}{\partial z}\left(\bar{k}\frac{dT}{dz}\right) + L_V\frac{\partial\varphi_A}{\partial t} = 0,\ T = T_M + m_B B + m_C C + n_C C^2,\ Vt + \delta_C < z < Vt + \delta, \quad (6)$$

$$D_B\frac{\partial}{\partial z}\left(\chi\frac{\partial B}{\partial z}\right) - \frac{\partial}{\partial t}\left(\chi^B\right) = 0,\ D_C\frac{\partial}{\partial z}\left(\chi\frac{\partial C}{\partial z}\right) - \frac{\partial}{\partial t}\left(\chi^C\right) = 0,\ Vt + \delta_C < z < Vt + \delta. \quad (7)$$

Here T_M is the phase transition temperature of the pure substance, m_B and m_C are the liquidus slope coefficients.

Let us write down the boundary conditions at the second phase interface between the cotectic and primary layers. These conditions, reflecting the heat and mass balance as well as the continuity of temperature and concentration fields, read [31,33–35]

$$L_V V\left[\varphi_A + \varphi_B\right]_-^+ = \left[\bar{k}\frac{\partial T}{\partial z}\right]_-^+,\ [T]_-^+ = [B]_-^+ = [C]_-^+ = 0, \quad (8)$$

$$B = B^C(T),\ C = C^C(T),\ z = Vt + \delta_C, \quad (9)$$

$$V\{B\left[\varphi_A\right]_-^+ + (B-1)\left[\varphi_B\right]_-^+\} = D_B\left[\chi\frac{\partial B}{\partial z}\right]_-^+,\ VC\left[\varphi_A + \varphi_B\right]_-^+ = D_C\left[\chi\frac{\partial C}{\partial z}\right]_-^+,\ z = Vt + \delta_C, \quad (10)$$

where φ_B indicates the solid phase fraction of component B.

Further, the heat and mass transfer equations in the cotectic mushy layer, where the phase transition undergo two components A and B ($\chi = 1 - \varphi_A - \varphi_B$), take the form

$$\frac{\partial}{\partial z}\left(\bar{k}\frac{\partial T}{\partial z}\right) + L_V\frac{\partial\left(\varphi_A + \varphi_B\right)}{\partial t} = 0,\ Vt < z < Vt + \delta_C, \quad (11)$$

$$B = B^C(T) = \frac{T_E + m_B^C B_E - T}{m_B^C},\ T = T_E^{AB} - m_C^C C + n_C^C C^2,\ Vt < z < Vt + \delta_C, \quad (12)$$

$$D_B\frac{\partial}{\partial z}\left(\chi\frac{\partial B}{\partial z}\right) - \frac{\partial}{\partial t}\left(\chi^B + \varphi_B\right) = 0,\ D_C\frac{\partial}{\partial z}\left(\chi\frac{\partial C}{\partial z}\right) - \frac{\partial}{\partial t}\left(\chi^C\right) = 0,\ Vt < z < Vt + \delta_C. \quad (13)$$

Here T_E, B_E and C_E are the known values of temperature and impurity concentrations at the eutectic point of ternary system and T_E^{AB} is the eutectic temperature of binary system. The boundary conditions at the interface between the solid phase and cotectic layer are

$$L_V V\left[\varphi_A+\varphi_B+\varphi_C\right]_-^+=\left[\bar{k}\frac{\partial T}{\partial z}\right]_-^+,\ z=Vt, \tag{14}$$

$$V\{B\left[\varphi_A\right]_-^+ + (B-1)\left[\varphi_B\right]_-^+ + B\left[\varphi_C\right]_-^+\}=D_B\left[\chi\frac{\partial B}{\partial z}\right]_-^+,\ z=Vt, \tag{15}$$

$$V\{C\left[\varphi_A\right]_-^+ + C\left[\varphi_B\right]_-^+ + (C-1)\left[\varphi_C\right]_-^+\}=D_C\left[\chi\frac{\partial C}{\partial z}\right]_-^+,\ z=Vt. \tag{16}$$

In the solid phase, we have a constant temperature gradient G_S, i.e.

$$\frac{\partial T}{\partial z}=G_S,\ z<Vt. \tag{17}$$

The model (1)-(17) represents a closed system of equations and boundary conditions and enables us to find the model solution, which describes the steady-state ternary melt crystallization.

3. Analytical solution of nonlinear model

Let us introduce the coordinate system which moves with the constant velocity V. The crystallization process is established in the new coordinate system where all unknown functions are independent of time. One can easily show that the diffusion equations (3) supplemented by the boundary conditions (1), have the following integrals

$$B(y)=B_\infty+B_1\exp\left(\frac{-Vy}{D_B}\right),\ C(y)=C_\infty+C_1\exp\left(\frac{-Vy}{D_C}\right),\ y>\delta=\delta_C+\delta_P, \tag{18}$$

where B_1 and C_1 are the constants of integration. Futher, we find the temperature and concentration derivatives performing the integration of heat and mass transfer equations (6) and (7) in the primary two-phase layer

$$\frac{dT}{dy}=\frac{(L_V V\varphi_A+A_1)}{\bar{k_P}(\varphi_A)},\ \bar{k_P}(\varphi_A)=k_L(1-\varphi_A)+k_S\varphi_A,\ \delta_C<y<\delta, \tag{19}$$

$$\frac{dB}{dy}=\frac{A_2-BV(1-\varphi_A)}{D_B(1-\varphi_A)},\ \frac{dC}{dy}=\frac{A_3-CV(1-\varphi_A)}{D_C(1-\varphi_A)},\ \delta_C<y<\delta, \tag{20}$$

where $A_1=\bar{k_P}(\varphi_{\bar{A}PL})G_{PL}-VL_V\varphi_{\bar{A}PL}$, A_2 and A_3 are the constants of integration, G_{PL} and $\varphi_{\bar{A}PL}$ are the temperature gradient and solid phase fraction at $y=\delta$. These unknowns will be found below. Furher, combining expressions (19), (20) and the liquidus equation (6), we determine the relationship between concentrations B and C in the primary two-phase region as

$$B(\varphi_A)=g(\varphi_A)-\frac{D_B m_C}{D_C m_B}C(\varphi_A)+\frac{2A_3 n_C D_B}{Vm_B D_C(1-\varphi_A)}C(\varphi_A)-\frac{2n_C D_B}{m_B D_C}C^2(\varphi_A),\ \delta_C<y<\delta,$$

$$g(\varphi_A)=\frac{D_B}{Vm_B}\left[\frac{m_B A_2}{D_B(1-\varphi_A)}+\frac{m_C A_3}{D_C(1-\varphi_A)}-\frac{\left(L_V V\left(\varphi_A-\varphi_{\bar{A}_{PL}}\right)+\bar{k_P}\left(\varphi_{\bar{A}_{PL}}\right)G_{PL}\right)}{\bar{k_P}(\varphi_A)}\right]. \tag{21}$$

Now we define $\frac{dB}{dC}$ in the primary mushy zone from expression (20)

$$\frac{dB}{dC} = \frac{(A_2 - VB(1-\varphi_A))\, D_C}{(A_3 - VC(1-\varphi_A))\, D_B}. \tag{22}$$

Next, we consider the case $D_B \neq D_C$ and substitute dB from (21) into (22). As a result, we obtain the expression

$$\frac{dC}{d\varphi_A} = \left[\frac{-2A_3 D_B n_C}{Vm_B(1-\varphi_A) D_C g'(\varphi_A)} + \frac{m_C D_B}{m_B D_C g'(\varphi_A)} + \frac{4n_C D_B C}{m_B D_C g'(\varphi_A)}\right.$$

$$\left. + \frac{D_C\left(A_2 - V(1-\varphi_A)\left[g(\varphi_A) + C\left(\frac{2A_3 n_C D_B}{Vm_B D_C(1-\varphi_A)} - \frac{m_C D_B}{m_B D_C}\right) - \frac{2n_C D_B}{m_B D_C}C^2\right]\right)}{D_B(A_3 - VC(1-\varphi_A))\, g'(\varphi_A)}\right]^{-1}. \tag{23}$$

The boundary condition for this equation can be obtained from expression (10). Combining formulas (20) and (23), we obtain

$$y(\varphi_A) = \delta_C + \int\limits_{\varphi^+_{ACP}}^{\varphi_A} \frac{dC}{d\varphi_A} \frac{D_C(1-\varphi_A)}{A_3 - VC(1-\varphi_A)} d\varphi_A, \tag{24}$$

$$\delta = \delta_C + \delta_P = \delta_C + \int\limits_{\varphi^+_{ACP}}^{\varphi^-_{APL}} \frac{dC}{d\varphi_A} \frac{D_C(1-\varphi_A)}{A_3 - VC(1-\varphi_A)} d\varphi_A. \tag{25}$$

Expressions (21)-(25) represent the parametric solution in the primary mushy layer (with parameter φ_A (or $\chi = 1 - \varphi_A$)).

Substituting these solutions into the boundary conditions (4), (5) we get

$$G_L = \frac{-m_B V B_1}{D_B} \exp\left(-\frac{V\delta}{D_B}\right) - \frac{m_C V C_1}{D_C} \exp\left(-\frac{V\delta}{D_C}\right)$$

$$+2n_C\left(C_\infty + C_1 \exp\left(-\frac{V\delta}{D_C}\right)\right)\left(\frac{VC_1}{D_C}\exp\left(-\frac{V\delta}{D_C}\right)\right), \tag{26}$$

$$G_{PL} = \frac{k_L G_L + L_V V \varphi^-_{APL}}{k^-_P\left(\varphi^-_{APL}\right)}, \; A_2 = VB_\infty, \; A_3 = VC_\infty, \tag{27}$$

$$B_1 = \exp\left(\frac{V\delta}{D_B}\right)\left[g\left(\varphi^-_{APL}\right) - B_\infty + \left(\frac{2A_3 n_C D_B}{Vm_B D_C\left(1-\varphi^-_{APL}\right)} - \frac{m_C D_B}{m_B D_C}\right)\right.$$

$$\left.\times\left(C_\infty + C_1\exp\left(-\frac{V\delta}{D_C}\right)\right) - \frac{2n_C D_B}{m_B D_C}\left(C_\infty + C_1\exp\left(-\frac{V\delta}{D_C}\right)\right)^2\right], \tag{28}$$

$$C_\infty + C_1 \exp\left(-\frac{V\delta}{D_C}\right) = C\left(\varphi^-_{APL}\right). \tag{29}$$

Next let us solve the problem in the cotectic zone. Integrating equations (11) and (13), we have

$$D_B\chi\frac{dB}{dy}+VB\chi+V\varphi_B=A_6,\ D_C\chi\frac{dC}{dy}+VC\chi=A_7, \tag{30}$$

$$\frac{dT}{dy}=\frac{\bar{k_C}\left(\chi_{SC}^{+}\right)G_{SC}-VL_V\left(1-\chi_{SC}^{+}\right)+VL_V\left(1-\chi\right)}{\bar{k_C}\left(\chi\right)}\equiv F_0\left(\chi\right). \tag{31}$$

Differentiating equation (12) one can obtain expressions connecting $\frac{dB}{dy}$, $\frac{dC}{dy}$ and $\frac{dT}{dy}$. Substituting them into (30) and (31), we obtain $\varphi_B\left(\chi\right)$, $\varphi_A\left(\chi\right)$ and $C\left(\chi\right)$

$$\varphi_B\left(\chi\right)=\frac{D_B\chi}{m_B^CV}F_0\left(\chi\right)+\frac{A_6}{V}-\chi B\left(\chi\right),\ \varphi_A\left(\chi\right)=1-\chi-\varphi_B\left(\chi\right), \tag{32}$$

$$C^2\left(\chi\right)\cdot\left(\frac{-2Vn_C^C}{D_C}\right)+C\left(\chi\right)\cdot\left(\frac{Vm_C^C}{D_C}+\frac{2n_C^CA_7}{D_C\chi}\right)-\left(\frac{m_C^CA_7}{D_C\chi}+F_0\left(\chi\right)\right)=0. \tag{33}$$

From (12), we get

$$B\left(\chi\right)=\frac{-n_C^C}{m_B^C}C^2\left(\chi\right)+m_{CB}^CC\left(\chi\right)+B_E+\frac{T_E-T_E^{AB}}{m_B^C}. \tag{34}$$

Note that only one of two solutions of equation (33) lies within the unit interval. Analyzing both solutions, we choose one of them (physically admissible)

$$C\left(\chi\right)=\frac{\frac{Vm_C^C}{D_C}+\frac{2n_C^CA_7}{D_C\chi}-\sqrt{\left(\frac{Vm_C^C}{D_C}+\frac{2n_C^CA_7}{D_C\chi}\right)^2-4\frac{2Vn_C^C}{D_C}\frac{m_C^CA_7}{D_C\chi}-4\frac{2Vn_C^C}{D_C}F_0\left(\chi\right)}}{\frac{4Vn_C^C}{D_C}}. \tag{35}$$

Furthermore, from expression (31), we come to the liquid phase fraction and thickness of the cotectic two-phase layer in the form

$$y\left(\chi\right)=\int\limits_{\chi_{SC}^{+}}^{\chi}\frac{dT}{d\chi}\frac{1}{F_0\left(\chi\right)}d\chi, \tag{36}$$

$$\delta_C=\int\limits_{\chi_{SC}^{+}}^{\chi_{CP}^{-}}\frac{dT}{d\chi}\frac{1}{F_0\left(\chi\right)}d\chi. \tag{37}$$

Substituting these solutions into the boundary conditions (14) - (16) and taking into account the temperature gradient (17), we arrive at the following expressions for constants

$$G_{SC}=\frac{k_SG_S-L_VV\chi_{SC}^{+}}{\bar{k_C}\left(\chi_{SC}^{+}\right)},\ V\left(\varphi_{BSC}^{-}-\varphi_{BSC}^{+}-B_E\chi_{SC}^{+}\right)+V\left(B_E\chi_{SC}^{+}+\varphi_{BSC}^{+}\right)=A_6, \tag{38}$$

$$V\left(\varphi_{CSC}^{-}-\varphi_{CSC}^{+}-C_E\chi_{SC}^{+}\right)+VC_E\chi_{SC}^{+}=A_7, \tag{39}$$

$$A_7=\frac{\left(G_{SC}D_C+2C_E^2Vn_C^C-C_EVm_C^C\right)\chi_{SC}^{+}}{2n_C^CC_E-m_C^C}, \tag{40}$$

where φ_{BSC}^{-} and φ_{BSC}^{+} are the solid phase fractions of component B on the left and right sides of the solid phase - cotectic layer boundary, φ_{CSC}^{-} and φ_{CSC}^{+} are the similar values for the component C. The boundary conditions (9) - (10) give

$$V = \frac{k_S G_S - k_L G_L}{L_V}, \ A_6 = A_2 = VB_\infty, \tag{41}$$

$$A_7 = A_3 = VC_\infty, \tag{42}$$

$$C_{CP} = \frac{\frac{Vm_C^C}{D_C} + \frac{2n_C^C A_7}{D_C \chi_{CP}^-} - \sqrt{\left(\frac{Vm_C^C}{D_C} + \frac{2n_C^C A_7}{D_C \chi_{CP}^-}\right)^2 - 4\frac{2Vn_C^C}{D_C}\frac{m_C^C A_7}{D_C \chi_{CP}^-} - 4\frac{2Vn_C^C}{D_C} F_0\left(\chi_{CP}^-\right)}}{\frac{4Vn_C^C}{D_C}}, \tag{43}$$

$$B_\infty - \left(1 - \varphi_{ACP}^+\right)\left[B^* - \frac{n_C^C}{m_B^C} C_{CP}^2 + m_{CB}^C C_{CP}\right]$$

$$= m_{CB}^C D_{BC}\left(C_\infty - C_{CP}\left(1 - \varphi_{ACP}^+\right)\right) - 2\frac{n_C^C}{m_B^C} D_{BC}\left(C_\infty C_{CP} - C_{CP}^2\left(1 - \varphi_{ACP}^+\right)\right), \tag{44}$$

$$C_{CP}^2\left(\frac{-2n_C D_B}{m_B D_C} + \frac{n_C^C}{m_B^C}\right)$$

$$+C_{CP}\left(\frac{2C_\infty n_C D_B}{m_B D_C\left(1 - \varphi_{ACP}^+\right)} - \frac{m_C D_B}{m_B D_C} - m_{CB}^C\right) + g\left(\varphi_{ACP}^+\right) - B^* = 0, \tag{45}$$

where $D_{BC} = \frac{D_B}{D_C}$, $m_{CB}^C = \frac{m_C^C}{m_B^C}$, $B^* = B_E + \frac{T_E - T_E^{AB}}{m_B^C}$. Now combining (38), (40) and (42), we obtain the following expressions

$$a_2\left(\chi_{SC}^+\right)^2 + a_1\left(\chi_{SC}^+\right) + a_0 = 0, \tag{46}$$

$$\chi_{SC}^+ = \frac{-a_1 \pm \sqrt{a_1^2 - 4a_2 a_0}}{2a_2}, \tag{47}$$

where

$$a_2 = -VL_V D_C + 2C_E^2 V n_C^C k_l - 2C_E^2 V n_C^C k_S - C_E V m_C^C k_l + C_E V m_C^C k_S,$$

$$a_1 = -Vk_l C_\infty\left(2n_C^C C_E - m_C^C\right) + C_\infty k_S V\left(2C_E n_C^C - m_C^C\right) + k_S G_S D_C + 2C_E^2 V n_C^C k_S - VC_E m_C^C k_S,$$

$$a_0 = -C_\infty k_S V\left(2C_E n_C^C - m_C^C\right).$$

Now from (38), (39) and (41), (42), one can get

$$\varphi_{B_{SC}}^- = B_\infty, \tag{48}$$

$$\varphi_{C_{SC}}^{-} = C_{\infty}, \tag{49}$$

$$\varphi_{C_{SC}}^{+} = 0. \tag{50}$$

Furthermore, we find $\varphi_{A_{CP}}^{+}$ and C_{CP} from expressions (44) and (45). Then we find χ_{CP}^{-} from (43) with allowance for C_{CP}. In addition, one can solve the differential equation (43) because $C_{CP} = C\left(\varphi_{A_{CP}}^{+}\right)$ is known. Next, combining expressions (26), (28), and (29), we obtain equations determining $\varphi_{A_{PL}}^{-}$, C_1 and B_1. The boundary values on the right side of interface between the solid and cotectic phases can be found from distributions (32) as $\varphi_{ASC}^{+} = \varphi_A\left(\chi_{SC}^{+}\right)$ and $\varphi_{BSC}^{+} = \varphi_B\left(\chi_{SC}^{+}\right)$. Furthermore, distribution (32) also determines the boundary values of solid phase fractions φ_{ACP}^{-} and φ_{BCP}^{-} of components A and B on the left side of interface between the primary and cotectic layers as $\varphi_{ACP}^{-} = \varphi_A\left(\chi_{CP}^{-}\right)$ and $\varphi_{BCP}^{-} = \varphi_B\left(\chi_{CP}^{-}\right)$.

4. Interdendritic spacing

Let us now calculate the primary interdendritic spacing on the basis of known impurity concentration as a function of spatial coordinate. In order to find the analytical dependence for interdendritic spacing λ, we use the previous theory [39]

$$\lambda = \sqrt{\frac{2\pi\rho}{0.86\frac{\partial(\varphi_A+\varphi_B)}{\partial y}}}, \tag{51}$$

where ρ is the radius of curvature corresponding to a stable mode of dendritic growth. The solvability theory [11, 40, 41] gives

$$\frac{2d_0 D_T}{\rho^2 V_d} = \sigma_0 \beta^{7/4}\left[\frac{1}{\left(1+a_1\sqrt{\beta}P_g\right)^2} + \frac{1}{\left(1+a_2\sqrt{\beta}P_g\frac{D_T}{D_C}\right)^2}\frac{2m_C^C C_i\left(1-k_0\right)D_T}{\frac{Q}{c_P}D_C}\right], \tag{52}$$

where

$$C_i = \frac{C_{CP}}{1-\left(1-k_0\right)\exp\left(P_g\frac{D_T}{D_C}\right)P_g\frac{D_T}{D_C}I_C\left(\infty\right)}, \tag{53}$$

$$I_C\left(\eta\right) = \int_1^{\eta}\exp\left[-P_g\frac{D_T}{D_C}h\right]\frac{dh}{\sqrt{h}}, \tag{54}$$

$a_2 = \sqrt{2}a_1 \approx 0.505\sqrt{\sigma_0}$, σ_0 is a fitting parameter, d_0 is the capillary length, V_d is the dendrite tip velocity, β is the anisotropic parameter, m_C^C is the liquidus slope, C_i is the concentration of impurity at the dendritic surface, k_0 is the equilibrium partition coefficient, Q is the latent heat per unit volume of the solid phase, c_P is the heat capacity, D_T is the thermal diffusivity, D_C is the solute diffusivity, and P_g is the growth Peclet number. Taking into account the theory under consideration one can calculate the explicit function $\lambda\left(P_g\right)$ on the basis of expressions (51)-(54).

5. Discussion

A new analytical method for solving the problem of heat and mass transfer in a ternary system is developed in this study. The distriburions of impurity concentration as well as of the solid and liquid fractions in the phase transition layer are obtained. The effect of quadratic term in the liquidus equation is studied as well.

Figure 2 shows the effect of deviation of the liquidus equation (coefficient n) from its linear dependence. So, its small deviations (at a fixed impurity concentration) are capable to change the reduced temperature several degrees. It should be noted that such variations often occur in practice. Fig. 2 shows that the temperature decreases with decreasing n and vice versa.

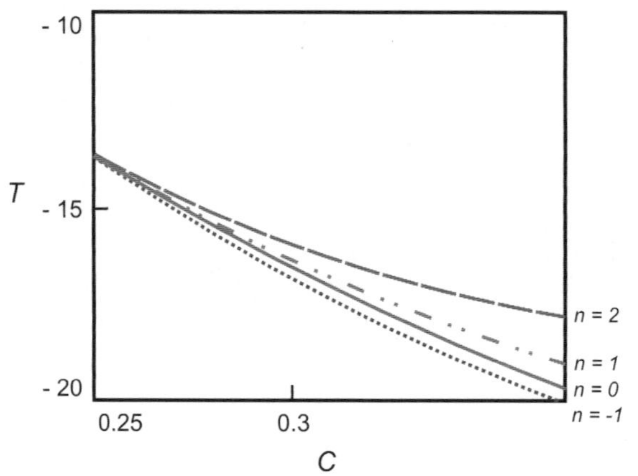

Figure 2. The influence of coefficient n on the deviation of liquidus equation from a linear function (temperature T is measured in Celcius).

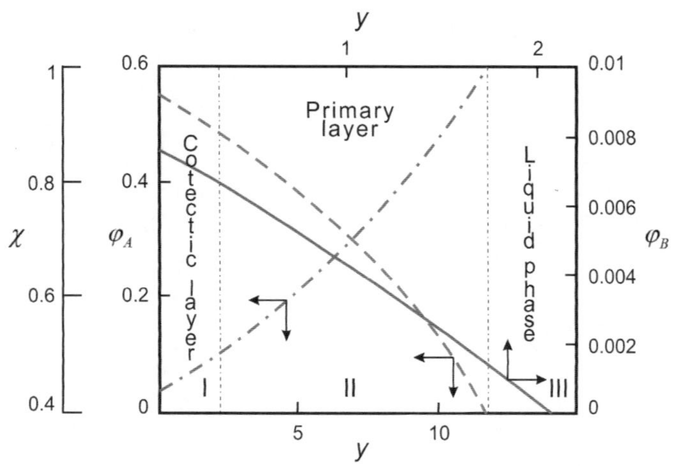

Figure 3. The solid fractions (solid and dashed lines), liquid fraction (dash-dotted line) as functions of the spatial coordinate. The cotectic, primary and liquid layers respectively correspond to regions I, II and III. The physical parameters are given in table 1 (y is measured in cantimeters).

Figure 3 illustrates the solid and liquid phase distributions in the phase transition layer

IOP Conf. Series: Materials Science and Engineering **192** (2017) 012007 doi:10.1088/1757-899X/192/1/012007

consisting of the primary and cotectic regions. The solid fraction φ_A of component A, which undergoes the phase transformation in regions I and II, decreases monotonically in the phase transition layer. In contrast, the solid fraction φ_B of component B, which undergoes the phase transition in region I, decreases in the cotectic layer only. The cotectic layer length (region I) therewith is lesser than the main layer (region II) of the phase transition. Accordingly to these dependencies, the liquid phase fraction χ increases monotonically in the total phase transformation domain (regions I and II). In addition, it is defined by the solid fractions φ_A and φ_B in region I, whereas it is described by fraction φ_A in region II. For this reason, the distribution $\chi(x)$ also has an inflection point in the primary layer as well as the solid fraction $\varphi_A(x)$. Our calculations show that the solid and liquid phase fractions at the boundaries between regions I and II, and II and III are continuous.

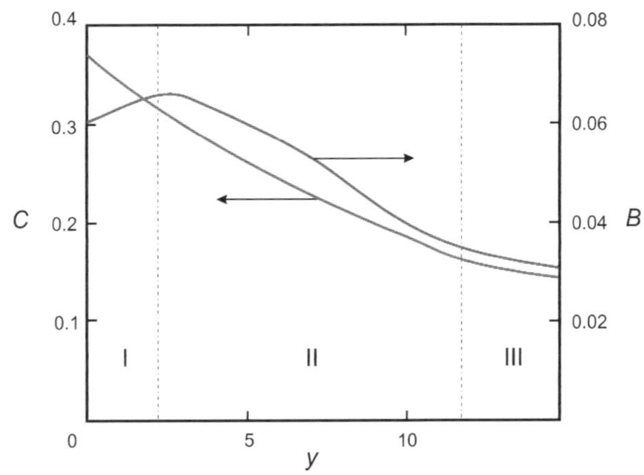

Figure 4. The impurity concentrations C and B as functions of the spatial coordinate $y = z - Vt$. All physical parameters correspond to Figure 2, $\delta_C = 2.168$ cm, $\delta = 11.803$ cm (y is measured in cantimeters).

Figure 4 demonstrates the distributions of impurity concentrations in the entire region of phase transition. The main impurity component, having a concentration $C(x)$, decreases monotonically in regions I and II due to the effect of impurity displacement by the growing solid phase. In contrast to this dependence, having a traditional behavior, the impurity concentration of the second component $B(x)$ increases in the cotectic layer, crosses the boundary between regions I and II, reaches a maximum in the primary mushy zone and then decreases to the initial concentration B_∞. This, at first sight, unusual behavior of the impurity concentration $B(x)$, occurs due to the fact that component B undergoes the phase transition in region I (this leads to a decreased concentration near the solid phase - cotectic layer boundary). Note that a similar behavior of the impurity concentration was obtained in the analysis of self-similar crystallization in references [13, 14, 40, 42, 43]. However, in these studies, the maximum point was found on the boundary of regions I and II. The reason explaining the maximum point displacement into region II lies in the fact that previous studies (see references [13, 14, 40, 42, 43]) have been carried out on the basis of approximate Sheil equations governing the impurity concentrations (equations without diffusion terms). Therefore, in a real ternary system, the maximum point displacement into the primary layer occurs due to the influence of diffusion transport of component $B(x)$.

It should be noted that the influence of coefficient n of the liquidus equation is seen in Figures 3 and 4. All dependencies have the same behavior, however the mushy region length increases significantly in comparison with the linear liquidus line equation.

Figure 5 shows the interdendritic lengh λ as a function of the growth Peclet number P_g. Note that the interdendritic spacing decreases with increasing P_g. This behavior is in consistent with known experimental data [44, 45] and the previously developed theory [46].

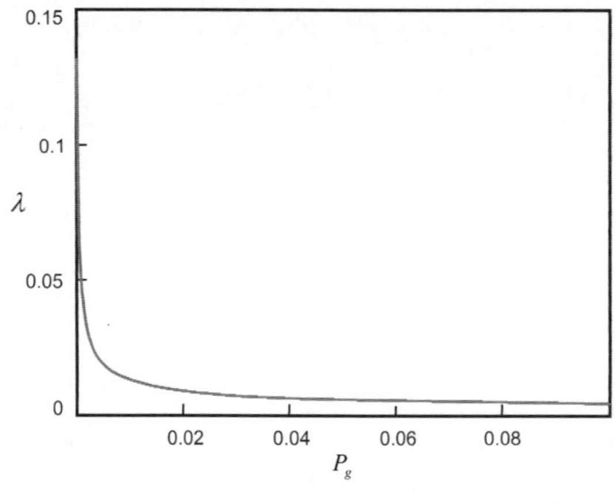

Figure 5. The interdendritic spacing λ (measured in cantimeters) as a function of the growth Peclet number P_g (thermophysical parameters are listed in Table 1).

Table 1. The physical parameters of three-component system $H_2O - KNO_3 - NaNO_3$ (B and C designate $NaNO_3$ and KNO_3).

B_E	0.06
C_E	0.37
B_E^{AB}	0.10
B_∞	0.035
C_∞	0.152
T_E (^{0}C)	-19
T_E^{AB} (^{0}C)	-5
T_M (^{0}C)	0
κ (cm^{2}s^{-1})	$1.1 \cdot 10^{-3}$
D_C (cm^{2}s^{-1})	$4.94 \cdot 10^{-6}$
D_B (cm^{2}s^{-1})	$4.94 \cdot 10^{-6}$
L_V/k_S (s^{0}C cm^{-2})	$1.52 \cdot 10^5$
k_L/k_S	0.25
k_S	$5.3 \cdot 10^{-3}$
G_S (^{0}C cm^{-1})	1
k_0	0
σ_0	2.1
β	0.195

The main result of this paper lies in the fact that even small deviations of the liquidus equation from a linear dependence can be responsible for significant changes of other parameters governing the ternary melt solidification process.

This work was partially supported by the Russian Foundation for Basic Research (grant no. 16-08-00932).

6. References

[1] Hills R N, Loper D E and Roberts P H 1983 *Q. J. Appl. Math.* **36** 505–539
[2] Fowler A C 1985 *IMA J. Appl. Math.* **35** 159–174
[3] Buyevich Yu A, Alexandrov D V and Mansurov V V 2001 *Macrokinetics of Crystallization* (New York: Begell House)
[4] Alexandrova I V, Alexandrov D V, Aseev D L and Bulitcheva S V 2009 *Acta Physica Polonica A* **115** 791–794
[5] Mullin J W 1972 *Crystallization* (London: Butterworth)
[6] Huppert H 1990 *J. Fluid Mech.* **212** 209-240
[7] Pelcé P 1988 *Dynamics of Curved Fronts* (Boston: Academic Press)
[8] Herlach D M 2008 *Phase Transformations in Multicomponent Melts* (Weinheim: Wiley-VCH)
[9] Herlach D M 2015 *Crystals* **5** 355–375
[10] Alexandrov D V, Galenko P K and Herlach D M 2010 *J. Crystal Growth* **312** 2122–2127
[11] Alexandrov D V and Galenko P K 2013 *Phys. Rev. E* **87** 062403
[12] Alexandrov D V and Malygin A P 2013 *J. Phys. A: Math. Theor.* **46** 455101
[13] Alexandrov D V and Nizovtseva I G 2014 *Proc. R. Soc. A* **470** 20130647
[14] Alexandrov D V and Malygin A P 2014 *Modelling Simul. Mater. Sci. Eng.* **22** 015003
[15] Alexandrov D V 2014 *Phys. Lett. A* **378** 1501–1504
[16] Alexandrov D V 2014 *J. Phys. A: Math. Theor.* **47** 125102
[17] Alexandrov D V 2014 *Phil. Mag. Lett.* **94** 786–793
[18] Shneidman V A 2013 *Phys. Rev. E* **88** 010401
[19] Worster M G 1986 *J. Fluid Mech.* **167** 481–501
[20] Alexandrov D V and Aseev D L 2005 *J. Fluid Mech.* **527** 57–66
[21] Alexandrov D V and Aseev D L 2006 *Comput. Mater. Sci.* **37** 1–6
[22] Aseev D L and Alexandrov D V 2006 *Acta Mater.* **54** 2401–2406
[23] Alexandrov D V, Aseev D L, Nizovtseva I G, Huang H-N and Lee D 2007 *Int. J. Heat Mass Trans.* **51** 5204–5208
[24] Alexandrov D V, Nizovtseva I G, Malygin A P, Huang,H-N and Lee D 2008 *J. Phys.: Condens. Matter* **20** 114105
[25] Alexandrov D V and Nizovtseva I G 2008 *Int. J. Heat Mass Trans.* **51** 5204–5208
[26] Alexandrov D V and Malygin A P 2011 *Int. J. Heat Mass Trans.* **54** 1144–1149
[27] Alexandrov D V and Malygin A P 2012 *Int. J. Heat Mass Trans.* **55** 3196–3204
[28] Aseev D L and Alexandrov D V 2006 *Int. J. Heat Mass Trans.* **49** 4903–4909
[29] Aseev D L and Alexandrov D V 2006 *Doklady Physics* **51** 291–295
[30] Alexandrov D V and Malygin A P 2011 *Phys. Earth Planet. Int.* **189** 134–141
[31] Anderson D M 2003 *J. Fluid Mech.* **483** 165–197
[32] Aitta A, Huppert H E and Worster M G 2001 *J. Fluid Mech.* **432** 201–217
[33] Alexandrov D V and Malygin A P 2012 *Int. J. Heat Mass Trans.* **55** 3755–3762
[34] Alexandrov D V and Ivanov A A 2009 *Int. J. Heat Mass Trans.* **52** 4807–4811
[35] Alexandrov D V and Ivanov A A 2009 *J. Exper. Theor. Phys.* **108** 821–829
[36] Jordan A S 1971 *Metall. Trans.* **2** 1965–1970
[37] Alexandrov D V, Rakhmatullina I V and Malygin A P 2010 *Russian Metallurgy (Metally)* **8** 745–750
[38] Herlach D M and Matson D M 2012 *Solidification of Containerless Undercooled Melts* (Weinheim: WileyVCH)
[39] Deguen R, Alboussiére A and Brito D 2007 *Phys. Earth Planet. Inter.* **164** 36-49
[40] Alexandrov D V and Galenko P K 2014 *Physics-Uspekhi* **57** 771-786
[41] Alexandrov D V and Galenko P K 2015 *Phys. Chem. Chem. Phys.* **17** 19149-19161
[42] Alexandrov D V 2003 *Physics-Doklady* **48** 481-486
[43] Gao J, Han M, Kao A, Pericleous K, Alexandrov D V and Galenko P K 2016 *Acta Materialia* **103** 184-191
[44] Somboonsuk K, Mason J T and Trivedi R 1984 *Metall. Trans. A* **15A** 967-975
[45] Trivedi R 1984 *Metall. Trans. A* **15A** 977-982
[46] Alexandrov D V and Britousova A V 2015 *AIP Conference Proceedings* **1648** 850101

Acknowledgments

This work was supported by the Russian Foundation for Basic Research (grant no. 16-08-00932) and the Ministry of Education and Science of the Russian Federation (project no. 1.9527.2017).

Influence of computational domain size on the pattern formation of the phase field crystals

Ilya Starodumov[1], Peter Galenko[2], Dmitri Alexandrov[1] and Nikolai Kropotin[3]

[1] Laboratory of Multi-Scale Mathematical Modelling, Department of Theoretical and Mathematical Physics, Ural Federal University, 620000 Ekaterinburg, Russia
[2] Friedrich Schiller University, Jena, D-07743, Germany
[3] MKM Ltd., Ilfata Zakirova st. 24, 426000 Izhevsk, Russia

E-mail: ilya.starodumov@urfu.ru

Abstract. Modeling of crystallization process by the phase field crystal method (PFC) represents one of the important directions of modern computational materials science. This method makes it possible to research the formation of stable or metastable crystal structures. In this paper, we study the effect of computational domain size on the crystal pattern formation obtained as a result of computer simulation by the PFC method. In the current report, we show that if the size of a computational domain is changed, the result of modeling may be a structure in metastable phase instead of pure stable state. The authors present a possible theoretical justification for the observed effect and provide explanations on the possible modification of the PFC method to account for this phenomenon.

1. Introduction

The original PFC model has been developed in a parabolic form [1–3]. Then it has been modified to the extended form described by partial differental equation of a hyperbolic type. Such the modified or hyperbolic PFC model (MPFC model) includes faster degrees of freedom in a form of inertia and as such leads to the description of both fast and slow dynamics. Fast front dynamics proceeds when the driving force for the phase transition is large. This occurs when the free energy difference between the (meta)stable periodic solid and initially unstable phase is very large which in general occurs when a system is quenched far below a transition point, or in this case far below the equilibrium temperature of phase transition [4]. These conditions lead to a fast phase transition when the velocity of the front is comparable to the speed of atomic diffusion or the speed of local structural relaxation. The movement of a phase transition front at such fast velocities can lead to bulk phases that are not in a local structural or chemical equilibrium. As a result, the modified (hyperbolic) PFC-model includes the description of both fast and slow dynamics of transition [5–11] by taking into account diffusion and inertial effects [7].

Using the free-energy functional

$$\mathcal{F}[\phi, \nabla\phi, \nabla^2\phi] = \int_\Omega \left[f(\phi) - |\nabla\phi|^2 + \frac{1}{2}(\nabla^2\phi)^2 \right] \mathrm{d}\Omega, \tag{1}$$

related to the domain Ω, one can define the chemical potential as

$$\mu(\phi) = \frac{\delta \mathcal{F}}{\delta \phi} = \frac{df}{d\phi} + 2\nabla^2\phi + \nabla^4\phi, \qquad (2)$$

where ϕ describes conserved order parameter. The free energy density $f(\phi)$ represents the homogeneous part of energy and in the present work it is chosen in the Landau-de Ginnes form:

$$f(\phi) = \frac{1-\epsilon}{2}\phi^2 + \frac{\alpha}{3}\phi^3 + \frac{1}{4}\phi^4. \qquad (3)$$

Here, $\epsilon = (T_c - T)/T_c$ is the dimensionless temperature T shows its deviation from T_c critical temperature of transition, and α is the measure of metastability. Finally, one can define the modified (hyperbolic) phase field crystal model, which describes a continuous and conserved field of atomic density $\phi(x,t)$ and it can be expressed by the following equation:

$$\tau\frac{\partial^2\phi}{\partial t^2} + \frac{\partial\phi}{\partial t} = \nabla^2\mu, \qquad (4)$$

where t is the time, τ is the relaxation time of the atomic flux to its stationary state. Using the chemical potential (2) with taking into account Eqs. (1) and (4), the PFC-equation (4) presents the second order in time and the six order in space equation. Its solution requires development special computational methods [11].

2. The results of simulations in computational domains of different size

In this section we present how the domain size influence on the steady solution of the MPFC task. This phenomenon can be illustrated by calculation results of the MPFC task with parameters of the average density $\phi = -0.34$ and the undercooling $\epsilon = 0.4$. This parameters correspond to the BCC crystalline pattern (Body Centered Crystal) from the three-dimension structure diagram of Jaatinen and Ala-Nissila [12]. The results were perfomed for the same task but for the cases of different cubic domains. Namely, consider domains with the size of the domain:

1) $25 \times 25 \times 25$,
2) $31 \times 31 \times 31$,
3) $32 \times 32 \times 32$,
4) $39 \times 39 \times 39$,
5) $45 \times 45 \times 45$.

All calculations are perfomed by the special computational method proposed in [13] which represented by the implicit finite element method (FEM) using regular hexagonal mesh. The computational efficiency of such algorithm was shown in [14, 15].

The ratio of the sizes of the computing domain and the mesh as shown in [16] should be taken into account for proper results. We have established this ratio to be equal to 0.78 for all tasks. Further in the section the figures with the visualization of stationary solutions of the described task are presented. The left side of each figure presents colored surface of computational domain in steady moment. The right side of each figure presents isocontours of the result into the computational domain it steady moment.

In figure 1, we can see pure BCC crystal pattern formed in the smallest domain 1. At the same time in domain 2 (figure 2) the stable solution is presented by curved rods or the "worms" pattern. Figure 3 presents that even small change of domain size can transform a stationary solution and instead of "worms" in the last case we can observe a mixed structure of domain 3. In figure 4, we can see pure BCC pattern again, this time in domain 4. But further increase

IOP Conf. Series: Materials Science and Engineering **192** (2017) 012008 doi:10.1088/1757-899X/192/1/012008

Figure 1. Stable state formation prepared in simulation of the task in domain 1.

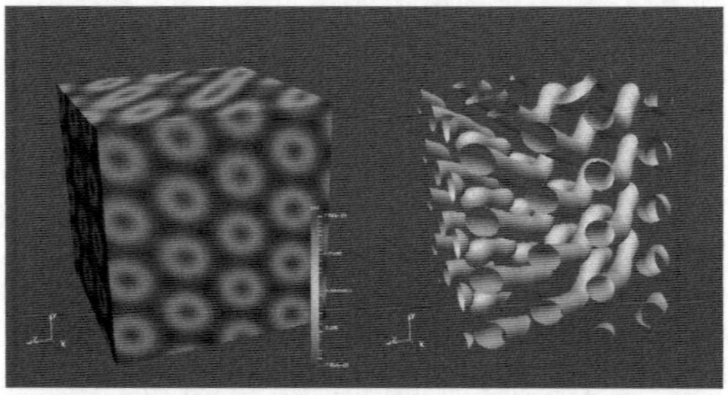

Figure 2. Stable state formation prepared in simulation of the task in domain 2.

Figure 3. Stable state formation prepared in simulation of the task in domain 3.

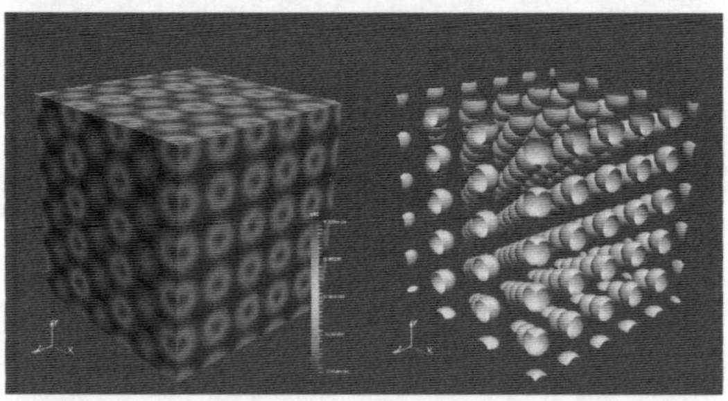

Figure 4. Stable state formation prepared in simulation of the task in domain 4.

Figure 5. Stable state formation prepared in simulation of the task in domain 5.

in the size of the domain again leads to an alternative stationary solution of a mixed type. It's presented in figure 5.

These results demonstrate the strong dependence of the stationary formation of the crystallization process on the sizes of the calculated domain. In two simulations the result is consistent with the expected BCC structure. In all other simulations the result could not be classified as known pattern type and probably is a metastable form. The reasons for this phenomenon may be a feature of the MPFC method. It consists in the fact that changes in the crystal structure occur due to minimization of the free energy functional (2). The local minima of the free energy functional can correspond to locally stable or metastable states of the crystal structure. With the increase in the size of the computational domain, the number of possible atomic density ϕ distributions in such a domain and in every moment of time is significantly increased. As a consequence, the probability of the system falling into the metastable region increases too. That, however, does not exclude the possibility of obtaining absolutely stable crystal formation.

3. Conclusions

The problem of crystal pattern formation in the Modified Phase Field Crystal Model has been discussed. To perform reliable computations, one should take into account the factors leading to instability. In the current report, we show that one of these factors is the size of computational domain. We believe that with increasing domain sizes, the probability of formation of metastable crystal patterns is substantially increased. In some cases, this can lead to a mismatch of the results with the predictions or to results that are unclassifiable. To improve the stability of the results we consider reasonable to add a stochastic vibrations in the atomic density field consistently with thermodynamic noise. It can be realized by modification of the PFC-model with taking into account the stochastic term. This should allow the system to come out of metastable states and increase the rate of formation of equilibrium and stable states consistent with diagrams of structures [12, 13].

References

[1] Elder K R, Katakowski M, Haataja M and Grant M 2002 *Physical Review Letters* **88** 245701

[2] Elder K R and Grant M 2004 *Physical Review E* **70** 051605

[3] Elder K R, Provatas N, Berry J, Stefanovich P and Grant M 2007 *Physical Review B* **75** 064107

[4] Herlach D, Galenko P and Holland-Moritz D 2007 *Metastable Solids from Undercooled Melts* (Elsevier)

[5] Stefanovic P, Haataja M and Provatas N 2006 *Physical Review Letters* **96** 225504

[6] Stefanovic P, Haataja M and Provatas N 2009 *Physical Review E* **80** 046107

[7] Galenko P K, Danilov D A and Lebedev V G 2009 *Physical Review E* **79** 051110

[8] Galenko P K and Elder K R 2011 *Physical Review B* **83** 064113

[9] Wang C and Wise S M 2011 *SIAM Journal on Numerical Analysis* **49** 945–969

[10] Baskaran A, Hu Z, Lowengrub J S, Wang C, Wise S M and Zhou P 2013 *J. Comput. Phys.* **250** 270–292 ISSN 0021-9991

[11] Galenko P K, Gomez H, Kropotin N V and Elder K R 2013 *Phys. Rev. E* **88**(1) 013310

[12] Jaatinen A and Ala-Nissila T 2010 *Journal of Physics: Condensed Matter* **22** 205402

[13] Bueno J, Starodumov I, Gomez H, Galenko P and Alexandrov D 2016 *Computational Materials Science* **111** 310 – 312 ISSN 0927-0256

[14] Starodumov I, Pavlyuk E, Klyuev L, Kovalenko M and Medyankin A 2015 *Ceur-WS Proceedings* **1513** 109–122

[15] Starodumov I, Pavlyuk E, Abramov S, Klyuev L, Galenko P and Alexandrov D 2016 *Bulletin of Udmurt University. Mathematics, Mechanics* **26** 445–450 URL http://dx.doi.org/10.1063/1.4959750
[16] Starodumov I and Kropotin N 2016 *AIP Conference Proceedings* **1759** URL http://dx.doi.org/10.1063/1.4959750

Acknowledgments
Authors achnowledge support from the Russian Scientific Foundation under project number 16-11-10095.

IOP Conf. Series: Materials Science and Engineering **192** (2017) 012009 doi:10.1088/1757-899X/192/1/012009

The evolution of structural and chemical heterogeneity during rapid solidification at gas atomization

V M Golod[1] and V Sh Sufiiarov[1]

[1] Peter the Great St.-Petersburg Polytechnic University, Institute of Metallurgy, Mechanical Engineering and Transport, Saint- Petersburg, 195251, Russia

e-mail: lpi2015@mail.ru

Abstract. Gas atomization is a high-performance process for manufacturing superfine metal powders. Formation of the powder particles takes place primarily through the fragmentation of alloy melt flow with high-pressure inert gas, which leads to the formation of non-uniform sized micron-scale particles and subsequent their rapid solidification due to heat exchange with gas environment. The article presents results of computer modeling of crystallization process, simulation and experimental studies of the cellular-dendrite structure formation and microsegregation in different size particles. It presents results of adaptation of the approach for local nonequilibrium solidification to conditions of crystallization at gas atomization, detected border values of the particle size at which it is possible a manifestation of diffusionless crystallization.

1. Introduction

A number of promising technological processes for manufacturing of the metal products are targeting to use micro-scale metallurgical blanks as raw materials, which are in contrast to the traditional large-scale ingots and castings having sizes $\approx 10^{-6}$ m. These primarily include selective laser melting, electron beam melting, direct metal deposition, cladding, spray forming, spark plasma sintering, extrusion, powder injection molding etc. Raw material for these processes usually produced by spinning processes, atomization (plasma, gas, water atomization), spray-based technologies et al [1-4]. Their appearance was due to a desire to get away from "birth defects" of the traditional technologies like coarse columnar structures in steel billets and castings, as well as macro- and micro segregation in combination with gaseous and nonmetallic inclusions. A transition to use micropowders provides a reduction in size microstructural parameters due to a sharp increase in crystallization rate and deep melts supercooling. A condition of local equilibrium at the liquid-solid phase interface is disrupted during rapid solidification as a result of increasing concentration and "trapping" of alloying components of the alloy, which leads to a sequent structural transition from dendrite to cellular and flat morphology with the establishment at the interface local nonequilibrium conditions, that in the limit gives diffusionless crystallization mode [3].

Atomization of the melt obtained through the fragmentation of alloy melt flow leads to an inhomogeneous array of small particles entrained by gas stream asynchronously and involved in complex heat exchange with it. Crystallization of particles is occurring under controlled rapid cooling conditions, ensures the formation of cellular-dendrite structures, local parameters of which are the subject of research in this article. Computer models for analysis of gas atomization process [1, 5-7] et al. exhibit almost complete agreement in describing the gas-dynamic and thermal parameters of the process and significantly differ in terms of the formation and growth of crystallites.

The nature of the redistribution of the alloying components between the phases is seen as the equilibrium one in a number of studies, i.e. occurring at full equalization of composition in liquid and solid phase, as opposed to nonequilibrium one, implemented under an additional condition of complete suppression of diffusion in the solid phase. In all cases, despite rapid crystallization conditions and deep supercooling of melt, it is using the local equilibrium conditions at phase interfaces.

Conditions of *local equilibrium* at the solid-liquid phase interface in the above-mentioned forced solidification modes get broken, when temperature, a composition of a liquid phase at the interface C_L* and composition of the crystallizing solid phase C_S* become dependent on the crystallization rate [3]. Values of phase diagrams parameters undergo the kinetically-dependent changes under *local nonequilibrium crystallization*, the nature of which depends on the relation of phase interface movement speed, diffusion rates of components in melt V_D and transfer of components across the phases interface V_{DI} [10-11].

In this paper, it is implemented a hybrid model, combining traditional analysis of heat transfer between molten particles and gas flow [8-9] with the thermodynamic and kinetic conditions at the front of the high-speed solidification [3, 10-11], which allows to identify the conditions of continuous transition from the local equilibrium diffusion-controlled dendrite structure to the local nonequilibrium diffusionless crystallization.

2. Formulation of a model

The models of microparticles solidification during gas atomization ([5-7] et al.) developed by several researchers are based on the description of equilibrium or local equilibrium crystallization of alloys.

The partition coefficient k_V in the local nonequilibrium model (subscript symbol *V*) [10-11] has the form:

$$k_V = \frac{k(1-\vartheta^2/V_D^2)+\vartheta/V_{DI}}{(1-\vartheta^2/V_D^2)+\vartheta/V_{DI}} \text{ at } \vartheta < V_D;\ k_V=1 \text{ if } \vartheta \geq V_D. \quad (1)$$

Under conditions of local equilibrium, i.e. $\vartheta \ll V_D$ and $\vartheta \ll V_{DI}$, $k_V = k$ – equilibrium partition coefficient. An analogous relation is written for the slope of the liquidus line at local nonequilibrium model:

$$m_V = \frac{m}{1-k}(1-k_V+\ln\left(\frac{k_V}{k}\right)+(1-k_V)^2\frac{\vartheta}{V_D}) \text{ at } \vartheta < V_D; \quad (2)$$

if $\vartheta \geq V_D$ then $m_V = m\frac{\ln(k)}{k-1}$. Under local equilibrium $m_V = m$, where m – equilibrium tangent of liquidus line angle.

Known values k_V (1) and m_V (2) allow to estimate corresponding change of non-equilibrium liquidus temperature for multicomponent alloys:

$$(T_L)_V = (T_L)_0 - \sum(m_V)_i[(C_L)_i - (C_0)_i], \quad (3)$$

as well as to reconstruct the geometry of nonequilibrium phase diagram, in particular, to determine the solidification range of a multi-component alloy Δt_{LS}:

$$(\Delta t_{LS})_V = \sum(m_V)_i(C_0)_i(1-(k_V)_i)/(k_V)_i \text{ at } \vartheta < V_D;$$

$$(\Delta t_{LS})_V = 0 \text{ at } \vartheta \geq V_D, \quad (4)$$

where $(T_L)_0$ – liquidus temperature of initial composition; $(C_0)_i$, $(C_L)_i$ – initial and current content of *i*-component in liquid phase.

The main components of the considered hybrid model are relations describing kinetics of formation of solid phase f_S and change in average composition of liquid phase, modified for conditions of local nonequilibrium crystallization of multicomponent alloys. The process of nucleation and growth of a new phase centers with specific density *N* describes by equation

$$\frac{df_S}{d\tau} = 4\pi(1-f_S)N\vartheta(\tau)R^2(\tau),$$

in which *R* – radius of growing crystallite and ϑ – growth rate is function of thermal supercooling melt $\Delta T(\tau)$. It's value is determined from the heat balance equation and used to estimate the speed of dendrite growth:

$$-V\rho\left(c\frac{dT}{d\tau} - L\frac{df_S}{d\tau}\right) = hS(T-T_g) + \sigma_0\varepsilon S(T^4 - T_g^4), \quad (5)$$

where T, T_g – absolute temperatures of melt and gas stream; c, L – volumetric heat capacity and heat of crystallization of the melt; τ – time of flight of the particle in the gas stream; V, S – volume and surface area of the particle; σ_0 – Stefan-Boltzmann constant; ε – emissivity factor of particle surface; h – convective heat transfer coefficient, estimated on the basis of gas-dynamic solving of task interaction of droplet with gas flow [8].

Linear crystallite growth rate determines the ratio $\vartheta = 2aP_T/R^*$ [12], where a – thermal diffusivity of melt; P_T – thermal Peclet number; $P_T = Iv^{-1}(\frac{c}{L}\Delta T)$; $Iv(P) = Pexp(P)Ei(P)$ – Ivantsov relation of Peclet number P; R^* – radius of curvature of the top of the dendrite cell-crystallite in the accelerated crystallization, determined by the equation [12]:

$$R^* = \frac{\Gamma}{\sigma^*}\left[\frac{L}{c}P_T - \sum\frac{2(m_V)_i(C_0)_i(P_C)_i(1-(k_V)_i)}{1-(1-(k_V)_i)Iv(P_C)_i}\right]^{-1}, \quad (6)$$

where Γ – Gibbs-Thomson coefficient; σ^* – stability coefficient ($\sigma^* \approx 0.025$); $(P_C)_i = \vartheta R^*/2(D_L)_i$ – diffusion Peclet number; $(D_L)_i$ – diffusion coefficient of *i*-component in the melt; summation sign marks the additive contribution of the alloy components.

Redistribution of components between phases is calculated using equation, which correlates content of the melt at the phases interface C_L^* with kinetically dependent crystallization conditions [12]:

$$(C_L^*)_i = (C_0)_i / [1-(1-k_V)\mathrm{Iv}(P_C)_i]. \quad (7)$$

This makes it possible to evaluate the average composition of the residual liquid phase $(C_L)_i$ by using the equation, transformed for the conditions of rapid crystallization, subject to the availability of cross-border enriched layer of small thickness $\delta = 2D_L/\vartheta$, having a content $(C_L^*)_i$, different from $(C_L)_i$ [12]:

$$\frac{d(C_L)_i}{df_s} = \frac{(C_L)_i - (k_V)_i(C_L^*)_i}{1-f_s(1-\sigma_i(k_V)_i)};\ \sigma_i = \frac{2\alpha_i}{2\alpha_i+1};\ \alpha_i = \frac{8(D_S)_i\tau_{LS}}{\lambda_2^2}, \quad (8)$$

where $(D_S)_i$ – diffusion coefficient of *i*–component in the solid phase; τ_{LS} – local solidification time; λ_2 – the size of microcells (dendrite arm spacing). Introduction $0 \le \sigma_i \le 1$ parameter in equation (8) includes a partial diffusion in the solid phase of *i*-component.

During high-speed crystallization, when k_V value increases to one, the thickness of border enriched layer $\delta \approx 2D_L/\vartheta$ becomes comparable with magnitude of interatomic distances in crystal lattice of solid phase [3], the conditions for entrainment and complete "trapping" of the components at the interface formed and the limiting state $(C_L^*)_i=(C_S^*)_i=(C_0)_i$ achieved.

Formed model allows us to analyze the evolution of the solidification conditions and the redistribution of the components of the alloy crystallization, covering a wide range of cooling rates, within which a change of the alloy particle size is initiated and completed the transition from cellular-dendrite structure to flat front of diffusionless crystallization.

3. Experimental procedure

To investigate the crystallization conditions at gas atomization and type of thus formed structure it was selected martensitic stainless steel X12CrMoV. Physical-chemical and thermal characteristics of the alloy in the liquid and solid state were determined using the thermodynamic modeling software of phase transformations in multicomponent alloys «Polytherm» [12], as well as with published data. Table 1-2 shows parameters and thermal characteristics of the investigated alloy relating to the phase diagram.

Table 1. Parameters of equilibrium phase diagram of steel 12CrMoV.

Element	Partition coefficient	Tangent of liquidus line angle, K/mass. %
C	0.14	92.6
Mn	0.75	4.00
Si	0.76	22.9
Cr	0.89	1.45
Ni	0.86	5.13
Mo	0.67	3.10

Note: $N = 7.44 \cdot 10^{16}$ m^{-3}; $(D_S)_C = 3.9 \cdot 10^{-9}$ m^2/s; $(D_S)_{Si} = 4.1 \cdot 10^{-12}$ m^2/s (δ-Fe, 1400 °C).

Table 2. Thermal characteristics of the studied steel.

Parameter	Dimension	Value
Liquidus temperature	°C	1488
Solidus temperature	°C	1399
Thermal conductivity of the melt	W/(m·K)	27
Thermal conductivity in the solid state	W/(m·K)	37
Melt density	kg/m^3	6800
Density in the solid state	kg/m^3	7270
Specific heat capacity of the melt	J/(kg·K)	690
Specific heat of solid	J/(kg·K)	775
Specific latent heat of crystallization	KJ/kg	196.5

Crystallization parameters required for solving the task in the analysis of thermal processes, was given on the basis of estimates made in the work [13]: the specific density of nucleation N in the melt ranges $9 \cdot 10^{14}$–$3 \cdot 10^{17}$ m^{-3}, the rate constant of nuclei growth k_ϑ–

from $2\cdot10^{-2}$ to $6\cdot10^{-3}$ m/(K·s). The experimental values of secondary dendrite arm spacing for steel X12CrMoV [14] showed are depended from local value of the cooling rate by equation $\lambda_2=54.38\,\vartheta_0^{-0,33}$.

Process conditions (initial melt temperature 1750 °C, the initial gas velocity of 1200 m/s, used gas - nitrogen) and the raw data to calculate the gas atomization process X12CrMoV alloy described in [8-9]; additionally specified V_D and V_{DI} values, which took characteristics: V_D = 17 m/s; V_{DI} = 3 m/s equal according estimations given in [10].

Investigations of the microstructure of the particles were carried out using an optical microscope Leica DMI5000 M at x50-x1000 magnifications, and scanning electron microscope (SEM) TESCAN Mira 3 LMU, operating at a magnification of x4 to x10^6 with accelerating voltage of from 200 V to 30 kV. SEM is equipped with Oxford Instruments for spectral energy-dispersive microanalysis (EDX) X-MAX^N based on nitrogen-free silicon drift X-ray detector.

To study dendrite structure of particles samples were pressed into an electrically conductive plastic, mechanically ground and polished, followed by etching with a solution containing 1.2 g NH_2SO_3H, 0.5-1 g NH_4FHF and 3 g of $K_2S_2O_5$, diluting with 100 ml of distilled water.

Research of local chemical heterogeneity and its quantitative analysis was carried out for particles with a diameter of 60 and 28.5 microns by constructing planar images distribution of elements with subsequent registration of content of test components in grid nodes, applied to an area the size 40x40 mm of particle area.

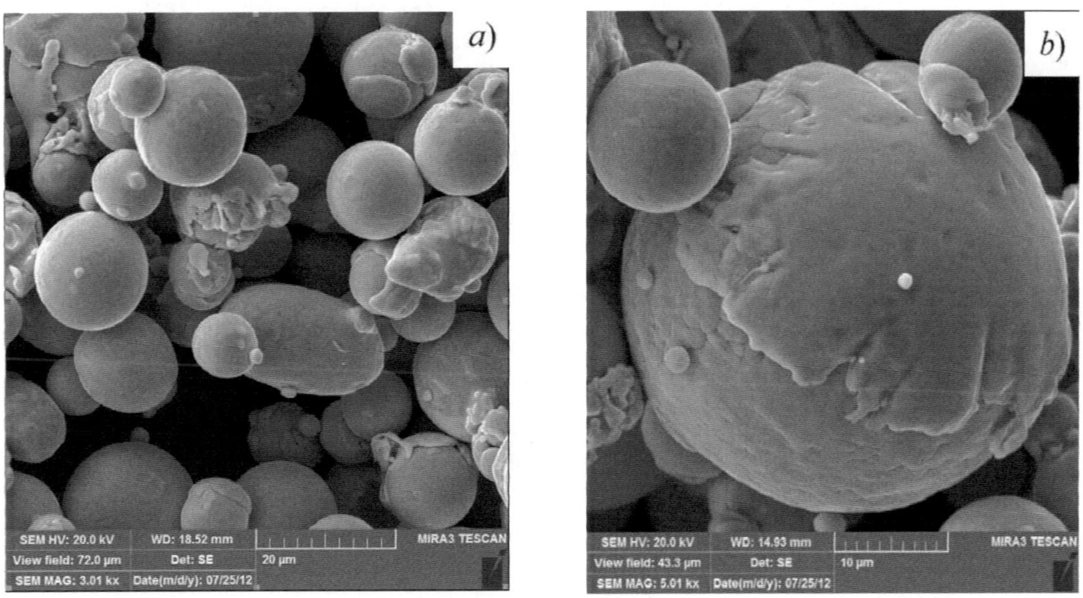

Figure 1. General view of X12CrMoV steel powder, received by gas atomization.

IOP Conf. Series: Materials Science and Engineering **192** (2017) 012009 doi:10.1088/1757-899X/192/1/012009

4. Characterization of atomized particles

General view of researched particles showed at Figure 1. Particles have near to spherical shape, some particles have satellites of smaller particles, which are attached to the surface of larger particles (Figure 1) or partially embedded therein (Figure 2,d) resulting in the collision (in solid state) and also during connection in liquid state (Figure 1,b).

Study of microsections at optical microscope showed that the particles have structure consisting of equiaxed dendrites (Figure 2,a-b) whose dimensions are on the periphery of the particle sizes do not differ from the central regions of dendrites. The collision of the particles, as shown by microsections (Figure 2,d), can initiate direction oriented cellular crystallites and their branching in the growth process.

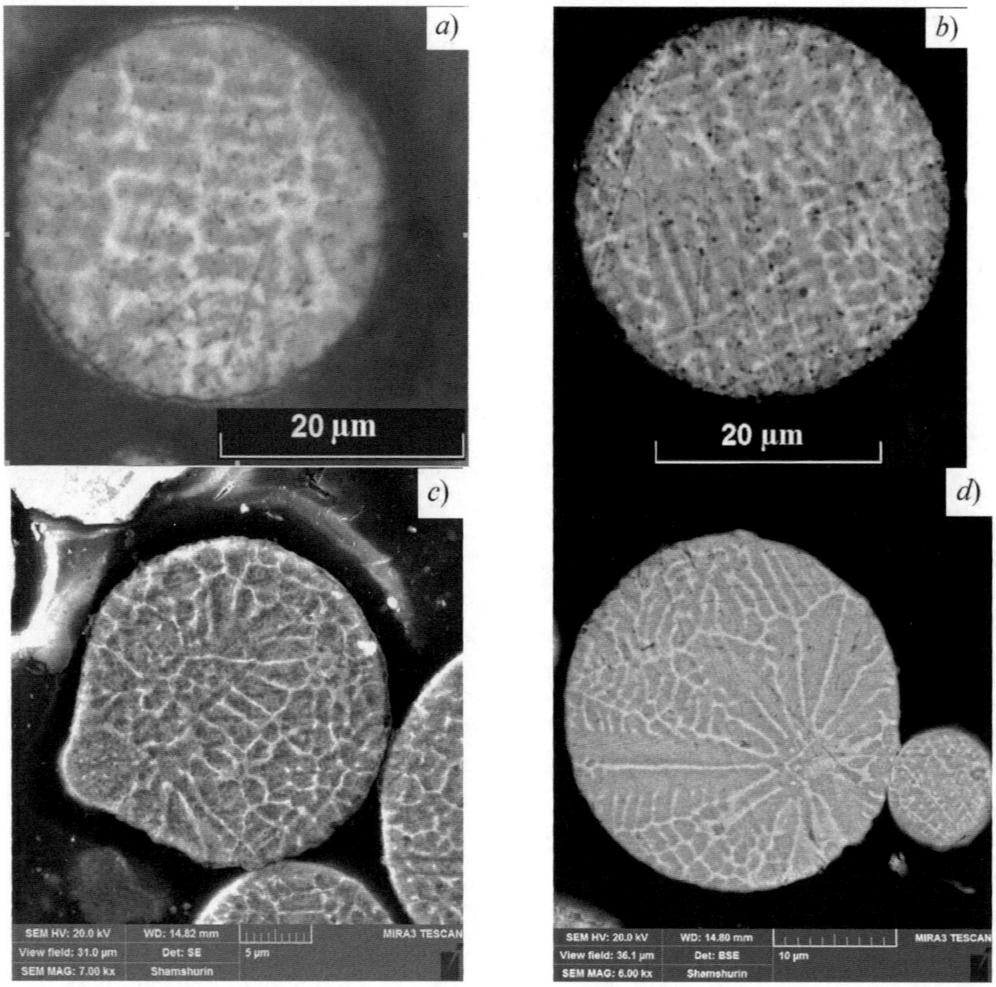

Figure 2. Cellular-dendrite microstructure of atomized powder Computer modelling results.

Computer analysis of thermal regime during solidification particles diameter of 10-100 microns alloys based on iron [8-9] shows that regular atomization modes provide substantial supercooling of the melt (50-120 K) and the cooling rate to 10^6-10^8 K/s.

Figure 3-4 shows the calculated solidification curves of steel X12CrMoV, which demonstrate the effect of most effective factors determining intensity of heat sink, supercooling of melt ΔT and average cooling rate during crystallization (Figure 4) at which the extrapolation gives for particle d = 1 mkm value V_0=10^8 K/s (Figure 4,c).

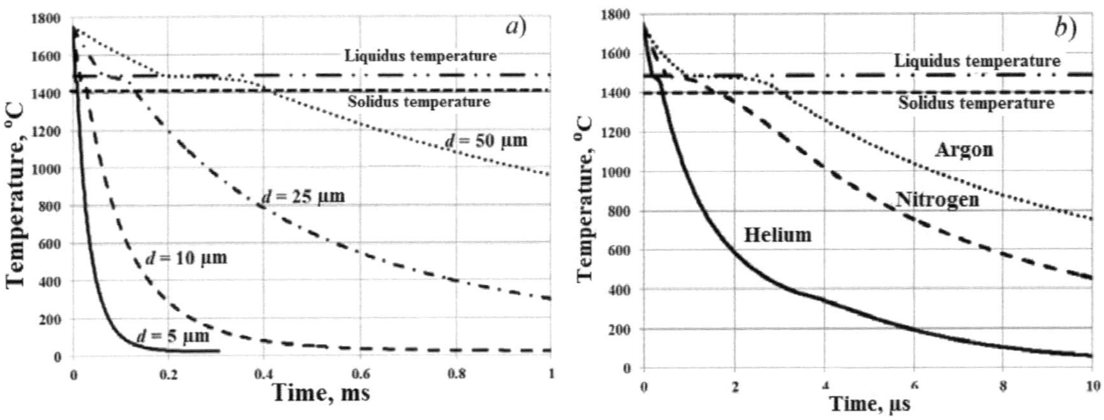

Figure 3. Thermal curves of solidification atomized particles depended of their size d (a) and (b) type of gas (for 50 mkm particle).

Summary results of the calculations (Table 3) show that decreasing particle size from 50 mkm to 1 mkm significantly increases the value of maximum supercooling ΔT_m and time of crystallization τ_{LS} is sharply reduced. This is due not only to the size of the particles, but also by increasing the heat transfer coefficient (from $1.2\cdot10^3$ to $800\cdot10^3$ W/m$^2\cdot$K) due to the fact that small particles are solidified in a short period when their speed and therefore the heat transfer coefficient have the greatest value, whereas the larger particles begin to solidify later and crystallize at a significantly lower flow rate.

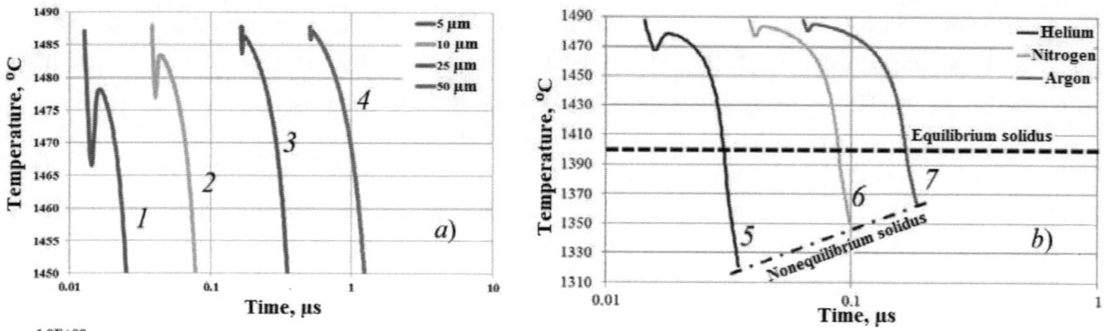

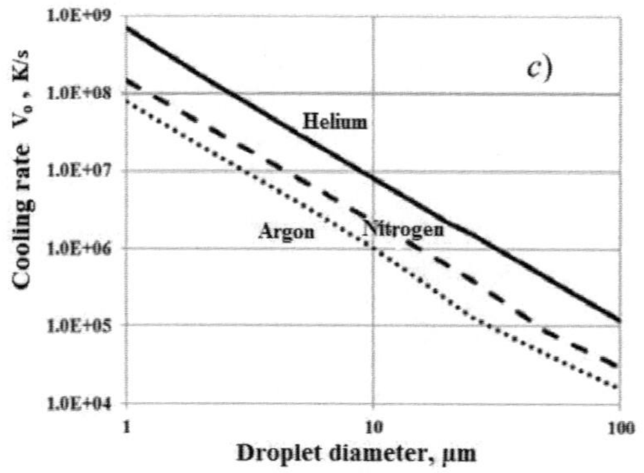

Figure 4. The dependence of initial supercooling ΔT (*a-b*) and average cooling rate V_0 (*c*) at solidification of particles different diameter (*a, c*) and different type of gas *d*, mkm: 5(*1*); 10(*2*); 25(*3*, *5 - 7*); 50(*4*) Gas: Helium (*5*); Nitrogen (*1- 4*, *6*); Argon (*7*).

Deep supercooling during the crystallization of small particles (up to 100-120 K) causes an increase in the maximum cooling rate during the crystallization $(V_0)_m$, which varies from 10^5 (for d=50 mkm) to 10^9 K/s (for d<1.0 mkm). At the moment of maximum supercooling also maximum crystallization speed ϑ_m achieved, calculated value of that varies from $4 \cdot 10^{-4}$ (for d=50 mkm) to 10^2 m/s respectively. Reducing specific nucleation density (from $7.5 \cdot 10^{16}$ to $1.5 \cdot 10^{15}$ m^{-3}) causes to significant increasing of supercooling ΔT_m and growth rate ϑ_m.

Table 3. Parameters of crystallization and redistribution of carbon and silicon in according to the particle size.

Parameter	Particle diameter *d*, mkm				
	1.0	2.0	4.0	10.0	50.0
ΔT_m, K	100	83	68	52	27
τ_{LS}, mkm	1.67	5.13	15.5	67.4	1575
$(V_o)_m$, 10^6 K/s	176	54.2	16.8	3.52	0.13
ϑ_m, m/s	7.92	2.65	0.902	0.218	0.011
Carbon					
$(k_V)_C$	0.805	0.549	0.332	0.191	0.145
$(m_V)_C$, K/%	209	197	165	119	94.2
C_L*, % C	0.248	0.363	0.595	1.00	1.19
C_S*, % C	0.200	0.199	0.198	0.192	0.174
$(C_L)_m$, % C	0.205	0.224	0.269	0.365	0.635

Silicon					
$(k_V)_{Si}$	0.918	0.811	0.753	0.664	0.641
$(m_V)_{Si}$, K/%	28.4	27.4	25.6	23.8	22.9
C_L*, % Si	0.269	0.304	0.341	0.370	0.381
C_S*, % Si	0.247	0.247	0.247	0.246	0.244
$(C_L)_m$, % Si	0.266	0.277	0.283	0.284	0.287

Designations: ΔT_m – maximum supercooling; τ_{LS} – local time of crystallization; $(V_0)_m$ – the maximum average cooling rate at crystallization; ϑ_m – maximum linear rate of crystallization speed; $(C_L)_m$ – the maximum content of a component in residual liquid phase.

The maximum value of Biot number in all cases is much less than unity (Bi=0.004-0.02), indicating non-important temperature difference over cross section of particles and characterizes bulk crystallization conditions. This is consistent with observed in Figure 2,a-b unoriented cell-dendrite structure with uniform microcells sizes λ, which are traced experimentally by varying the particle diameter of 5 to 130 mkm (Figure 5).

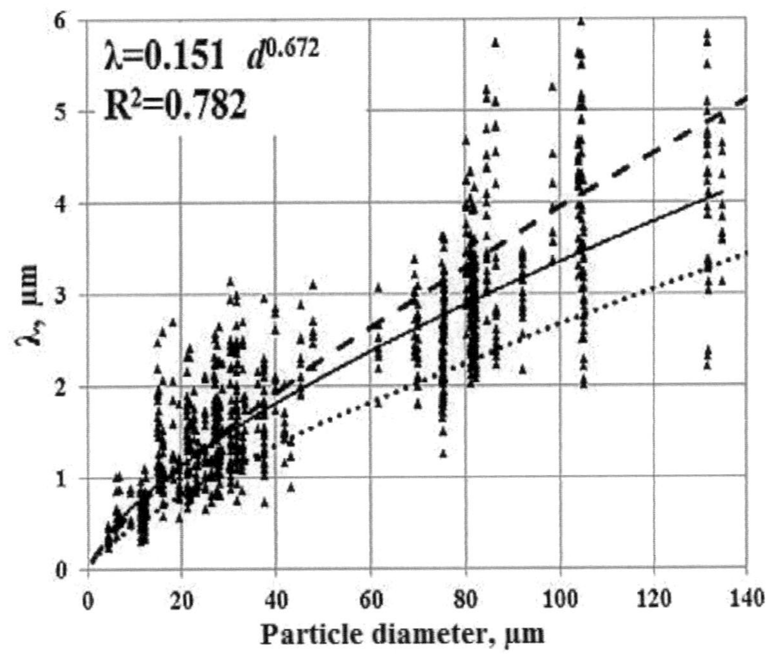

Figure 5. Dependence of dendrite secondary spaces λ_2 on particles diameter d in comparison with the calculation of [9] at initial gas velocity V_g = 100 (dotted line) and 1200 m/s (points). Line – statistical approximation of dependence $\lambda(d)$ [9].

To identify boundary values of parameters that determine changes in conditions redistribution of components at the phases interface in Table 3 and Figure 6 a number of characteristics of the crystallization process is mapped depending on the diameter of particles on example carbon.

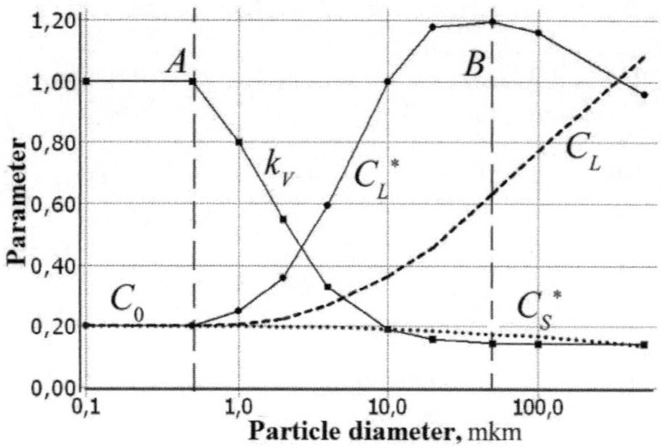

Figure 6. The effect of particle diameter d on the partition coefficient k_V, the carbon content C^*_L and C^*_S at the phases interface and the average composition of the liquid phase C_L (dotted line) during crystallization in nitrogen.

With increasing of particle diameter values of the partition coefficients are reduced from limit values $k_V = 1$ (at $d \leq A$) to equilibrium values (at $d \geq B$). In the same range of particle size (points A and B in Figure 6) there is an intensive change of liquid phase composition C^*_L at phases interface and solid phase composition C^*_S from the limits ($C_L^*=C_S^*=C_0$) at diffusionless crystallization (point A) to values that meet conditions of local equilibrium ahead of solidification front (point B).

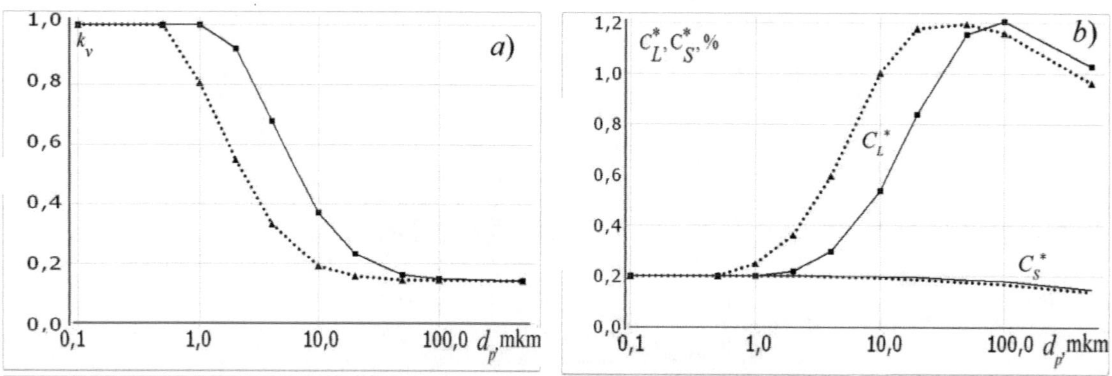

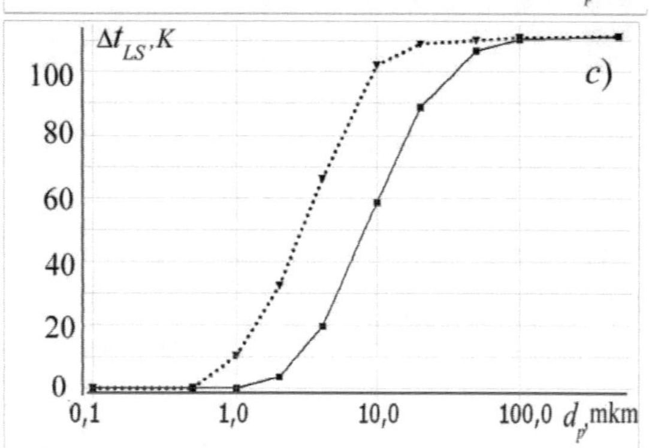

Figure 7. Dependency of distribution coefficient k_V (a), carbon concentration C_L^* and C_S^* in liquid and solid phase at the interface (b) and the solidification temperature interval $(\Delta t_{LS})_V$ (c) from particle diameter d_p and the type of working gas (line – He, dotted – Ar).

When used as a vehicle gas of argon and helium (Figure 6 - 7) a comparison of the curves k_V, C_L^*, C_S^* and $(\Delta t_{LS})_V$, depending on the diameter of the particles reveals a noticeable boundaries shift in the region of self-similarity mode at absolute stability (left to point A in Figure 6, where the change in diameter of particles has no effect at k_V =1, $(\Delta t_{LS})_V$ = 0,

$C_L^* = C_0$, $C_S^* = C_0$). A similar shift is occurring in the field of the dominant character of local equilibrium mode on the interface (right to point *B* with $k_V \approx k$, $\Delta t_{LS} \approx mC_0(1-k)/k$, $m_V \approx m$). This effect is due to the impact of more intensive heat sink, which is characteristic for helium with intensive heat transfer coefficient in the gas flow (see Figure 3,b and 4,b-c).

In order to assess microsegregation in solid phase it was made comparison of averaged data of silicon distribution across section of dendrite cells obtained by EDX, with the results of calculation (see Figure 8) [15], which takes into account diffusion suppression solid phase D_S in process of crystallization with local time τ_{LS} on the cross section of secondary dendrite space with size λ_2, whose value changes as a result of diffusion coalescence, described by the empirical equation $\lambda = K\,\tau_{LS}^{\,n}$ at $n = 0.33$.

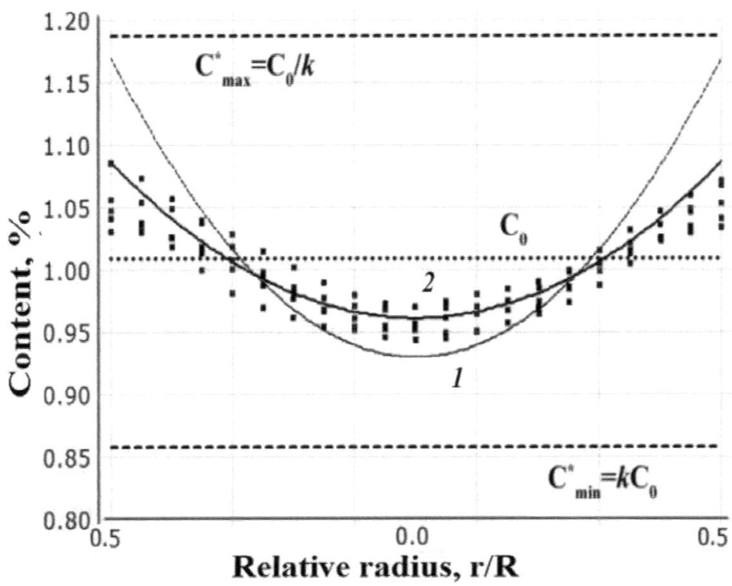

Figure 8. The silicon content distribution over the dendrite space cross section, according EDX (points) in comparison with simulation results (lines) at $k = 0.76$ (*1*) and $k_V = 0.85$ (*2*).

For atomized powder particles the agreement with the experimental data of the dendrite segregation of silicon is achieved (Figure 8) by using the partition coefficient $k_V = 0.85–0.87$ significantly exceeding the equilibrium value (k=0.76), which indicates the need to consider a significant contribution of nonequilibrium "trapping components" in the crystallization of particles diameter of 28-60 mkm, for which λ value is within 1.5-2.4 mkm and local time of crystallization from 0.55 to 1.85 mks, respectively.

5. Conclusion

As an important summary of the presented analysis, we can conclude that the studied thermal-physical conditions of gas atomization of melt (type and velocity of gas, supercooling of liquid, size of droplet and it's solidification rate et al.) not only indicate the influence on the formation of microstructure and chemical inhomogeneity of material but the ability to radical change the mode of components redistribution near the interface – from microsegregation to diffusionless type. This results indicate the ways and scopes for use of this technology in order to implement the transient regimes of local nonequilibrium crystallization, providing the required level of structural and chemical homogeneity of atomized powder and articles made of them.

6. References

[1] Lavemia E and Srivatsan T 2010 The rapid solidification processing of materials: science, principles, technology, advances, and applications *J. Mater. Sci.* **45** 287-325

[2] Glezer A and Permyakova I 2012 *Nanocrystals, tempered from melt* (Moscow: Fizmatlit)

[3] Herlach D M, Galenko P and Holland-Moritz D 2007 *Metastable solids from undercooled melts* (Amsterdam: Elsevier)

[4] Laheniht F et al. 2010 Jet compaction - advanced steelmaking process for high quality alloys *Metallurgist* **10** 36-43

[5] Grant P S, Cantor B and Katgerman L 1993 Modelling of droplet dynamic and thermal histories during sprayforming Pt. 1 *Acta Metall. Mater.* **41** 3097-3108

[6] Hattel J H et al. 1999 A quasi-stationary numerical model of atomized metal droplets Pt. 1 *Mater. Sci. Eng.* **7** 413-430

[7] Vedovato G, Zambon A and Ramons E 2001 A simplified model for gas atomization *Mater. Sci. Eng. A* **304-306** 235-239

[8] Sufiiarov V Sh, Golod V M and Gjulihandanov E L 2012 Modelling of thermal and crystallization processes in the gas atomizatsin of iron-based alloys Proc. 9th Int. Conf. Foundry today and tomorrow 478-494

[9] Sufiiarov V Sh, Golod V M and Gjulihandanov E L 2013 Simulation the conditions of forming a dendritic structure iron-based alloys by gas atomization *Powder metallurgy and functional coatings* **2** 14-19

[10] Danilov D A and Galenko P K 2008 Diagram of morphological stability in the crystallization of a binary alloy *Bull. Udmurt University Physics. Chemistry* **1** 88-100

[11] Galenko P K et al. 2009 Dendritic solidification in undercooled Ni-Zr-Al melts: Experiments and modeling *Acta Mater.* **57** 6166-6175

[12] Golod V M, Savel'ev K D and Basin A S 2008 Modeling and computer analysis of crystallization of multicomponent iron-based alloys (Saint Petersburg: Polytechnic Univ Press)

[13] Pryds N H and Pedersen A S 2002 Rapid solidification of martensitic stainless steel atomized droplets *Metall. Mater. Trans.* **33** A 3755-61

[14] Volkova O, Heller H-P and Janke D 2003 Microstructure and cleanliness of rapidly solidified steels *ISIJ Int.* **43** 1724-1732

[15] Shin Y H et al. 2001 An analytical model of microsegregation in alloy solidification *ISIJ Int.* **41** 158-163

IOP Conf. Series: Materials Science and Engineering **192** (2017) 012010 doi:10.1088/1757-899X/192/1/012010

Microstructure based simulations for prediction of flow curves and selection of process parameters for inter-critical annealing in DP steel

M J Deepu[a], H Farivar[b], U Prahl[b] and G Phanikumar[a*]

[a]Integrated Computational Materials Engineering Laboratory, Department of Metallurgical and Materials Engineering, Indian Institute of Technology Madras, Chennai, India - 600036
[b]Integrated Computational Materials Engineering group, Steel Institute (IEHK), RWTH Aachen University, Intzestraße 1, 52072 Aachen, Germany

E-mail: *gphani@iitm.ac.in

Abstract. Dual phase steels are versatile advanced high strength steels that are being used for sheet metal applications in automotive industry. It also has the potential for application in bulk components like gear. The inter-critical annealing in dual phase steels is one of the crucial steps that determine the mechanical properties of the material. Selection of the process parameters for inter-critical annealing, in particular, the inter-critical annealing temperature and time is important as it plays a major role in determining the volume fractions of ferrite and martensite, which in turn determines the mechanical properties. Selection of these process parameters to obtain a particular required mechanical property requires large number of experimental trials. Simulation of microstructure evolution and virtual compression/tensile testing can help in reducing the number of such experimental trials. In the present work, phase field modeling implemented in the commercial software Micress® is used to predict the microstructure evolution during inter-critical annealing. Virtual compression tests are performed on the simulated microstructure using finite element method implemented in the commercial software, to obtain the effective flow curve of the macroscopic material. The flow curves obtained by simulation are experimentally validated with physical simulation in Gleeble® and compared with that obtained using linear rule of mixture. The methodology could be used in determining the inter-critical annealing process parameters required for achieving a particular flow curve.

1. Introduction

Dual phase steels(DP steel) are advanced high strength steels that consists of hard martensite and soft ferrite phases in their microstructure. DP steel sheets are used in automobile industry due to the combination of high strength and formability, high crashworthiness, high tensile to yield strength ratio, high work hardening rate and good ductility and hence can help in automobile weight reduction [1, 2]. In future cars, it may occupy up to 80% of the total weight of car [3]. The DP steels used for sheet metal applications typically consists of around 0.2 and 0.8 volume fractions of martensite and ferrite respectively [4]. Thus, it contains hard martensite islands in soft ferrite matrix. DP steels also have high potential for application in bulk components like gear. DP steel has shown improved fatigue strength and decrease in heat treatment distortion of carburized gears and hence reduce the noise in automotive applications.

It also has added advantages that the existing manufacturing lines can be used without major modifications [3]. The DP steel that is to be used in such an application should typically have 0.8 and 0.2 volume fractions of martensite and ferrite respectively. Thus it contains soft ferrite islands in a hard martensite matrix. The present work focuses on this type of DP steel, that could be used for automotive gear applications.

Microstructure plays a major role in governing the properties of dual phase steels, especially the phase fractions, morphology and distribution of the phases [4]. The phase transformation happening during inter-critical annealing in DP steel has a major role in governing the microstructure of the material. Phase field modeling is a versatile tool that can be used to simulate the microstructure evolution [5–7]. It had been used previously to perform the simulation of phase transformation during inter-critical annealing in DP steel [1, 9]. The ability to predict mechanical properties from microstructure is of high demand as it can lead to predictive capability that can help in reducing the number of experimental trials required or to help in designing careful experiments. Among the various approaches that can perform this, Finite Element Method (FEM) is of special interest, as it takes into account the morphology of the phases [2, 4]. Ramazani et.al had used virtual tensile test using FEM to predict the macroscale flow curve from the microstructure [3]. This methodology can also be used to predict the high temperature flow curves of a material to judge the capability of a material for high temperature application. Attempts have also been made to couple phase field modeling and FEM to analyse the microstructure evolution and deformation behaviour [10–14].

In the present work, phase field simulation is used to simulate the microstructure evolution during the heat treatment process. Virtual tests are performed on the simulated microstructure to obtain the effective macroscopic flow curves and the obtained flow curve are compared with the ones obtained by using linear rule of mixture(ROM). Physical simulation in Gleeble® is used to validate the virtual test results.

2. Phase field simulation of microstructure evolution

The austenite to ferrite transformation during inter-critical annealing in DP steel was simulated using the multi phase field model(1) implemented in the commercial software Micress® [15].

$$\dot{\phi_\alpha} = \sum_{\beta=1,\dots,N} \mu_{\alpha\beta} \left\{ \sigma_{\alpha\beta} \left[\phi_\beta \nabla^2 \phi_\alpha - \phi_\alpha \nabla^2 \phi_\beta + \frac{\pi^2}{2\eta^2} (\phi_\alpha - \phi_\beta) \right] + \frac{\pi}{\eta} \sqrt{\phi_\alpha \phi_\beta} \Delta g_{\alpha\beta} \right\}, \quad (1)$$

where, ϕ_α and ϕ_β are the order parameters, $\mu_{\alpha\beta}$ is the interface mobility, $\sigma_{\alpha\beta}$ is the surface energy, $\Delta g_{\alpha\beta}$ is the thermodynamic driving force and η is the thickness of the diffuse interface. The multi phase field equation is coupled with the diffusion equation(2), since the driving force($\Delta g_{\alpha\beta}$) depends on the local concentration [9].

$$\dot{c} = \nabla \sum_{i=1}^{N} \phi_i D_i \nabla c_i \ , \quad (2)$$

where, D_i is the multi-component diffusion coefficient matrix. The thermodynamic and the diffusion data required for the simulation were obtained from the ThermoCalc® databases TCFE7 and MOBFE1 using the Micress® TQ interface. The heat treat treatment cycle, the chemical composition, simulation conditions and the experimental validation of the Micress® simulation is discussed in one of the work by the authors [16].

3. Physical simulation of high temperature compression test using Gleeble®

In order to obtain the flow curve, a compression test was performed on the material using Gleeble 3800®. A cylindrical specimen of 6mm radius and 9mm length was used for this purpose. The specimen was prepared using Electrical Discharge Machining(EDM) and thermocouples were welded on to it. Then it was solutionized and heat treated before the start of the compression test, so as to obtain the required dual phase microstructure in the material. Solutionizing was performed at 950 °C followed by cooling to the inter-critical annealing temperature of 790 °C at a cooling rate of 4 °C/min. It was then held at this temperature for 30 minutes, following which the compression test was performed at a strain rate of 0.001 s^{-1}. Figure 1 shows the overall thermal cycle that was used in Gleeble® and figure 2 shows the flow curve obtained.

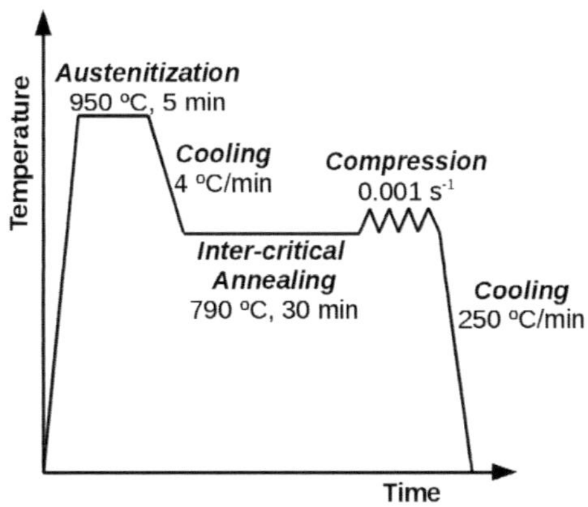

Figure 1. Overall heat treatment cycle.

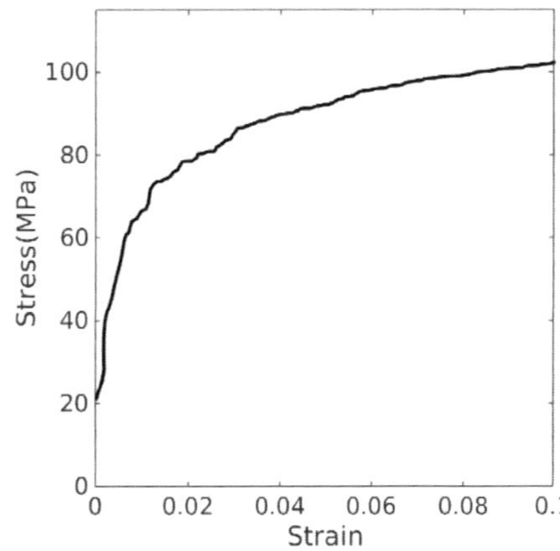

Figure 2. Flow curve obtained from Gleeble®.

4. Virtual testing for flow curves

In order to obtain the effective macroscale flow curve, virtual compression tests were performed on the phase field simulated microstructure using FEM implemented in the commercial software Abaqus®. The phase field simulated microstructure was assumed to be the Representative Volume Element (RVE). Figure 4 shows a phase field simulated microstructure in which the white colour represents the ferrite phase and the grey colour represents austenite phase. The RVE was meshed with C3D8 elements and periodic boundary conditions were applied. The properties of the individual phases, ferrite and austenite were obtained from JMatPro® database, for an austenitization temperature of 950 °C and an initial grain size of 10 μm. The initial grain size input was based on the experimentally observed initial austenite grain size, taken by averaging the data from several optical micrographs. Figure 3 shows one such optical micrograph of the initial austenite grains. The RVE was uniaxially loaded at 790 °C and the average von Mises stress and effective strain of all the elements in the microstructure were recorded to obtain the effective macroscopic flow curve. Figure 5 shows the von Mises stress distribution in the microstructure for an applied strain of 2%.

Linear rule of mixture (ROM) is one of the methods for obtaining the macroscale properties from the properties of individual phases and their phase fractions. The flow curve for the material can be calculated using ROM with the following equations:

$$\sigma = \sigma_\gamma V_\gamma + \sigma_\alpha V_\alpha \ , \tag{3}$$

$$\varepsilon = \varepsilon_\gamma V_\gamma + \varepsilon_\alpha V_\alpha \ , \tag{4}$$

where σ_γ and σ_α are the true stress, ε_γ and ε_α are the true strain and V_γ and V_α are volume fraction for austenite and ferrite phase respectively. This approach is also used in some of the

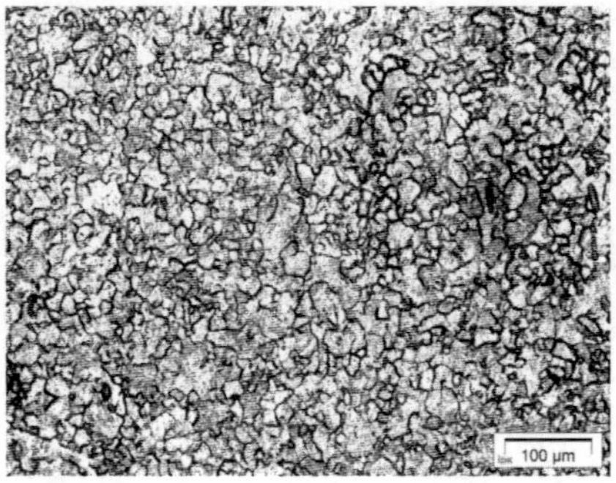

Figure 3. Representative light optical micrograph showing the prior austenite grains austenitized at 950 °C .

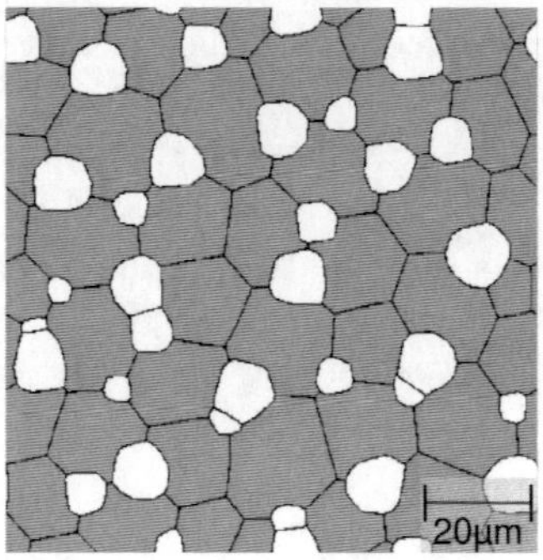

Figure 4. Phase field simulated microstructure in Micress®.

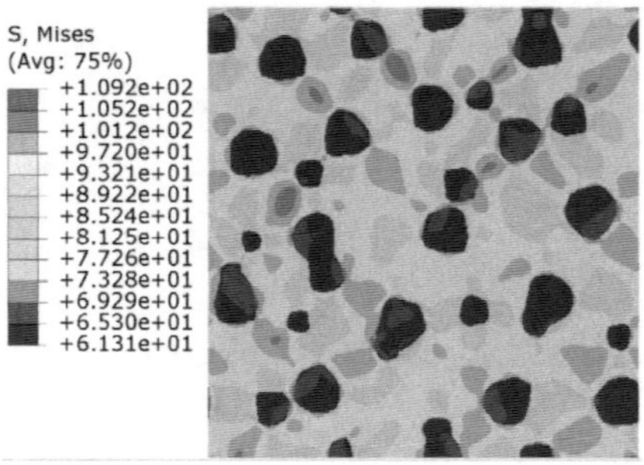

Figure 5. von Mises stress distribution during virtual test, for an applied strain of 2%.

commercial softwares for the calculation of properties. Hence, in the present work the results obtained from virtual tests are also compared with that obtained using ROM.

4.1. Simulations with and without consideration of grain orientations

While performing the virtual test simulations, the orientation of the grains are important, especially to distinguish two grains of same phase. Hence, a study was conducted to see the effect of the grain orientation assignment on the virtually obtained flow curve. Figure 6 shows the flow curves obtained when the orientation of the grains are not taken into account, along with the flow curve of austenite and ferrite phases, flow curve obtained with ROM and the macroscopic flow curve obtained from experiment. The microstructure was loaded in X and Y directions and both yielded same results. When the orientation of the grains are not taken into account, the virtually simulated flow curve matched closely with that of ROM, but both of these showed considerable deviation from the experimentally obtained flow curve. In order to consider the grain orientations, a random orientation was assigned to the grains using Micress® and these orientations were exported to the FEM software. Though the orientations were assigned to the grains, the material properties assigned were isotropic. Figure 7 shows the virtually obtained flow curve, along with the experimentally obtained flow curve and the one obtained by ROM. The virtually simulated flow curve deviated from ROM and moved closer to the experimentally obtained flow curve.

4.2. Simulations with varying initial grain size in JMatPro®

For obtaining the flow curves for the individual phases(austenite and ferrite) from JMatPro®, the austenitization temperature and the initial austenite grain size are to be provided as the inputs. The austenitization temperature in the present study is 950 °C and the initial austenite grain size is 10 μm. But, the model used in JMatPro® predicted an initial austenite grain size of 30 μm for an austenitization temperature of 950 °C. Hence, a study was conducted to see the effect of varying the initial grain size in JMatPro® on the final flow curves obtained by virtual test. The grain size was varied from 10 μm to 30 μm. Figure 8 shows the flows curves obtained for initial grain sizes of 10, 20 and 30 μm. The difference in the flow curves is due to the difference in the properties of the individual phases predicted by JMatPro®, when the initial grain size is varied. It was observed that for the grain size predicted by JMatPro®, the simulated flow curves matched very closely to that of the experiment. Though the matching with experiment is better for the grain size predicted by JMatPro®, in the following studies a grain size of 10 μm is used, since it is the actual grain size in the material.

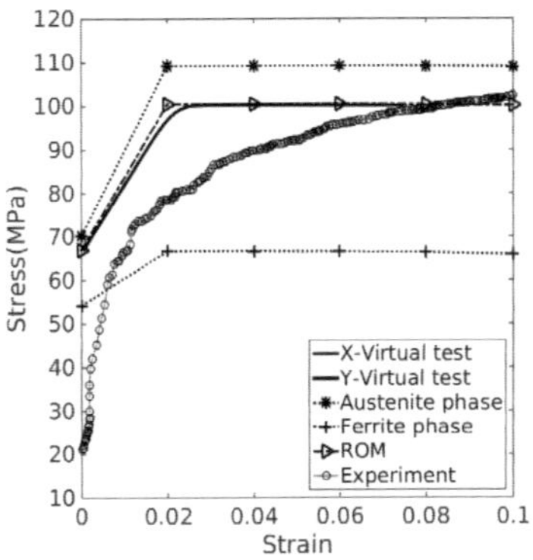

Figure 6. Virtual test without considering grain orientations.

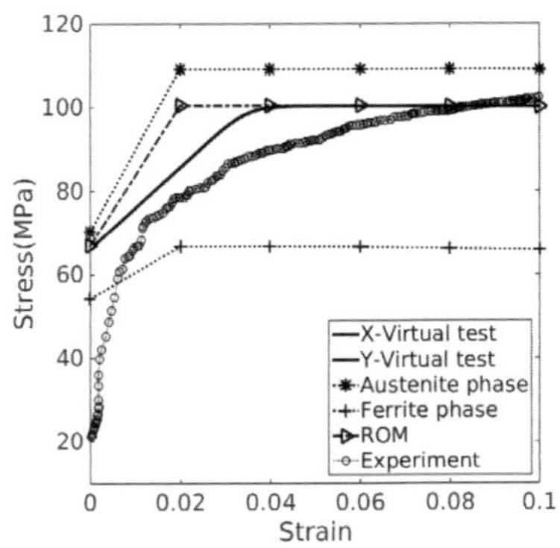

Figure 7. Virtual test with random orientations assigned to grains.

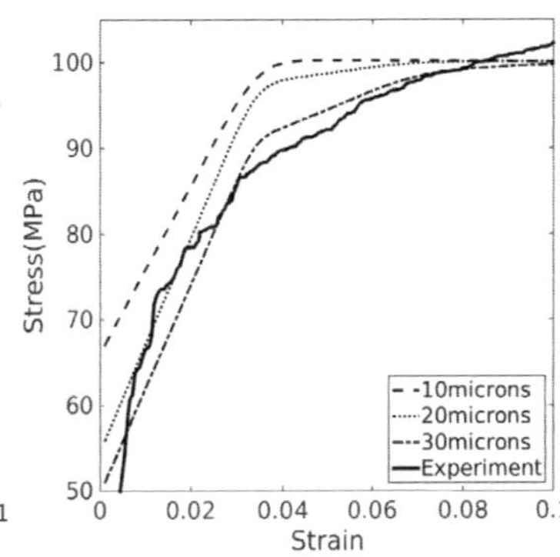

Figure 8. Virtual test with varying initial grain size in JMatPro®.

5. Towards the use of microstructure simulations for selection of process parameters for inter-critical annealing

The above simulation chain was repeated for different process parameters. The inter-critical annealing temperature and time was varied and the effect of these parameters on the virtually simulated flow curves was studied.

5.1. Flow curve variation with inter-critical annealing temperature

The inter-critical annealing temperature was varied between 720 °C and 780 °C in steps of 20 °C, for a constant holding time of 30 minutes. For each of these temperature, a phase field simulation was performed to predict the microstructure evolution during the cooling and holding after the solutionizing at 950 °C. Figure 9 shows the final microstructure obtained for various temperatures. White colour represents ferrite phase and grey colour represents austenite. With decrease in holding temperature, the final volume fraction of ferrite increased.

Virtual compression test was performed on each of the above simulated microstructure to obtain the flow curve. The properties of austenite and ferrite phases were obtained from JMatPro® for an initial grain size of 10 μm and austenitization temperature of 950 °C. Figure 10 shows the variation of flow curves with annealing temperature. With increase in temperature,

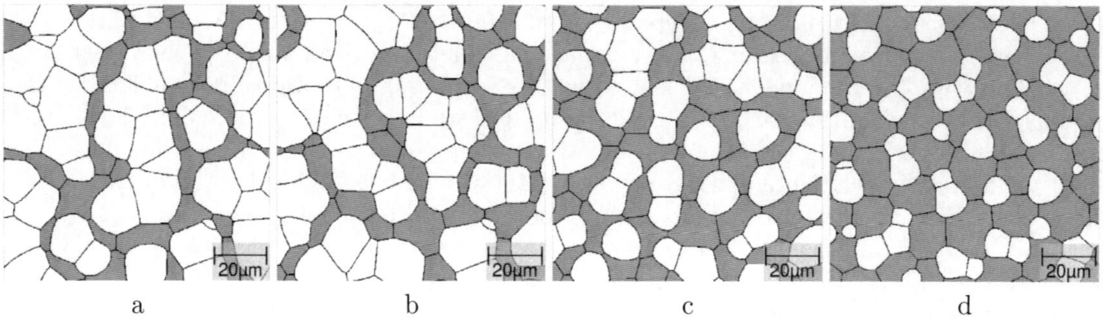

Figure 9. Final microstructure after 30 minutes holding at various temperatures: (a) 720 °C (b) 740 °C (c) 760 °C (d) 780 °C.

the flow curve shifted down. This decrease in the flow property is due to the combined effect of change in phase properties with temperature, change in volume fraction of phases and change in morphology. These combined effects are captured in the present methodology for obtaining the flow curves.

5.2. Flow curve variation with inter-critical annealing time

The inter-critical annealing time was varied for a constant holding temperature of 740 °C. The phase field simulated microstructure for each of the holding time was used to perform the virtual test for obtaining the flow curve. Figure 11 shows the flow curves obtained for various holding time. With increase in holding time, the flow curves shifted down. The primary reason for this decrease in flow property is the variation in the volume fraction of phases with time. There was only a slight change in the morphology with time during the isothermal holding.

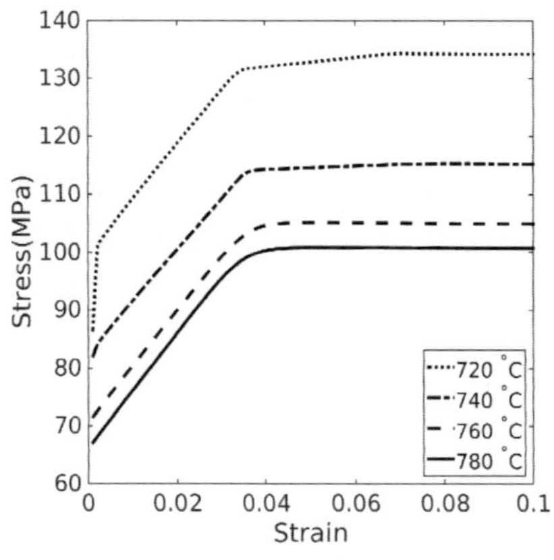

Figure 10. Flow curve variation with inter-critical annealing temperature for a constant holding time.

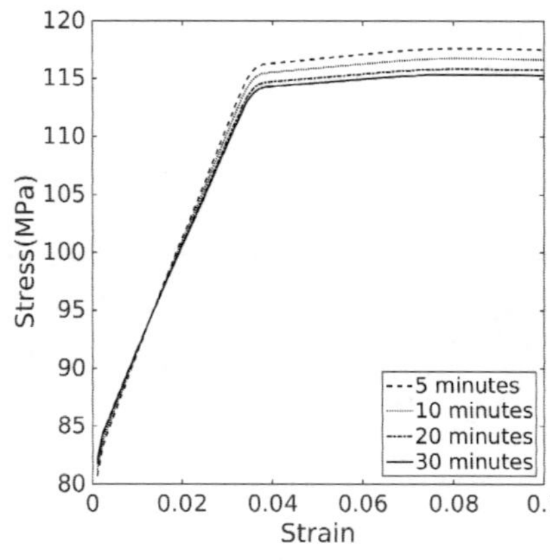

Figure 11. Flow curve variation with inter-critical annealing time for a constant holding temperature.

6. Conclusions

The microstructure evolution in dual phase steel during heat treatment was simulated using phase field simulation. Virtual test was conducted on the simulated microstructure to obtain the flow curve and a Gleeble® simulation was performed to validate the virtual test results. It was observed that the orientation of the grains do play a role in obtaining an accurate prediction of flow curve using virtual test. On comparing the obtained flow curves with ROM, it was observed that when the orientation of grains were not considered, the virtual test results came very close to ROM, but both deviated from experimental results. On assigning random orientation to grains, it was observed that the virtual test results deviated from the ROM and came closer to the experimental results. The properties of the individual phases required for the virtual test were obtained from JMatPro®. These properties varied on varying the initial grain size input to JMatPro®. On using the initial grain size predicted by the model in JMatPro® it was observed that the virtual test results came vary close to the experimentally obtained flow curve. The simulations were repeated for various inter-critical annealing temperatures and times. The methodology can be used in determining the inter-critical annealing process parameters required for achieving a particular flow property in a material. The present methodology is able to capture the combined effects of change in phase properties, change in volume fraction of phases and change in morphology with the variation in process parameters, within the limits of a two dimensional microstructure simulation.

7. References

[1] Zhu B and Militzer M 2014 *Metallurgical and Materials Transactions A: Physical Metallurgy and Materials Science* **46** 1073-1084
[2] Marvi-Mashhadi M, Mazinani M and Rezaee-Bazzaz A 2012 *Computational Materials Science* **65** 197 - 202
[3] Ramazani A, Mukherjee K, Quade H, Prahl U and Bleck W 2013 *Materials Science and Engineering A* **560** 129-139
[4] Ramazani A, Mukherjee K, Prahl U and Bleck W 2012 *Computational Materials Science* **52** 46-54
[5] Konter A W A, Farivar H, Post J and Prahl U 2016 *JOM* **68** 59-69
[6] Rudnizki J, Böttger B, Prahl U and Bleck W 2011 *Metallurgical and Materials Transactions A: Physical Metallurgy and Materials Science* **42** 2516-2525
[7] Ramazani A, Li Y, Mukherjee K, Prahl U, Bleck W, Abdurakhmanov A, Schleser M and Reisgen U 2013 *Computational Materials Science* **68** 107-116
[8] Fukuoka Kazuaki, Tomita Kunikazu S T 2010 *JFE Technical Report* **15** 17-23
[9] Jenny Rudnizki, Prahl U, and Bleck W 2012 Phase-field modelling of microstructure evolution during processing of cold-rolled dual phase steels *Integrating Materials and Manufacturing Innovation* **1:3**
[10] Yamanaka A and Takaki T 2012 *International Journal of Automation Technology* **7** 16-23
[11] Yoshimoto C and Takaki T 2014 *ISIJ International* **54** 452-459
[12] Cho Y, Kim J, Cho H, Cha P, Suh D, Lee J and Han H 2012 Analysis of transformation plasticity in steel using a finite element method coupled with a phase field model *PloS one* **7** e35987
[13] Jamshidian M, Thamburaja P and Rabczuk T 2016 *Computational Physics* **327** 779-798
[14] Vasconcelos P, Giebmann A, Oliveira J and Andrade-Campos A 2014 *Key Engineering Materials* **611-612** 117-124
[15] Steinbach I 2009 *Modelling Simul. Mater. Sci. Eng* **17** 073001
[16] Deepu Mathew John, Hamidreza Farivar, Gerald Rothenbucher, Ranjeet Kumar, Pramod Zagade, Danish Khan, Aravind Babu, Gautham BP, Ralph Bernhardt, Phanikumar G, Ulrich Prahl 2017 An attempt to integrate software tools at microscale and above towards an ICME approach for heat treatment of a DP steel gear *Proceedings of the 4th World Congress on Integrated Computational Materials Engineering (ICME 2017)* (under review)

Acknowledgements

The authors would like to acknowledge Mr. Allan Roy, Mr. Siddharth Kumar, Mr. Aaditya Lakshmanan and Mr. Rahul M. R. of Department of Metallurgical and Materials Engineering, IIT Madras for their help in this work. The financial support from German Federal Ministry of Education and Research (BMBF) and Department of Science and Technology (DST), India under the context of Indo-German Science and Technology Center (IGSTC) is gratefully acknowledged.

Multistability and stochastic phenomena in a randomly forced thermochemical system

I Bashkirtseva[1], A N Pisarchik[2] and L Ryashko[1]

[1] Ural Federal University, Lenina, 51, Ekaterinburg, 620000, Russia

[2] Center for Biomedical Technology, Technical University of Madrid, 28223 Pozuelo de Alarcon, Madrid, Spain

E-mail: lev.ryashko@urfu.ru

Abstract. A model of the thermochemical reactor with well mixing is considered. An impact of random noise on the attractors of this model is studied in zones of saddle-node bifurcations with transitions from mono- to bimodal dynamics. Phenomena of the stochastic excitability and noise-induced transitions are demonstrated. Critical values of the noise intensities corresponding to the onset of generation of the large-amplitude stochastic oscillations are estimated.

1. Introduction

It is well known that in nonlinear systems even a weak noise can generate the wide variety of stochastic phenomena [1-4]. Such effects as the stochastic resonance, noise-induced chaos, stochastic bifurcations are observed in various nonlinear models in different domains of science [5-9]. In particular, the mathematical models of chemical kinetics [10, 11] demonstrate the complex dynamic regimes extremely sensitive to noise. A nonlinear model of thermochemical reactor [12-14] is one of such examples with excitable stochastic dynamics.

In the present paper, we study this model in zones of saddle-node bifurcations where the transitions from mono- to bimodal behavior are observed. An impact of random noise on the attractors of this model is investigated. We focus on two variants of the saddle-node bifurcation when the additional attractor (equilibrium or cycle) appears.

In Section 2, the dynamic regimes of the deterministic model of the thermochemical reactor with well mixing are shortly discussed.

In Section 3, an influence of random noise on attractors of this model is studied both in the mono- and bistable zones. In zones of stable equilibria near the saddle-node bifurcation, a phenomenon of the stochastic excitability is shown. We demonstrate how increasing noise transforms the system to the regime of large-amplitude stochastic oscillations. The noise-induced transitions between coexisting attractors (equilibrium–equilibrium and equilibrium–cycle) are discussed.

2. Deterministic thermochemical model

Consider a deterministic system

$$\begin{aligned} \dot{x} &= \sqrt{y}\left(-x\exp\left(-\frac{\delta}{y}\right) + p\,(1-x)\right) \\ \dot{y} &= \frac{2}{3}q\sqrt{y}\left(x\exp\left(-\frac{\delta}{y}\right) + r\,(1-y)\right) \end{aligned}, \qquad (1)$$

where x and y are the dimensionless concentration and temperature, respectively; parameters p, q, r and δ are positive. The system (1) models the dynamic processes in the thermochemical reactor with well mixing [12].

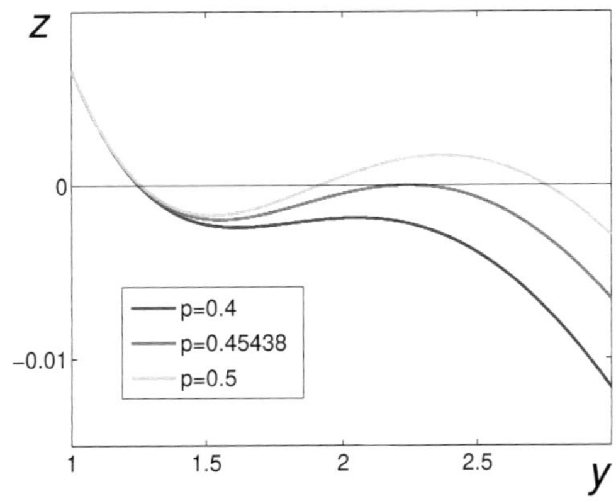

Figure 1. Plots of the function $z(y)$.

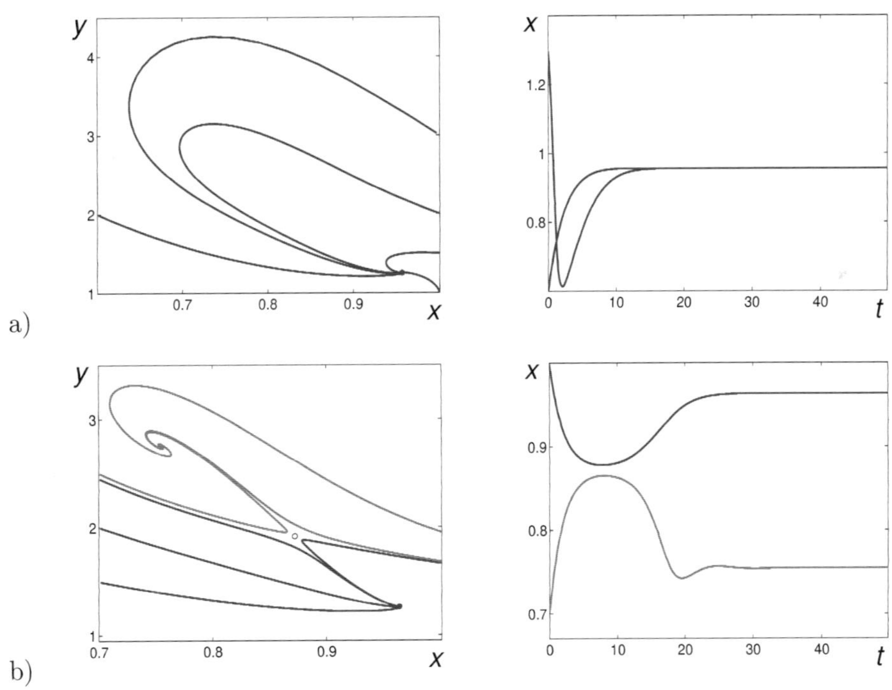

Figure 2. Phase trajectories and time series of deterministic system (1) with $\delta = 5$, $r = 0.07$, $q = 40$ and a) $p = 0.4$, b) $p = 0.5$.

In the present paper, we study a behavior of this system in zones of the transition from mono- to bistability. Here, we consider a case when an appearance of the additional attractor occurs as a result of the saddle-node bifurcation. In system (1), two variants of the saddle-node bifurcation can be observed. One of them (variant A) is connected with the birth of the pair of the stable and unstable equilibria. The second one (variant B) is a result of the appearance of stable and unstable cycles.

At first, consider the variant A of the saddle-node bifurcation for the fixed $\delta = 5$, $r = 0.07$, $q = 40$ and varying p.

The coordinates of equilibria of model (1) satisfy the following system

$$z(y) = 0, \qquad x = \frac{p}{p + \exp\left(-\frac{\delta}{y}\right)},$$

where

$$z(y) = \frac{p}{p\exp\left(\frac{\delta}{y}\right) + 1} + r(1 - y).$$

In Fig. 1, plots of the function $z(y)$ are shown for three values of parameter p. As can be seen, system (1) has a single equilibrium for $p = 0.4$, and has three equilibria for $p = 0.5$. The value $p_* = 0.45438$ where the plot of $z(y)$ is tangent to the y-axis corresponds to the saddle-node bifurcation. In Fig. 2, the phase portraits and time series of system (1) for $p = 0.4$ and $p = 0.5$ are presented.

The system (1) with $p = 0.4$ is monostable: all trajectories tend to the single stable equilibrium $\bar{x} = 0.9561$, $\bar{y} = 1.2508$ (see Fig. 2a).

For $p = 0.5$, system (1) is bistable with two stable equilibria $M_1(0.7544, 2.7544)$ and $M_2(0.9634, 1.2612)$, separated by the saddle equilibrium $M_3(0.8719, 1.9153)$ (see Fig. 2b). The basins of attraction of equilibria M_1 and M_2 are separated by the stable manifold of the saddle M_3.

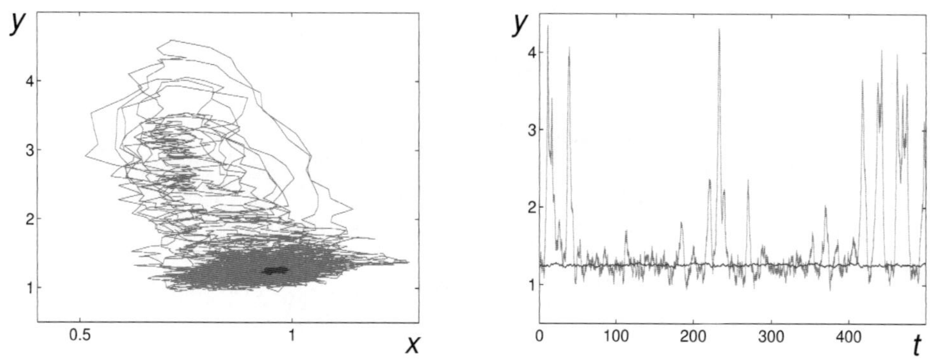

Figure 3. Phase trajectories and time series of stochastic system (2) with $\delta = 5$, $r = 0.07$, $q = 40$, $p = 0.4543$ for $\varepsilon = 0.01$ (blue) and $\varepsilon = 0.1$ (red).

3. Stochastic model

To study an impact of random disturbances on the dynamics of the thermochemical system, we will use the following stochastic model:

$$\begin{aligned} \dot{x} &= \sqrt{y}\left(-x\exp\left(-\tfrac{\delta}{y}\right) + p\,(1 - x)\right) + \varepsilon_1\xi_1(t), \\ \dot{y} &= \tfrac{2}{3}q\sqrt{y}\left(x\exp\left(-\tfrac{\delta}{y}\right) + r\,(1 - y)\right) + \varepsilon_2\xi_2(t). \end{aligned} \tag{2}$$

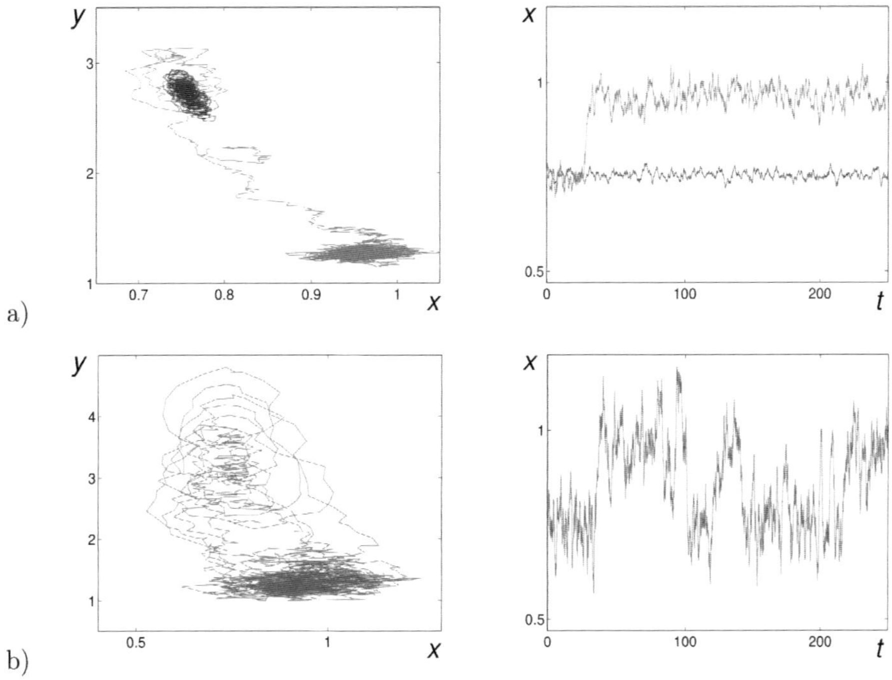

Figure 4. Phase trajectories and time series of the stochastic system (2) with $\delta = 5$, $r = 0.07$, $q = 40$, $p = 0.5$ for a) $\varepsilon = 0.01$ (blue), $\varepsilon = 0.03$ (red); b) $\varepsilon = 0.08$ (green).

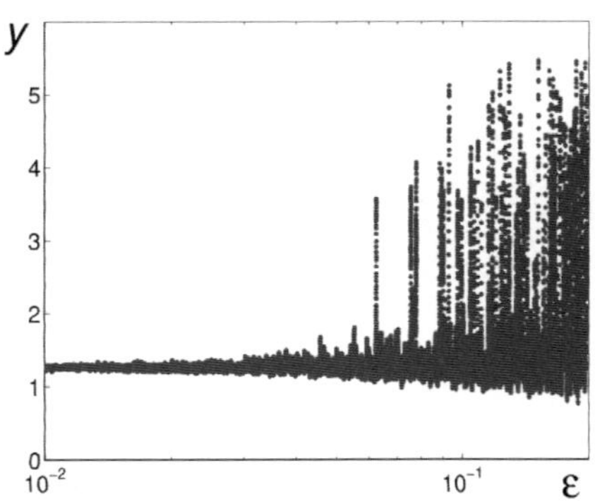

Figure 5. Random states of the stochastic system (2) $\delta = 5$, $r = 0.07$, $q = 40$, $p = 0.5$.

Here, $\xi_{1,2}(t)$ are the standard Gaussian uncorrelated processes, and $\varepsilon_{1,2}$ are the noise intensities. In what follows, we put $\varepsilon_1 = \varepsilon_2 = \varepsilon$.

3.1. Stochastic excitability

Consider an influence of noise on the monostable system with $p = 0.4543$ that is close to the saddle-node bifurcation point p_*. In Fig. 3, for two values of the noise intensity, the random trajectories of system (2) starting from the stable equilibrium are shown.

For a weak noise ($\varepsilon = 0.01$), the random trajectories are localized in a small vicinity of the equilibrium, and system (2) exhibits the small-amplitude stochastic oscillations (SASO). When the noise intensity increases and exceeds some threshold value, along with SASO, the large-amplitude stochastic oscillations (LASO) appear (see Fig. 3 for $\varepsilon = 0.1$). So, an alternation of SASO and LASO specifies a stochastic excitability of the stochastic system (2) close to p_*.

3.2. Bistability and noise-induced transitions

Consider now an impact of noise on the system in the bimodal zone where two stable equilibria M_1 and M_2 coexist.

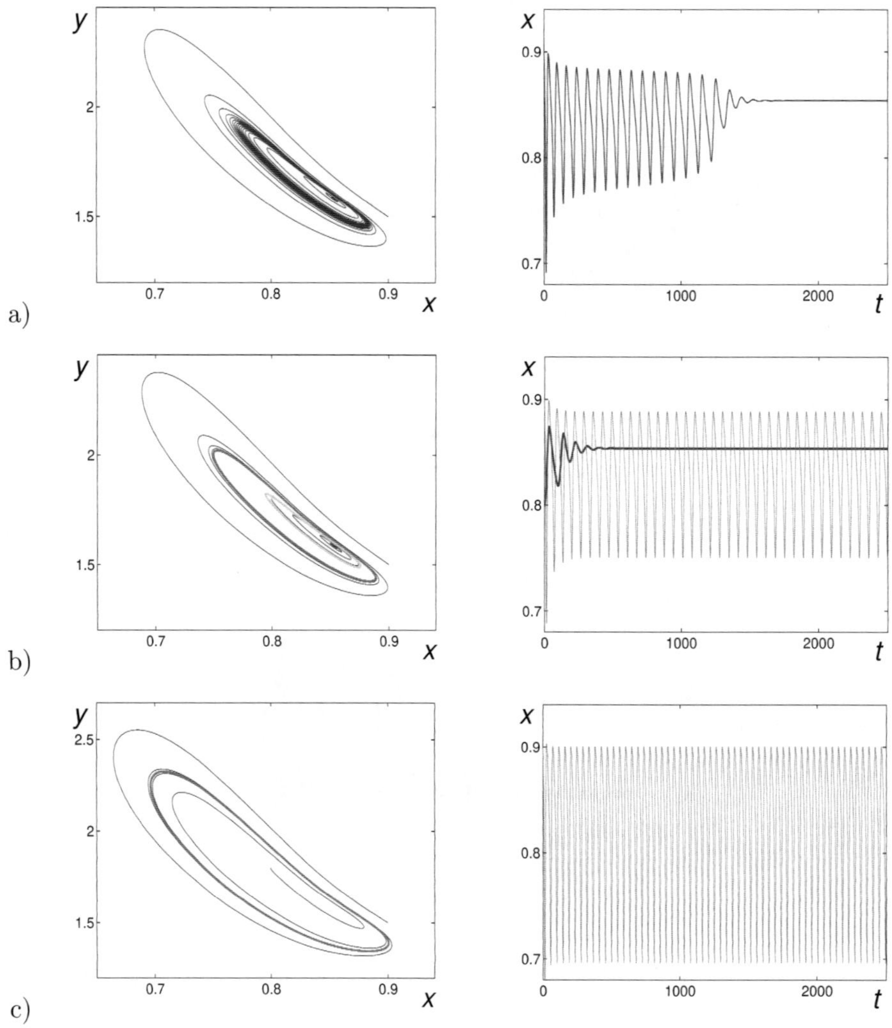

Figure 6. Phase trajectories and time series of the deterministic system (2) with $\delta = 5$, $r = 0.0622$, $p = 0.2475$ for a) $q = 39.6$, b) $q = 40$, c) $q = 43.5$.

In Fig. 4, for three values of the noise intensity, the random trajectories of system (2) starting from the stable equilibrium M_1 are shown.

For a weak noise ($\varepsilon = 0.01$), the random trajectories are still reside in a small vicinity of the equilibrium M_1. When the noise intensity increases, a transition from the basin of attraction of

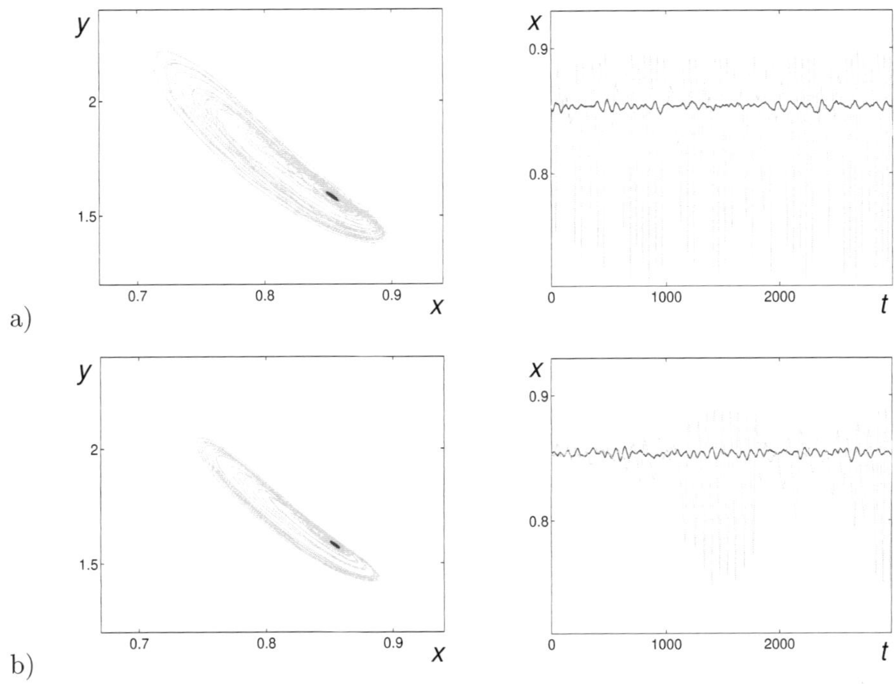

Figure 7. Phase trajectories and time series of the stochastic system (2) with $\delta = 5$, $r = 0.0622$, $p = 0.2475$ for a) $q = 39.6$ and $\varepsilon = 0.0001$ (blue), $\varepsilon = 0.001$ (green); b) $q = 40$ and $\varepsilon = 0.0001$ (blue), $\varepsilon = 0.0003$ (green).

M_1 to M_2 occurs (see Fig. 4a for $\varepsilon = 0.03$). So, the noise induces a transition of the considered thermochemical system to another alternative noisy equilibrium regime.

A further increase of noise generates the repetitive transitions between basins of attraction of M_1 and M_2. As a result, the large-amplitude oscillations are observed (see Fig. 4b for $\varepsilon = 0.08$). Some details of such changes in stochastic dynamics are well seen in Fig. 5 where the random states of the stochastic system (2) versus noise intensity ε are plotted. Here, the critical values of noise intensities corresponding to the onset of generation of the large-amplitude stochastic oscillations can be estimated.

3.3. Stochastic dynamics near the bifurcation of the cycle birth

Consider now the variant B of saddle-node bifurcation with the appearance of stable and unstable cycles.

Following [14], we put $\delta = 5$, $r = 0.0622$, $p = 0.2475$ and consider the dynamics depending on parameter q.

In Fig. 6, for three values of parameter q, the phase portraits and time series of deterministic system are shown. For $q = 39.6$, the system is monostable with a stable focus as a single attractor, and all trajectories tend to it (see Fig. 6a). For $q = 40$, the system is bistable and demonstrates a coexistence of the stable equilibrium and stable cycle. Their basins of attraction are separated by the unstable cycle (see Fig. 6b). For $q = 43.5$, the system is again monostable with a stable cycle as a single attractor (see Fig. 6c).

It is interesting that near the saddle-node bifurcation, in the zone of monostability where a stable equilibrium is a single attractor, system (2) is excitable as well. Indeed, for a small noise,

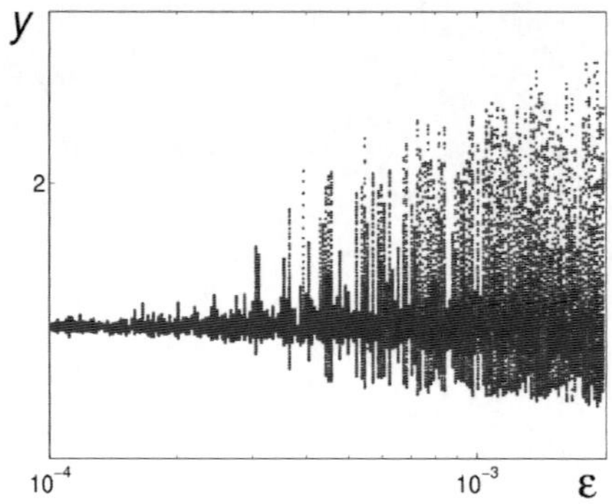

Figure 8. Random states of the stochastic system (2) with $\delta = 5$, $r = 0.0622$, $p = 0.2475$, $q = 39.6$.

system (2) exhibits SASO with random trajectories located near the stable equilibrium (see Fig. 7a) for $\varepsilon = 0.0001$). With increasing noise, an alternation of SASO and LASO is observed (see Fig. 7a) for $\varepsilon = 0.001$).

Such alternation in the zone of bistability (see Fig. 7b) is also observed.

Some details of such noise-induced transitions to LASO are shown in Fig. 8 where the random states of stochastic system (2) versus noise intensity ε are plotted for $q = 39.6$.

As a conclusion, we have to note that the nonlinear model under consideration demonstrates a wide variety of noise-induced effects which must be taken into consideration in the engineering problems.

4. References

[1] Moss F and McClintock P V E 1989 *Noise in Nonlinear Dynamical Systems* (Cambridge: Cambridge University Press)
[2] Horsthemke W and Lefever R 1984 *Noise-Induced Transitions* (Berlin: Springer)
[3] Fedotov S, Bashkirtseva I and Ryashko L 2002 *Phys. Rev. E* **66** 066310
[4] Bashkirtseva I, Chen G and Ryashko L 2010 *Int. J. Bifurcat. Chaos* **20** 1439
[5] Anishchenko V S, Astakhov V V, Neiman A B, Vadivasova T E and Schimansky-Geier L 2007 *Nonlinear Dynamics of Chaotic and Stochastic Systems. Tutorial and Modern Development* (Berlin/Heidelberg: Springer-Verlag)
[6] Lindner B, Garcia-Ojalvo J, Neiman A and Schimansky-Geier L 2004 *Phys. Rep.* **392** 321
[7] Bashkirtseva I, Ryashko L and Schurz H 2009 *Chaos Solitons Fractals* **39** 72
[8] Ryashko L and Bashkirtseva I 2011 *Phys. Rev. E* **83** 061109
[9] Pisarchik A N and Feudel U 2014 *Phys. Rep.* **540** 167
[10] Field R J and Burger M 1985 *Oscillations and Traveling Waves in Chemical Systems* (New York: Wiley)
[11] Jakobsen H A 2014 *Chemical Reactor Modeling: Multiphase Reactive Flows* (Berlin: Springer)
[12] Nowakowski B and Kawczynski A L 2005 *J. Phys. Chem. A* **109** 3134
[13] Kawczynski A L and Nowakowski B 2008 *Phys. Chem. Chem. Phys.* **10** 289
[14] Kolbus A, Nowakowski B and Kawczynski A L 2013 *Eur. Phys. J. B* **86** 262

Acknowledgments
This work was supported by RFBR (16-08-00388).

IOP Conf. Series: Materials Science and Engineering **192** (2017) 012012 doi:10.1088/1757-899X/192/1/012012

On the existence of metastable microheterogeneities in metallic melts

P Popel[1], **U Dahlborg**[2] **and M Calvo-Dahlborg**[2]

[1] Ural State Pedagogical University, Ekaterinburg, Russia

[2] GPM-UMR6634-CNRS, University of Rouen Normandie, Saint-Etienne-duRouvray, France

E-mail: pspopel@mail.ru

Abstract. Whether a molten alloy after melting is homogeneous or not on the atomic scale has been an issue for many experimental investigations, which have used a manifold of techniques. In most of these investigations, the anomalies in measured property/temperature curves, which were found at temperatures well above the liquidus temperature, have been interpreted as the indications for the existence of a dissolution process of alloy microheterogeneities which were inherited from the precursor solid alloy. In this report new results on the structural and transport properties of Al-Si and Al-Cu alloys are presented. In particular, new accurate measurements of the viscosity and the density of Al-Cu liquid alloys are reported. A large difference between the quantities measured during heating and the quantities obtained after the subsequent cooling was observed in narrow composition ranges corresponding to the stoichiometric compositions AlCu and Al_2Cu. Furthermore, the ultrasound velocity and viscosity measurements on alloys, which are suggested to be used as heat carriers in nuclear power plants (Pb-Bi, Pb-Sn and Ga-In alloys of near-eutectic concentrations), are presented. The results show that microheterogeneities exist in the melts at their planned temperatures of use.

1. Introduction

Many experimental results, which were obtained since 1960, demonstrate non-homogeneity of liquid eutectic alloys, i.e. existence of microscopic domains enriched with different components inside them. Basing on the results of his densimetric experiments on Pb-Sn melts, one of the authors of the present paper (P.P) had formulated a hypothesis of metastable microheterogeneity of liquid alloys of an eutectic type [1]. According to him, the thermodynamically stable state of eutectic melts at all temperatures of their existence is the state of a homogeneous solution. Their micro-inhomogeneity is determined by a prolonged existence of microdomains of various composition, which were inherited from the original heterogeneous ingot. Relying on their size (about 10 nm and larger), these domains can be regarded as dispersed particles, and the melt itself – as a microheterogeneous (colloid) system consisting of both dispersed and continuous phases. After the melting is over the eutectic melt relaxes into a thermodynamically stable state. However, this process can pass kinetically very slow and come to an end with the establishment of a metastable equilibrium between the dispersed particles and the surrounding melt. Such a microheterogeneous structure gets irreversibly destroyed when it undergoes the overheating above the liquidus up to temperatures, which depend on the composition of the melt or because of the other external factors.

Later many experimental results have indirectly confirmed the validity of the concept of metastable microheterogeneity of eutectic melts (see, for example, [2,3]). These ideas were theoretically substantiated in the context of Cahn-Hilliard thermodynamics of inhomogeneous systems [4]. However, it was only the end of the 1990s, when the small angle neutron scattering (SANS) measurements were represented as a direct confirmation of the existence of a microheterogeneity state in a molten eutectic Sn-26.1 at.% Pb alloy [5]. It was shown that the melt contains the dispersed particles, which greatly vary in size (from 1 to 80 nm) and also that their size depends on the heat treatment of the melt. Irreversible changes in their size were detected after heating the melt up to the maximum temperature of the experiment.

It is the aim of this paper to present some new experimental data developing the idea of metastable microheterogeneity of melts and useful for widening the field of technological applications of alloys. We concentrate on two kinds of alloys: two different Al-based ones and some liquid alloys, which are considered suitable to be applied as the coolants in nuclear and thermonuclear reactors. Some of the results have already been partly reported earlier, but here they are represented together in a general context.

2. Results of investigating Al-Si melts by neutron scattering techniques

With regard to AlSi alloys, there exists a large bibliography on the influence of superheat on the properties of solidified binary alloys, e.g. [6-9], but also for industrial alloys such as A356 [10, 11]. The reported effects generally deal with the fact that a finer and/or more uniform microstructure, as well as an increase of yield strength, tensile strength and ductility, occurs in the samples, which were produced with superheating of the melt before the solidification [12-15]. Moreover, even if on an atomic scale the structure for different compositions and at different temperatures, as well as the effect of the addition of small amounts of a third element, has been investigated by neutron and X-ray scattering techniques and by microscopy [16-23, 24 and references therein], the knowledge of the microstructure, i.e. the structure on a length scale from about 1 to about 100 nm, of the alloys in the molten state, is marginal. It may thus be concluded that although AlSi alloys are used in a manifold of technological applications, the understanding of the influence of synthesis conditions, especially superheat, on the different physical properties of the solidified alloy is incomplete.

The structure and microstructure of molten AlSi alloys were recently studied by neutron diffraction [23] and by neutron small angle scattering (SANS) [24]. It was found that at temperatures more than 100 degrees above the liquidus, the silicon-rich particles or clusters exist in the eutectic melt. For such a kind of particles, they must have a size larger than about 10 nm in order to be visible in a diffraction experiment. The originating from these particles diffraction peaks also indicate that they have a rather definite structure even in case if they had a large width.

The microstructure of molten hypoeutectic $Al_{93}Si_7$, the eutectic $Al_{88}Si_{12}$ and the hypereutectic $Al_{80}Si_{20}$ alloys was recently studied at different temperatures in a series of SANS experiments [24]. The length scale in the measurements ranged from about 1 to 100 nm and they have undoubtedly demonstrated the existence of particles in a wide particle size range for all the alloys at the three investigated melt temperatures (973K, 1223K and 1473K).

Measured SANS intensities at three different Q values for the three alloys during continuous heating and subsequent cooling are shown in figure 1. The neutron wavevector Q is given by

$Q=4\pi/\lambda\sin(\theta)$ where λ is the neutron wavelength and 2θ is the scattering angle. The liquidus temperatures for the three alloy compositions are about 900K, 859 and 970K, respectively.

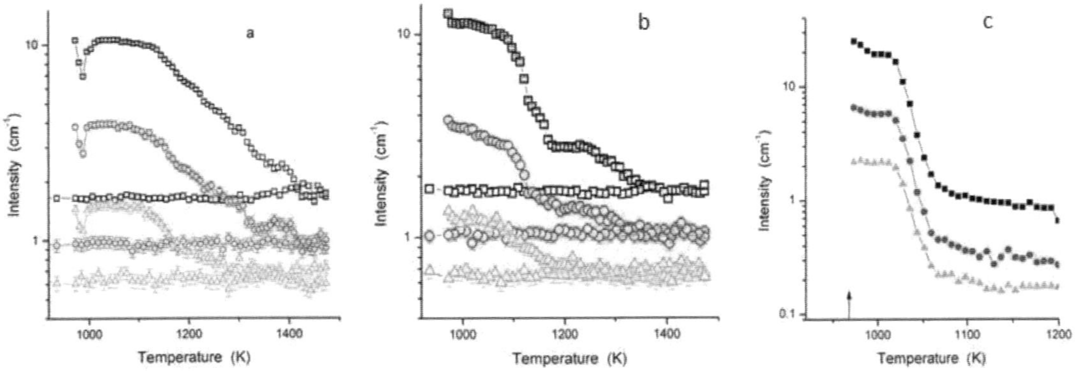

Figure 1. Measured intensities at three different Q-values during continuous heating and subsequent cooling for the a) $Al_{80}Si_{20}$, b) $Al_{88}Si_{12}$ and c) $Al_{93}Si_7$ alloys. Q=0.0042 $Å^{-1}$ (squares), Q= 0.011 $Å^{-1}$ (circles), and Q= 0.017$Å^{-1}$ (triangles).

The drop in intensity at about 973K exists most probably due to the breaking up of the oxide layer around the ingot and the molten alloy now fills the entire sample container. Afterwards the intensities stay for both alloys more or less constant up to about 1100K indicating that the microstructure does not change remarkably in this temperature range. For the hypoeutectic alloy the measured intensities drop monotonically during the heating up to 1473K in a very similar way for all three Q values. For all the alloys the value of the temperature, which is required for complete dissolving of the larger particles, is higher for smaller Q. During the subsequent cooling to the solidification temperatures the intensities do not change, showing that the change in microstructure that have occurred during heating is irreversible and that it corresponds to thermodynamically metastable or even stable states of the melts. For the eutectic alloy the change in microstructure during heating from 1100K to 1473K occurs in two steps, first is a rapid change from 1100K to about 1173K and the following is a slow change between 1173K and 1373K. During the cooling from 1473K the measured intensities stay constant.

Further information of the melt microstructure can be obtained from particle size distribution functions (PSDF) that may be derived from the measured SANS patterns during a stepwise heating-cooling sequence. For numerical reasons the volume weighted PSDF's were derived from these. The obtained PSDF's are shown in figure 2. It can be seen that after melting the hypoeutectic alloy contains particles with sizes up to more than 150 nm and their number is decreasing drastically during the heating up to 1473K. The break down to a smaller particles seems to be irreversible. Middle sized particles, perhaps resulting from an incomplete breaking up of the large ones, seem to exist at the lowest temperature (973K) after cooling and there are indications that small particles recombine in order to form the large ones, at least to some extent, during the cooling. This partly reversible dissolution of the particles takes place, probably, due to the fact that the alloy has not been sufficiently heated to reach a fully homogeneous state.

In the case of the hypereutectic alloy the large particles, which still can be observed after melting, are fully dissolved during the heating to 1473K. Small particles, which average size was of about 7 nm at 973K, have at 1473K decreased their average size to about 3 nm. For the eutectic alloy (see figure 2b) three different size distributions are present after the melting is over. The middle-sized particles with maximum size of about 15 nm rapidly dissolve during the heating process and the number of large particles decreases significantly, when the sample is heated to 1473K. When the sample is subsequently cooled again to 973K, the particles increase in their number, while the middle-sized ones do not recombine. The small particles increase in number during heating and recombine partly during the subsequent cooling.

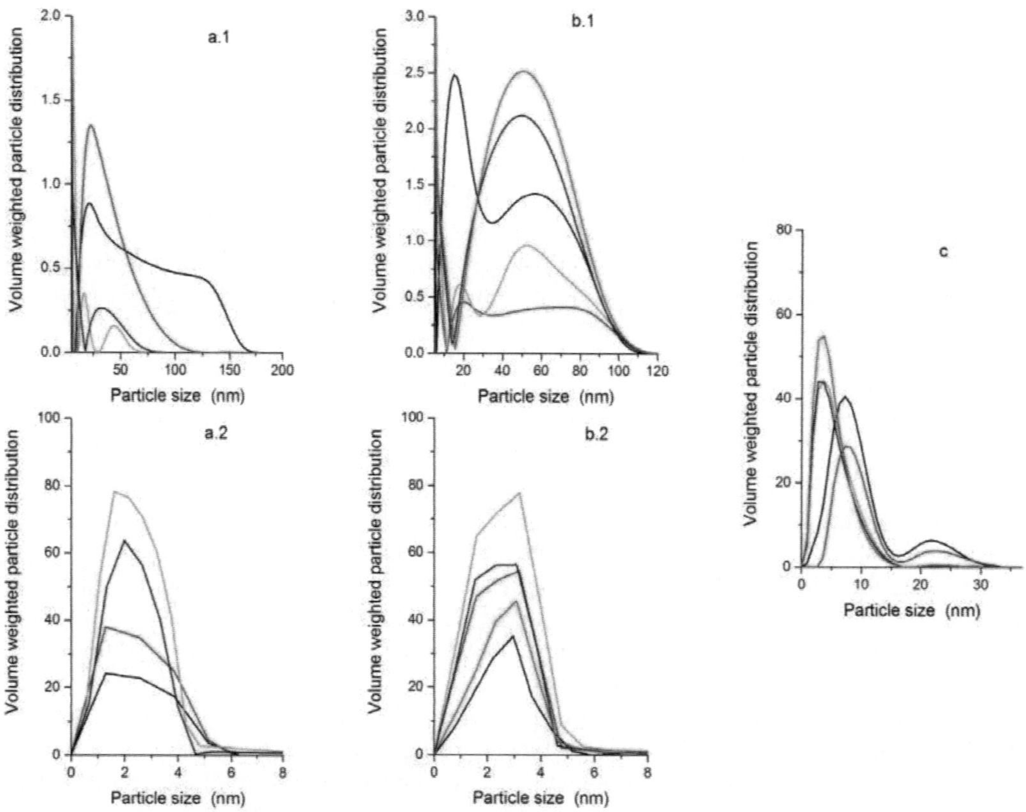

Figure 2. Derived volume weighted particle size distribution functions (PSDF) for the a) $Al_{80}Si_{20}$, the b) $Al_{88}Si_{12}$ and the c) $Al_{93}Si_{7}$ alloys at 973K (black curves), at 1223K (red curves) and 1473K (green curves) during stepwise heating and at 1223K (blue curves) and at 973K (magenta curves) during subsequent stepwise cooling.

The number of small clusters of particles varies significantly during the heating/cooling cycle. The dimensional distribution parameters are nearly the same for the particular temperature, which indicates that the size distribution changes in an almost reversible manner. However, in neutron diffraction measurements on the eutectic AlSi alloy, a significant difference in shape between the static structure factors measured at 973K before and after the heating of the alloy to 1473K was observed [23]. Thus, the reversibility is not complete and it can be conjectured that the eutectic AlSi alloy is considerably more inhomogeneous than both the hypo-and hypereutectic ones irrespective of its thermal history. This is in agreement with viscosity measurements.

It is interesting to note that the size of the small particles after heating to 1473K is about 3 nm and it is very similar for all the alloys. As the size of the large particles increases with Si concentration, one can assume that the 1000 to 4000 atoms in a small particle is in thermodynamic equilibrium with the surrounding melt. From the present SANS experiments it is not possible to determine accurately the composition of the particles, but they have to be enriched in Si in order to have a contrast large enough to be detected in a SANS measurement.

3. Metastable microheterogeneities in Cu-Al melts of different compositions

Some results of measuring the density ρ of Cu-Al melts starting at room temperature and up to 1300-1400°C by penetrating gamma-radiation method have been presented in [25]. A divergence of the gamma-ray beam attenuation curves corresponding to heating and cooling branches of liquid samples *(hysteresis)* was registered [26]. Here we present a detailed study of this phenomenon and its interpretation proceeding from the idea of metastable microheterogeneity of melts.

We used the absolute version of gamma-method for density measurement based on the well-known law of attenuation of a gamma-quanta beam by the substance. All the necessary details as well as the error estimate (~0.2% for Cu-Al alloys) are given in [27] that also contains the block diagram of the experimental setup. A gamma-quanta beam about 3 mm in diameter passed within about 10 mm from the bottom of the crucible, the height of the sample being 35-40 mm.

The melting of the sample was registered visually through the window of the furnace. Then the thermocouple was immersed into the melt, the last was slowly cooled below solidus temperature and the measurements of the passing beam intensity in the process of heating up to 1300-1400°C and a subsequent cooling at the rate of about 1 K/s have began.

According to the opinion of the authors, the formation of a micro-heterogeneous melt of micro-emulsion or micro-suspension type is possible after melting the samples that are heterogeneous in their crystal state. The formed melt consists of the dispersed particles enriched with one of the components and suspended in a dispersion medium rich in the other one. In the system whose components differ so greatly in their density, it was of a high probability that there would be observed the sedimentation of denser particles enriched with copper in a less dense environment rich in aluminium, and, vice versa, rising of the particles rich in aluminium in a copper-rich environment. As it has been stated above, in our experiments the gamma-beam has penetrated the investigated samples near the bottom of the crucible where the mean copper concentration could be considerably higher than the calculated one assigned at mixing. Therefore, it was decided to build temperature dependences of the expression $\rho\mu$ and not of the density ρ at the first stage of the interpretation of the obtained results. The expression in question describes the degree to which γ-radiation is absorbed by the melt, but it does not depend on the mass attenuation factor μ that, in turn, depends on the chemical composition of the zone, which is subjected to radiation.

Nineteen Cu-Al samples of various compositions have been investigated, including pure copper and aluminium. Due to the complexity of the constitution diagram of the system, their concentrations were chosen in the way to have different phase compositions of the samples before the melting (table 1).

The most general common feature of the temperature dependencies of $\rho\mu$ in the zone, which was subjected to radiation, is the divergence of heating and cooling curves *(hysteresis)*, which indicates the irreversible changes in the structure of the most of investigated binary melts after their heating above the curves $\rho\mu(T)$ branching points. Hysteresis of $\rho\mu$ product of the zone, which was subjected to radiation, is especially pronounced for the majority of samples, that are enriched with aluminium (figure 3). In all such a cases the temperature dependence, which is obtained at heating, lies above the cooling curve. Already for the 55 at.% Cu alloy there is a pronounced divergence of $\rho\mu(T)$ dependencies which exceeds 9% near the liquidus temperature and disappears only near by 1300°C.

IOP Conf. Series: Materials Science and Engineering **192** (2017) 012012 doi:10.1088/1757-899X/192/1/012012

With a further increase of aluminium content this branching becomes less pronounced and almost disappears at 65 at. % Al. However, it grows again with the further increase of aluminium content and turns out to be in the range of 10% to 16% for an Al content being in the range 65% to 85%. The maximum temperature, at which a maximum difference between the heating and cooling curves for these alloys takes place, exceeds 1350°C.

Table 1. Investigated in the current work *Cu - Al* melts and their phase composition before the beginning of melting.

Al content (at.%)	Phase composition of the sample before the beginning of melting
5	Single-phase solid solution based on copper α(*Cu*).
10	Single-phase solid solution based on copper α(*Cu*).
18	Eutectic composition: α(*Cu*) + *β*-phase (Cu_3Al).
25	Strongly hypereutectic alloy: α(*Cu*) + *β*-phase (Cu_3Al), passing into a monostructured *β*-phase at heating above 1032°C.
30	Hypereutectic alloy: primary dendrites of *β*-phase and eutectic colony structure α(*Cu*) + *β*.
34	Two-phase eutectoid – solid solutions based on copper and Cu_2Al.
40	γ_1-phase (solid solution based on Cu_2Al + γ_2 (solid solution based on *Cu*
45	Area of homogeneity, probably, a solid solution based on Cu_3Al_2..
50	η_2–phase (*CuAl*) + *θ*-phase ($CuAl_2$ along the borders of dendritic cells.
55	Eutectoid – solid solution based on *CuAl* and $CuAl_2$.
60	Eutectoid – solid solution based on *CuAl* and $CuAl_2$.
65	Solid solution based on $CuAl_2$ + precipitation of *CuAl* along the grain boundaries
67.8	*θ* –phase ($CuAl_2$) + eutectics (α(*Al*) + $CuAl_2$) of liquation origin.
75	Primary *θ*-phase +eutectics (α(*Al*) + $CuAl_2$) .
82.9	Eutectic structure (α(*Al*) + $CuAl_2$).
90	Primary dendrites of α-solid solution based on *Al* + eutectics (α(*Al*) + $CuAl_2$).
95	Primary dendrites of α- solid solution based on *Al* + eutectics (α(*Al*) + $CuAl_2$

Characteristic features of the temperature dependencies of ρμ complex for Cu-Al samples of various compositions are systematized in table 2. The results obtained indicate considerable and irreversible changes in the structure of the zone, which was subjected to radiation in the majority of the investigated samples. It should be pointed out that these kind of effects have been registered before at measuring the kinematic viscosity ν of Cu-Al samples [28]. The similarity of the effects described in [26] and our experiments leads to the conclusion about their common nature. Apparently, in densitometric experiments a microheterogeneous melt is formed first, when the original crystal sample was heterogeneous.

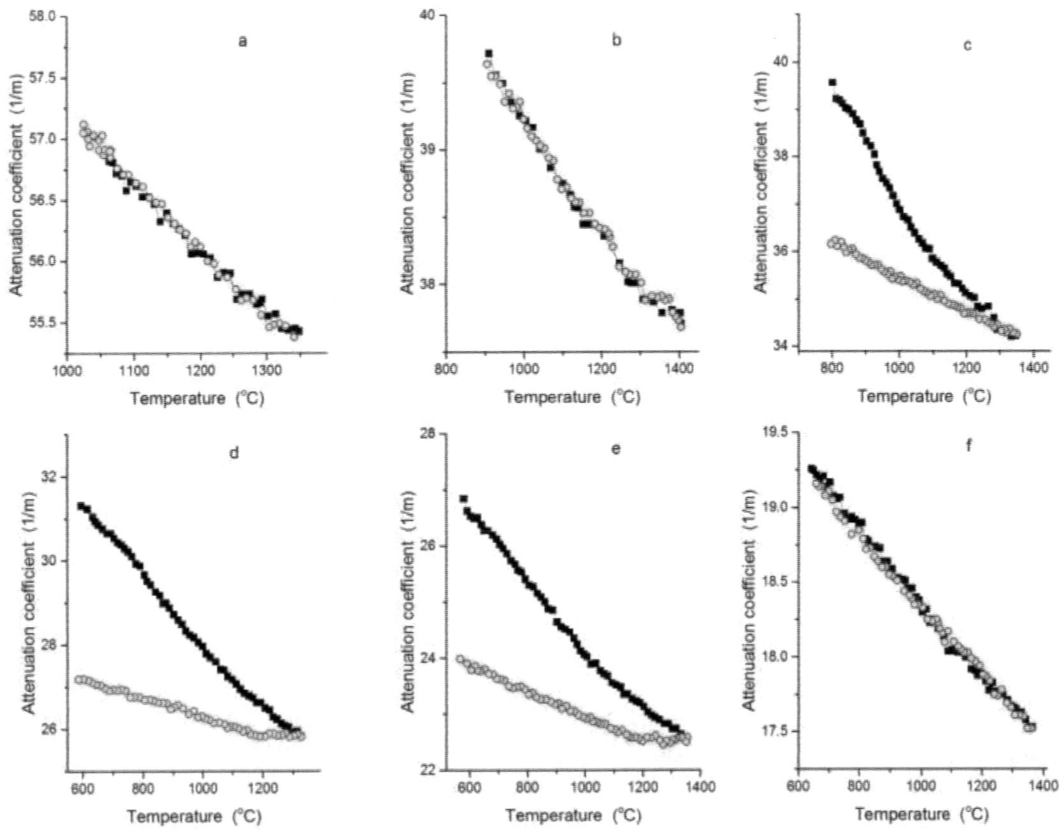

Figure 3. Some examples of measured absorption coefficient ρμ for some AlCu compositions. a) Al_5Cu_{95}, b) $Al_{45}Cu_{55}$, c) $Al_{55}Cu_{45}$, d) $Al_{75}Cu_{25}$, e) $Al_{83}Cu_{17}$, f) $Al_{95}Cu_5$ during heating (full symbols) and subsequent cooling (open symbols).

The enriched with one of the components dispersion medium, as well as the dispersed phase of such a metastable system, rich in the other component, were formed from the fragments of the phases of the original sample. Upon reaching the temperatures, close to the branching points of the temperature dependences of the ρμ quantity at heating and a subsequent cooling, there was an irreversible breakdown of this microheterogeneity and the system passed into a thermodynamically stable true solution state which it retained at the subsequent cooling down to the start of crystallization. Anomalously large differences in ρμ values, which were registered at heating and cooling, can only be explained by the sedimentation of denser dispersed particles in the melt of lower density (it should be recalled that in our experiments the gamma-ray beam have penetrated the melt in its lower part).

The most unexpected result of our densimetric experiments was the measured hysteresis of the ρμ value of the melts, obtained by melting homogeneous crystal samples of stoichiometric compositions CuAl (50 at.% Al) and $CuAl_2$ (67.8 at.% Al). Earlier, while studying temperature dependences of density in several molten refractory compositions (borides of 3d-transition metals), such a phenomena were not revealed: obtained at the heating and cooling of the melt $\rho(T)$ curves coincided, which we considered to be another proof of the inherited nature of the microheterogeneity in eutectic melts [27]. As for the compositions discussed in the present paper, there was hysteresis of ρμ quantity of the penetrated zone of stoichiometric melts, moreover, the divergence of heating and cooling curves reached its maximum in the case of stoichiometric concentrations and grew smaller with distance from them.

Table 2. Characteristic properties and branching temperatures of ρμ(T) dependences of the investigated samples, obtained at their heating after melting and at a subsequent cooling.

Al content (at.%)	Existence of branching	Largest divergence	Branching temeperature (°C)
5	?	-	-
10	no	-	-
18	yes	0.6	1130
25	yes	0.7	1280
30	yes	0.9	1060
34	yes	0.3	1310
40	yes	0.5	1300
45	no	-	-
50	yes	0.9	1280
55	yes	5	1300
60	yes	5	1150
65	yes	0.7	870
67.8	yes	16	1320
75	yes	15	1310
82.9	yes	11	1330
90	yes	1.5	900
95	?	-	-

For a qualitative explanation of this phenomenon the authors rely on the works by L. D. Son and R. E. Ryltsev [29,30]. In their opinion, when relatively low-melting intermetallic compounds, such as *CuAl* and $CuAl_2$, are melted, the melt can retain strong interatomic bonds characteristic of the most refractory intermetallic compounds of the system under investigation (in our case, Cu_3Al). We suppose that these bonds can represent the basis of the formation of corresponding dispersed intermetallic

IOP Conf. Series: Materials Science and Engineering **192** (2017) 012012 doi:10.1088/1757-899X/192/1/012012

phases. Excessive aluminum forms a dispersion medium of low density in this colloid system, and in this medium the particles of the refractory intermetallic compound gravitate to the bottom.

An indirect proof of the fact that dispersed particles of a refractory intermetallic compound remain in aluminum-rich melts can be found in the closeness of their homogenizing temperatures to the temperature of the branching of $\rho\mu(T)$ curves obtained at heating and a subsequent cooling of liquid Cu_3Al. Indeed, in the process of enriching copper with aluminum, the homogenizing temperature of about 1300°C appears for the first time in the case of this compound and later repeats itself in the cases of 6 more alloys of various concentrations (table 2).

4. Metastable microheterogeneity in liquid metal coolants for nuclear reactors

Liquid mercury and sodium have been used as coolants of nuclear power plants for a long time. Meanwhile more researchers nowadays have recently started using molten lead, bismuth, tin, gallium, indium and their mutual solutions (most often of eutectic concentration, as they are the easiest to melt [31-33]). It should be noted that nuclear power plants for submarines with liquid lead-bismuth eutectic have been designed before [32].

As far as we know, designers of reactors do not take into consideration the possible metastable microheterogeneity of Pb-Bi, Sn-Pb and Ga-In binary melts with the composition close to eutectic, even though this phenomenon is the most characteristic for such a solutions. Suffice it to recall that the hypothesis of metastable microheterogeneity takes its origin from the results of measuring the density of Sn-Pb liquid eutectic [1].

As it is known, in cases of emergencies the temperature in the nuclear reactor core can vary over a wide range. Both reversible and irreversible changes may occur in the inner structure of the coolant as a result of exposure to high temperatures. The properties of the coolant may determine the course and consequences of a possible accident. New eutectic coolants were not investigated systematically before this work, and there were even no suppositions about the changes that take place in their inner structure when they are exposed to high temperatures. Thus, the possibility for irreversible transitions of Sn-Pb, Pb-Bi and In-Ga melts from microheterogeneous conditions inherited from the original solid samples into a true solution state was investigated [34,35].

To this purpose, the measurements of the speed and attenuation of ultrasound were performed. The amplitude attenuation coefficient α was calculated by exponential approximation of the dependence of the amplitude of the ultrasound signal on the distance between the transmitting and receiving waveguide at various temperatures. The experiments were conducted at heating the sample after melting the components to 1370 K for approximately 12 hours and at subsequent cooling at the same rate.

Figure 4 shows the results of this experiment. It is noticeable that there is a considerable (approximately fivefold) irreversible decrease of α at cooling the melt after it was heated up to the above mentioned temperature. Despite a relatively large experimental error, especially at heating, the discovered hysteresis is not accidental. It evidently indicates that there are large-scale inhomogeneities in the heated melt, which are comparable with the length of the ultrasound wave (about 10 micron), which get irreversibly destroyed after heating up to the maximum temperature of the experiment. In view of the fact that the presence of the above-stated microheterogeneity should cause a considerable enrichment of the lower part of the melt with a heavier component (in this case – lead), a series of experiments was conducted, that represented itself the measurement of the speed of ultrasound υ_S at the various distances h from the bottom of the crucible.

In order to measure the speed of ultrasound in the experiments with molten *lead-bismuth eutectic,* the initial components were mixed directly in the cell. The ultrasound signal first appeared at 550 K. After that the sample was heated at the rate not exceeding 50 K/hour, and the dependence of the speed of ultrasound υ_S on the vertical coordinate h (the distance to the bottom of the cell) was measured. The above-stated dependences were found at temperatures between 582 K (the lowest temperature at which these measurements were revealed) and 822 K. At higher temperatures the $\upsilon_S(h)$ dependence disappeared. As it was shown already in [26], such an inhomogeneity of a macroscopic kind over the height of the sample about several centimeters in size could be found only if the atoms of the denser component (in this case, the lead) were united into particles containing hundreds or thousands of these atoms. Therefore, the conducted experiments clearly show the microheterogeneity of the Pb-Bi eutectic alloy after its melting, which disappears at a temperature close to 850 K.

To measure the homogenization temperature more accurately we used the method, based on the extrapolation of the temperature dependence of the difference in the speed of ultrasound for two different distances h from the bottom of the crucible $\Delta\upsilon_S(T) = \upsilon_S(h_2,T) - \upsilon_S(h_1,T)$ to the intersection

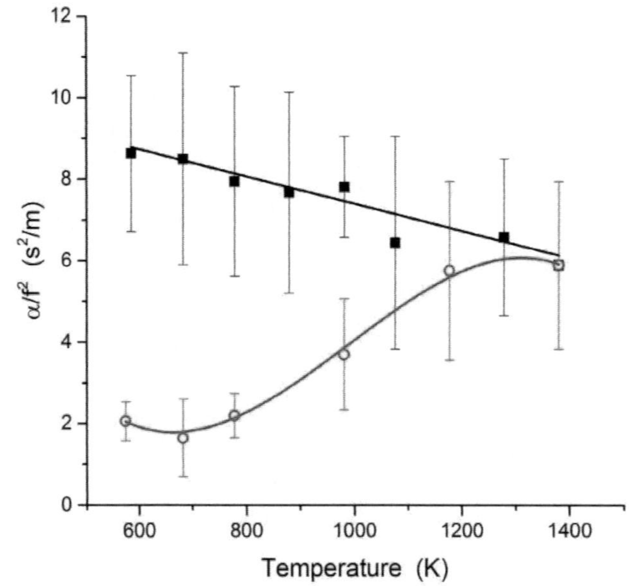

Figure 4. Temperature dependence of amplitude attenuation coefficient of ultrasound α in the melt of lead-bismuth eutectic: • – heating, ○ – cooling. *f* is the frequency of ultrasound.

with the temperature axis. Figure 5 shows the diagram of $\Delta\upsilon_S(T)$ dependence. It is clear from the diagram, that the temperature, at which the large-scale inhomogeneity over the height of the sample disappears, is (950 ± 30) K, which is within the error of the measurement of ultrasound attenuation coefficient and coincides with the branching point of its temperature dependence which was obtained at heating the sample after melting and at its subsequent cooling.

Experiments on Sn-Pb and Ga-In melts showed a very similar variation of the sound velocity over the height. However, unlike in the above described experiments, for equiatomic Pb-Sn alloy its melt *was first heated above homogenization temperature* (up to 1270 K), thoroughly mixed at this temperature, then crystallized and then the sample was cooled down to 320 K. The temperature dependence of the speed of ultrasound was measured all the way down to crystallization. Then the sample was melted again and the dependence of the speed of ultrasound on the height was measured. After measuring this dependence at the first temperature since melting, the melt was held at this temperature for one hour, then the $\upsilon_S(h)$ dependence was measured again. Same as in the previous experiments with Pb-Sn melts, the prolonged holding did not lead to the establishment of a homogenous distribution of the speed of ultrasound over the height of the sample. The sample have passed into a homogenous state only at the temperature of about 1365 K. This temperature is close to the homogenization points of Pb-Sn melts and it have melted without the homogenizing overheating of the original melt. These results show that more or less similar metastable microheterogeneous states of Pb-Sn melts appear at melting of the original heterogeneous crystal samples, regardless of the method of their preparation.

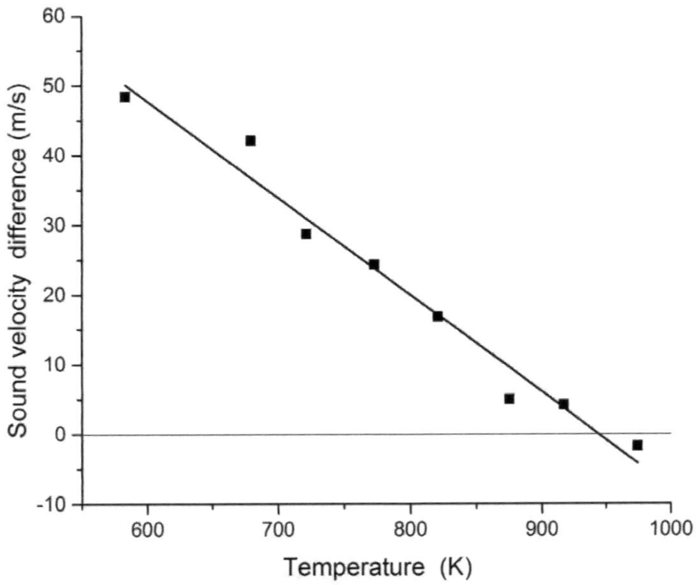

Figure 5. Dependence of the difference of the speed of ultrasound near the bottom of the cell and near the surface of the Pb-55.2 at.% Bi melt on temperature.

The values of homogenization temperatures were calculated from the points of intersection between $\Delta \upsilon_S(T)$ dependences and the abscissa. For Pb-Sn and Ga-In alloys in figures 6,7 these values are fitted in the phase diagrams of these systems. It should be noted that regardless of the way of preparing the original melts (whether the components were mixed immediately after melting of their mixture, or whether a heterogeneous crystalline sample was previously prepared from a homogenized or non-homogenized melt), the above-mentioned temperatures are close to each other. The corresponding points can be fitted with a satisfactory precision in the general dome-like curve which may be viewed as the boundary of existence of metastable microheterogeneous states of the corresponding melts. Above this curve the acoustic properties of the melt correspond to the thermodynamically stable true solution state. At cooling after heating above this curve the homogenous state persists until its crystallization.

It should be noted, that the temperatures, at which a liquid metal coolant is used, vary greatly. In the modern reactors even under the normal conditions its temperature rises by 100-120°C when it passes through the reactor, which means that in emergencies the temperatures of the above-described structural change (from 310 to 770 °C) can be reached in the reactor with such a coolant. Therefore, the possibility of an irreversible change in the structure of the melt and its thermal physical properties should be taken into consideration when nuclear power plants are designed and used. It is especially important to take this into account when choosing the operating modes for the coolant during the first days of its usage in the heat-transfer loop of the reactor.

5. Conclusion

We performed detailed analysis of SANS results obtained earlier on hypo-, hyper- and eutectic Al-Si melts. In addition to earlier results, it was shown that the eutectic melt is more inhomogeneous than both the hypo-and hypereutectic ones irrespective of its thermal history. It is interesting to note that the size of the small particles after heating to 1473K is about 3 nm and very similar for all alloys. As the size of the large particles increases with Si concentration one can assume that 1000 to 4000 atoms in a small particle is in thermodynamic equilibrium with the surrounding melt.

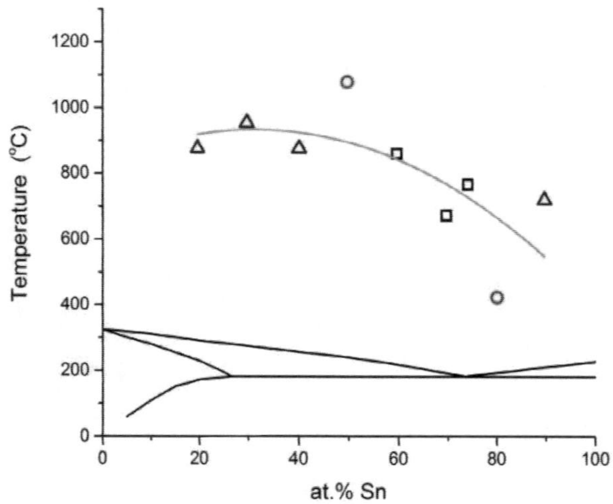

Figure 6. Boundary of existence of metastable microheterogeneity on the phase diagram of the Pb-Sn system constructed from the results of acoustic experiments with melts prepared at heating: ▲ – the mixture of original components; ● – a crystallized heterogeneous ingot obtained from a non-homogenized melt; ○ – a crystallized heterogeneous ingot obtained from the melt overheated above the homogenization temperature.

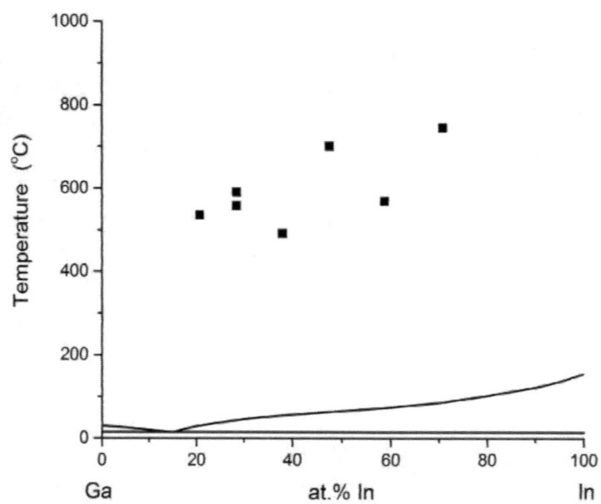

Figure 7. The same as in figure 6 for the Ga-In alloys heating: ● – the mixture of original components; ○ – a crystallized heterogeneous ingot obtained from the melt overheated above the homogenization temperature.

While performing gamma-ray attenuation measurement on Al-Cu melts, the obvious signs of their metastable microheterogeneity where fixed too. The most unexpected result of the experiments is hysteresis of the attenuation value of the melts obtained by melting homogeneous crystal samples of stoichiometric compositions CuAl and $CuAl_2$. Earlier, while studying temperature dependences of density in several molten refractory compositions (borides of 3d-transition metals), such a phenomena were not found. And, finally, the signs of a metastable microheterogeneity of liquid eutectics considering as a perspective heat carriers for nuclear reactors where discovered. Therefore, the possibility of an irreversible change in the structure of the melt and its thermal physical properties should be taken into consideration when nuclear power plants are designed and used.

6. References

[1] Popel P S 1985 *Izvestija vuzov, Chernaja Metallurgija* **5** 34-41
[2] Manov V P, Manukhin A B and Popel P S 1985 *Doklady AN SSSR* **281** 107-109
[3] Popel P S, Djomina E L, Arkhangel'sky E L and Baum B A 1987 *Teplophyzika Vysokikh Temperatur* **25** 487-491
[4] Popel P S, Chikova O A and Matvejev V M 1995 *High Temp. Mat. Proc.* **14** 219-234
[5] Dahlborg U, Calvo-Dahlborg M, Popel P S and Sidorov V E 2000 *Eur. Phys. J. B* **14** 639-648
[6] Gupta M and Lavernia E J 1995 *J. Mater. Proc. Tech.* **54** 261-270
[7] Gupta M and Ling S 1999 *J. Alloys Compd.* **287** 284-294
[8] Dai H S and Liu X F 2008 *Mater. Charact.* **59** 1559-1563
[9] Li P, Nikitin V I, Kandalova E G and Nikitin K V 2002 *Mater. Sci. Eng. A* **332** 371-374
[10] Venkataramani R, Simpson R and Ravindran C 1995 *Mater. Charact.* **35** 81-92
[11] Wang J, He S, Sun B, Guo Q and Nishio M 2003 *J. Mater. Proc. Tech.* **141** 29-34
[12] Wang W M, Bian X F, Qin J Q and Syliusarenko S I 2000 *Metall. Mater. Trans. A* **31** 2163-2168
[13] Hegde S and Prabhu K 2008 *J. Mater. Sci.* **43** 3009-3027
[14] Wang J, Qi J, Du H and Zhang Z 2007 *J. Iron Steel Res. Int.* **14** 75-78
[15] Zhang R, Cao O, Pang S, Wei Y and Liu L 2001 *Sci. Technol. Adv. Mater.* **2** 3-5
[16] Wang R, Lu W and Hogan L M 2003 *Mater. Sci. Eng. A* **348** 289-298
[17] Haque M M and Ismail A F 2005 *J. Mater. Proc. Tech.* **162-163** 312-316
[18] Bian X, Wang W and Qin J 2001 *Mater. Charact.* **46** 25-29
[19] Kobayashi K F and Hogan L M 1985 *J. Mater. Sci.* **20** 1961-1975
[20] Song X, Bian X, Zhang J and Zhang J 2009 *J. Alloys Compd.* **479** 670-673
[21] Xu C and Jiang Q C 2006 *Mater. Sci. Eng. A* **437** 451-455
[22] Srirangam P, Kramer M J and Shankar S 2011 *Acta Mater.* **59** 503-513
[23] Dahlborg U, Kramer M J, Besser M, Morris J R and Calvo-Dahlborg M 2013 *J. Non-Cryst. Solids* **361** 63-69
[24] Calvo-Dahlborg M, Popel P S, Kramer M J, Besser M, Morris J R and Dahlborg U 2013 *J. Alloys Compd.* 9-22
[25] Kurochkin A R, Borisenko A V, Popel P S, Yagodin D A and Okhapkin A V 2013 *High Temperature* **51** 197-205
[26] Kurochkin A R, Popel P S, Borisenko A V and Yagodin D A 2015 *High Temperature – High Pressure* **44** 265-283
[27] Brodova I G, Popel P S and Eskin G I 2002 *Series Part: Advances in Metallic Alloys 1. Liquid metal processing* (London and New York: Taylor & Francis)
[28] Konstantinova N Yu and Popel P S 2007 *J. Phys: Conf. Ser.* **98** 062022
[29] Son L, Ryltcev R, Sidorov V and Sordelet D 2007 *Mat. Sci. Eng. A* **449-451** 582-585

[30] Ryltcev R E and Son L D 2011 *Physica B* **406** 3625-3630
[31] Takibajev Zh S 2006 *Vestnik NJaC RK* **4** 5-6
[32] Naumov V V 2006 *Atomnaja Strategija XXI* **6** 8-9
[33] Burns S, Site MTI (http://www.technologyreview.com/14625/page1/)
[34] Subbotin V I, Arnoldov M N, Kozlov F A and Shimkevich A L 2002 Atomnaja Energija **92** 31-42
[35] Popel P, Stankus S, Mozgovoy A Khairulin R Pokrasin V Yagodin D, Konstantinova N, Borisenko A and Guzachev M 2011 *Eur. Phys. J., Web of Conf.* **15** 01014

Acknowledgements
This work was supported by the Grants obtained from the Ministry of Education and Science of the Russian Federation (state task 2014/392, projects No. 2391 and 1177) and the Government of the Russian Federation (program No.211, agreement No.02.A03.21.006).

IOP Conf. Series: Materials Science and Engineering **192** (2017) 012013 doi:10.1088/1757-899X/192/1/012013

Transport phenomena in non-uniform gas subjected to laser radiation

I V Chermyaninov and V G Chernyak

Department of Physics, Ural Federal University, 620083, Yekaterinburg, Russia

vladimir.chernyak@urfu.ru

Abstract. The paper discusses the theory of transport processes in one-component gas located in capillary subjected to resonant laser radiation and both temperature and pressure gradients. The equations for the kinetic coefficients determining heat- and mass transport in the gas are derived on the basis of modified Boltzmann equations for the excited and unexcited particles. The cross kinetic coefficients satisfy the Onsager reciprocity for all Knudsen numbers and laws of gas particles interaction with each other and with boundary surface of the capillary. Analysis of possible non-equilibrium stationary states of first and second order for the one-component gas in the capillary has been developed on the basis of the Prigogine theorem of stationary states. Equations describing the stationary states in Knudsen limit ($Kn >> 1$) and slip-flow regime ($Kn << 1$) were derived.

1. Introduction

Thermodynamic description of transport phenomena is based on the laws of conservation of mass, momentum and energy and the principle of increasing entropy. According to the thermodynamics of irreversible processes[1] the entropy production in the system is determined as a sum of products of generalized fluxes and generalized forces bound to those fluxes. These forces are generally associated with spatial inhomogeneity of thermodynamic parameters of the system or with a deviation of some of the internal variables from their equilibrium values [1].

Specific feature of light-induced transport phenomena is that they take place at homogeneous thermodynamic parameters of gas and under certain conditions and can generate temperature and pressure gradients [2]. These light-induced gradients stimulate their own fluxes of mass and heat in the gas. Theoretical description of these phenomena at the phenomenological level is impossible since the mechanisms of light-induced transport phenomena are associated only with the molecular-kinetic properties of the gas. Development of Onsager kinetic theory of these processes is the main objective of this study.

Laser radiation under resonant interaction with the gaseous medium behaves as the «two-faced Janus». The radiation can simultaneously violate the gas equilibrium by internal and translational degrees of freedom. This means that radiation stimulates two types of fluxes, «scalar» flux, characterizing change of population of the particles energy levels compared to their equilibrium values and vector fluxes of mass and heat of the gas.

Disequilibrium due to translational degrees of freedom is the result of asymmetry in the distribution functions of the excited and unexcited particles related to zero- component of the particles velocity vector in the direction of radiation propagation (Figure 1).

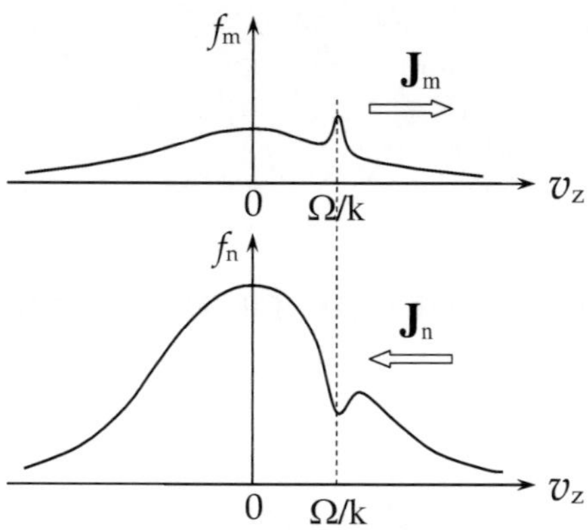

Figure 1. The distribution functions of excited and unexcited particles.

In accordance with the Doppler effect, radiation interacts only with those particles having Doppler shift **kv** (**k** is the wave vector, **v** is the velocity vector of the particle) compensates detuning Ω of radiation frequency ω relatively to ω_{mn}, the $m-n$ frequency transition from the ground state of particle n to excited state m. In linear approximation, when the magnitude of the particle velocity is much less than light speed, Doppler shift is determined by the expression

$$\mathbf{kv} = \Omega = \omega - \omega_{mn}, \qquad |\Omega| << \omega, \omega_{mn} \tag{1}$$

As a result of light absorption, distributions of excited and unexcited particles near the resonant velocity $v_p = \Omega/k$ are distorted due to the appearance of Bennett's peak and dip [3] (Figure 1). The asymmetry of velocity distributions at $\Omega \neq 0$ means the existence of fluxes of excited $\mathbf{J}_m$ and unexcited $\mathbf{J}_n$ particles having opposite directions. The particles absorbing light change their transport characteristics. If buffer gas [2, 4] or boundary surface with which m- and n-particles interact differently [5-7] are available in the system, these fluxes do not prevent each other and gas as a whole moves. The flux of the absorbing gas $\mathbf{J} = \mathbf{J}_n + \mathbf{J}_m$ is known as the light-induced drift – LID [4-7]. Simultaneously with LID, resonant laser radiation stimulates light-induced heat transfer in the gas, designated as LIHT [8].

The problem becomes much more complicated if the state of the gas is governed by the pressure and temperature gradients in addition to the resonant radiation. Each of these gradients creates its own fluxes, interacting with the light-induced mass and heat fluxes. In the closed gaseous systems under certain conditions this interaction leads to occurrence of various nonequilibrium steady states.

Cross-transport processes in one-component gas in the presence of laser radiation and both pressure and temperature gradients were overviewed in [9]. The paper was focused on the discussion of light-induced baro- and thermo- effects that take place in closed heat insulated systems. However the proof of an Onsager reciprocity for the cross kinetic coefficients was not given in this reference. In addition, the entropy production due to the interaction of gas particles with the surface of the capillary was only stated, but not analyzed.

The objectives of this work are as follows.

First, based on the kinetic theory of gases we derive the expression for the entropy production in one-component gas in the field of the laser radiation in presence of the pressure and temperature gradients; and determine the generalized thermodynamic forces and corresponding fluxes;

Second, we prove Onsager's reciprocity for the cross kinetic coefficients at arbitrary Knudsen numbers and any nature of gas particles interaction with each other and boundary surface.

Third, using theorems of the non-equilibrium thermodynamics we consider the possible stationary states of the gas.

2. Statement of the problem

We consider one-component gas in the capillary. Let capillary length L is much greater than its radius r_0 (L >> r_0). The resonant optical radiation is uniform over the cross section of the capillary and propagates along the axis of the capillary as a traveling monochromatic wave of ω frequency. In addition, the state of the gas is disturbed by the longitudinal temperature and pressure gradients.

We will use the two-level approximation according to which the particles can be either in the ground state n, or in the excited state m. Particles absorbing radiation transit to an excited state. They change their transport properties, in particular, the collision cross-sections and accommodative characteristics. The process of radiative decay of the excited m level with frequency Γ_m occurs simultaneously with the induced transitions. Thus, the gas can be considered as a mixture consisting of particles of equal mass m_0 but with different interaction cross-sections and accommodative properties. The components continuously exchange the particles.

The state of gas is described by distribution functions of excited f_m and unexcited f_n particles that satisfy the Boltzmann kinetic equations including terms responsible for the interaction of particles with radiation [10]:

$$\frac{\partial f_i}{\partial t} + \mathbf{v}\nabla f_i = a_i R + I_i, \tag{2}$$

$$R = \frac{\Gamma_m \chi(\mathbf{v})}{2}(f_m - f_n) + \Gamma_m f_m, \qquad \chi(\mathbf{v}) = \frac{4|G_{mn}|^2 \Gamma}{\Gamma_m\left[\Gamma^2 + (\Omega - \mathbf{kv})^2\right]}, \qquad G_{mn} = \frac{E d_{mn}}{2\hbar},$$

$$a_i = \begin{cases} 1, & i = n \\ -1, & i = m \end{cases}, \qquad I_i = I_{ii} + I_{ij}, \qquad i, j = n, m.$$

Here Γ is homogeneous half-width of the absorption line, E is the electric field amplitude of the light wave, d_{mn} is the dipole moment of m–n transition, $\mathbf{k}$ is the wave vector, and I_{ij} are Boltzmann integrals of collisions between the particles of i-th and j-th kind. Rabi frequency G_{mn} determines the rate of induced transitions between the levels of particles per unit time and depends on the intensity of radiation. The saturation parameter $\chi(\mathbf{v})$ characterizes the probability of radiation absorption per unit time by the particles which are moving with speed $\mathbf{v}$ and determines the width of the Bennett's peak.

The boundary conditions for the distribution functions of particles have the form [11]:

$$|\mathbf{v}\cdot\mathbf{n}| f_i(\mathbf{v}) = \sum_{j=n,m} \int_{(\mathbf{v}'\mathbf{n})<0} B(j,\mathbf{v}' \to i,\mathbf{v}) f_j(\mathbf{v}') |\mathbf{v}'\cdot\mathbf{n}| d\mathbf{v}' \tag{3}$$

$$(\mathbf{v}\cdot\mathbf{n}) > 0 \qquad (i,j) = m,n,$$

where $\mathbf{n}$ is inward normal to the surface of the capillary, $f_i(\mathbf{v})$ and $f_i(\mathbf{v}')$ are distribution functions of the reflected and incident particles on the wall of i-th kind, respectively, $B(j,\mathbf{v}' \to i,\mathbf{v})$ is the scattering kernel (the probability density is such that the internal energy E_j of the state j and the velocity $\mathbf{v}'$ of the incident particle on the wall after reflection is equal to E_i and $\mathbf{v}$). The scattering kernel satisfies the following conditions [11]:

normalization

$$\sum_{j=n,m} \int_{(\mathbf{vn})>0} B(j,\mathbf{v}' \to i,\mathbf{v})d\mathbf{v} = 1 \tag{4}$$

reciprocity relation expressing the principle of detailed balance on the gas—surface border

$$|\mathbf{v}' \cdot \mathbf{n}| f_{j0}(\mathbf{v}') G_j B(j,\mathbf{v}' \to i,\mathbf{v}) = |\mathbf{v} \cdot \mathbf{n}| f_{i0}(\mathbf{v}') G_i B(i,-\mathbf{v} \to j,-\mathbf{v}') \tag{5}$$

where f_{i0} is the equilibrium Maxwell-Boltzmann distribution function, G_i is the degree of degeneracy.

Integrating (5) over velocity space $(\mathbf{v}' \cdot \mathbf{n}) < 0$ and summing over the states *j*, taking into account the normalization (4), we obtain

$$|\mathbf{v} \cdot \mathbf{n}| f_{i0}(\mathbf{v}) = \sum_{j=n,m} \int_{(\mathbf{v}'\mathbf{n})<0} B(j,\mathbf{v}' \to i,\mathbf{v}) f_{j0}(\mathbf{v}') |\mathbf{v}' \cdot \mathbf{n}| d\mathbf{v}' \tag{6}$$

We choose a cylindrical coordinate system (*r*, ϕ, z), so that the polar axis z is directed along the axis of the capillary.

In the weak-field approximation ($\chi(\mathbf{v}) << 1$) and at small pressure and temperature gradients, the distribution functions of excited and unexcited particles are insignificantly different from local equilibrium Maxwell-Boltzmann distributions:

$$f_i = f_{i0}\left[1 + h_i(\mathbf{r},\mathbf{v})\right] \tag{7}$$

The local equilibrium Maxwell-Boltzmann distribution looks like

$$f_{i0} = n_i(z)\left(\frac{m_0}{2\pi k_B T(z)}\right)^{3/2} \exp\left(-\frac{m_0 v^2}{2k_B T(z)}\right),$$

$$n_i(z) = n(z)\frac{\exp(-E_i/k_B T(z))}{\sum_l \exp(-E_l/k_B T(z))}. \tag{8}$$

Here $T(z), n(z), n_i(z)$ are the local values of the gas temperature, the total particle number density, and the population of the *i*-th level, respectively; k_B is the Boltzmann constant.

Due to linearization (7) and equation (6), the boundary conditions (3) on the wall of the capillary for the perturbation functions take the following form:

$$|\mathbf{v} \cdot \mathbf{n}| f_{i0} h_i(\mathbf{v}) = \sum_{j=n,m} \int_{(\mathbf{v}'\mathbf{n})<0} B(j,\mathbf{v}' \to i,\mathbf{v}) f_{j0} h_j(\mathbf{v}') |\mathbf{v}' \cdot \mathbf{n}| d\mathbf{v}' \tag{9}$$

To make a separate analysis for vector fluxes directed along the capillary and scalar fluxes associated only with the change of the level population we distinguish even and odd parts of the longitudinal component of the particles v_z velocity vector in perturbation functions and in saturation parameter

$$\chi(v_z) = \chi^+(v_z) + \chi^-(v_z), \qquad h_i(v_z) = h_i^+(v_z) + h_i^-(v_z), \tag{10}$$

$$\chi^{\mp}(v_z) = \frac{\chi(v_z) \mp \chi(-v_z)}{2}, \qquad h_i^{\mp}(v_z) = \frac{h_i(v_z) \mp h_i(-v_z)}{2}.$$

Based on the assumptions made the kinetic equations (2) are linearized with respect to perturbation functions h_i and for steady gas state are written as:

$$\mathbf{v}_\perp \frac{\partial h_i^\mp}{\partial \mathbf{r}} = \sum_{j=n,m} L_{ij}\left(h_i^\mp\right) + b_i h_m^\mp + g_i^\mp \tag{11}$$

where the following notations are used:

$$g_i^- = -v_z \frac{1}{p}\frac{\partial p}{\partial z} - v_z\left(\frac{m_0 v^2}{2kT} - \frac{5}{2} + \varepsilon_i - \bar{\varepsilon}\right)\frac{1}{T}\frac{\partial T}{\partial z} + \frac{\Gamma_m}{2}\chi^-\left(v_z\right)\left(\frac{n_{j0}}{n_{i0}} - 1\right),$$

$$g_i^+ = \frac{\Gamma_m}{2}\chi^+\left(v_z\right)\left(\frac{n_{j0}}{n_{i0}} - 1\right) + b_i, \quad v^2 = v_z^2 + v_\perp^2 \quad \left(i, j\right) = n, m \quad i \neq j \tag{12}$$

$$b_i = \begin{cases} \Gamma_m \dfrac{n_{m0}}{n_{n0}}, & i = n \\ -\Gamma_m, & i = m \end{cases}, \quad \bar{\varepsilon} = \frac{\sum_l \varepsilon_l exp(-\varepsilon_l)}{\sum_l exp(-\varepsilon_l)} = \frac{\bar{E}}{k_B T_0}, \quad v_\perp^2 = v_r^2 + v_\phi^2, \quad \varepsilon_i = \frac{E_i}{k_B T_0}.$$

Here $\mathbf{r}$ and $\mathbf{v}_\perp$ are two-dimensional radius vector and vector component of the velocity of particles in the cross-section of the capillary, $L_{ij}\left(h_i\right)$ are the linearized collision integrals for particles of i-th and j-th kinds, n_{n0}, n_{m0} are the equilibrium populations of the ground and excited states at the equilibrium temperature of the gas T_0, $\bar{E}$ is the average internal energy of the particles in the equilibrium state. One can see [12] that the linearized collision integral $L_{ij}(h_i)$ off the odd function h_i over velocity v_z is an odd function, i.e. $L_{ij}(h_i(-v_z)) = -L_{ij}(h_i(v_z))$.

The linearized kinetic equations (11) are separated into odd ("minus") and even ("plus") relatively velocity v_z. The first equations describe vector fluxes that are associated with transfer of mass, heat and entropy of the gas along the capillary, second – with the scalar fluxes describing the pressure anisotropy and radiation-collisional heating or cooling gas [10].

3. The equation of entropy balance

Consider a volume V of gas contained in the capillary tube of length L. We define its entropy in conventional manner [1]:

$$S_e = -k_B \sum_{i=m,n} \int_V dV \int f_i \ln f_i d\mathbf{v} \tag{13}$$

The rate of entropy change in the respective volume of gas is equal to:

$$\frac{\partial S_e}{\partial t} = -k_B \sum_i \int_V dV \int \left(\ln f_i + 1\right)\frac{\partial f_i}{\partial t} d\mathbf{v} \tag{14}$$

Taking into account the kinetic equation (2) we obtain the entropy balance equation in the form:

$$\frac{\partial S_e}{\partial t} + \int_V div\mathbf{j} dV = P_R + P_c \tag{15}$$

where $\mathbf{j}$ is vector flux density of entropy:

$$\mathbf{j} = -k_B \sum_i \int \mathbf{v} f_i \ln f_i d\mathbf{v} \tag{16}$$

P_R is the total entropy production in the considered volume due to radiation interaction with the gas:

$$P_R = -k_B \int dV \int R \ln\left(\frac{f_n}{f_m}\right) d\mathbf{v} \tag{17}$$

P_c is the total entropy production due to the interaction of gas particles with each other:

$$P_c = -k_B \sum_i \int dV \int I_i \ln f_i d\mathbf{v} \tag{18}$$

Consider the stationary state of the system when the entropy of the gas volume does not change in time. In this case, the entropy balance equation (15) can be written as:

$$\int div\mathbf{j} dV = P_R + P_c \tag{19}$$

Using linearization of distribution functions (7) we obtain the entropy balance equation for stationary weakly nonequilibrium state of the system.

Taking into account (11) and (12) the equation (19) is divided into equations with vector («-») and scalar («+») fluxes:

$$\int (div\mathbf{j})^{\mp} dV = P_R^{\mp} + P_c^{\mp} \tag{20}$$

We consider only vector fluxes. To simplify the equations hereafter we omit the superscript «-» in all the physical quantities, believing that function $h_i(v_z)$ is odd relatively to velocity v_z.

Using the Gauss theorem and decomposition $\ln(1+h_i) \approx h_i - h_i^2/2$, we get:

$$\int div\mathbf{j} dV = -k_B \sum_i \int dV \int h_i \mathbf{v} \nabla f_{i0} d\mathbf{v} + k_B \sum_i \int_{S_w} dS \int (\mathbf{v}\cdot\mathbf{n}) f_{i0} \frac{h_i^2}{2} d\mathbf{v} \tag{21}$$

When writing the second integral in the right-hand side of equation (21) the integration is performed only over the lateral surface S_w of the capillary, since the integrals over face sections S in volume V in the stationary state are equal in magnitude and are opposite in sign.

The entropy production due to the interaction of radiation with the gas can be written as the sum of the entropy production due to the radiative decay of the excited level P_r and by induced transitions:

$$P_R = P_r + k_B \int dV \int \frac{\Gamma_m \chi^-(v_z)}{2} (f_{n0} - f_{m0})(h_n - h_m) d\mathbf{v} \tag{22}$$

$$P_r = k_B \frac{\Gamma_m}{2} \int_V dV \int f_{m0} (h_m - h_n)^2 d\mathbf{v} \tag{23}$$

Entropy production due to intermolecular collisions after linearization takes the form:

$$P_c = -k_B \sum_{i,j} \int dV \int f_{i0} h_i L_{ij}(h_i) d\mathbf{v} \tag{24}$$

To find the meaning of the second term of the right side of expression (21) we calculate the entropy flux from wall to gas due to collisions of the gas particles with the surface of the capillary:

$$P_w = -k_B \sum_i \int_{S_w} dS \int (\mathbf{v}\cdot\mathbf{n}) f_i \ln f_i d\mathbf{v} = -k_B \sum_i \int_{S_w} dS \int (\mathbf{v}\cdot\mathbf{n}) f_{i0} \frac{h_i^2}{2} d\mathbf{v} + \int_{S_w} \frac{(\mathbf{q}\cdot\mathbf{n})}{T_w} dS \quad (25)$$

Here $(\mathbf{q}\cdot\mathbf{n}) = q_n^{gas}$ is the normal component of the heat flux density in gas on the wall of the capillary, T_w is the gas temperature at the surface of the capillary. The condition of continuity of the normal flux of heat in the steady state is performed at any point of the surface: $q_n^{gas} = q_n^{cap}$, where $q_n^{cap} = (\mathbf{q}^{cap}\cdot\mathbf{n})$ is the normal component of the heat flux density in the wall of the capillary.

The contribution of the second term in the right side of (25) in the entropy flux P_w depends not only on the thermal conductivity of the material of the capillary but also from the ratio of the radius of the capillary r_0 to its length L. In the limit $L >> r_0$ its contribution becomes negligible under any coefficient of thermal conductivity of the wall. In this case in any cross section of the capillary in stationary conditions the state of local equilibrium is established when the normal component of vector of the heat flux density is zero and the temperature of the wall is equal to the temperature of the gas at any point of section. Therefore further, the second term in the right side of (25) we omit. Hence:

$$P_w = -k_B \sum_i \int_{S_w} dS \int (\mathbf{v}\cdot\mathbf{n}) f_{i0} \frac{h_i^2}{2} d\mathbf{v} \quad (26)$$

Substituting (21) - (24) and (26) into (20) we obtain the equation of entropy balance in the gas for the stationary state:

$$P = P_c + P_r + P_w = -k_B \sum_{i=m,n} \int_V dV \int h_i \mathbf{v} \nabla f_{i0}\, d\mathbf{v} + k_B \int dV \int \frac{\Gamma_m \chi^-(v_z)}{2} (f_{m0} - f_{n0})(h_n - h_m) d\mathbf{v}. \quad (27)$$

The entropy production inside the allocated volume of gas is caused by presence of the following sources of irreversible processes: intermolecular collisions (P_c), radiative decay of the excited level (P_r) and gas particles interaction with the capillary surface (P_w).

From property of symmetry of the collision integral [13] follows that $P_c \geq 0$. According to equation (23), we have $P_r \geq 0$. Using reciprocity for the scattering probability density (5) and Jensen's inequality for the continuous, strictly convex down functions [13], we get $P_w \geq 0$. Thus, the entropy production in the gas due to vector fluxes is nonnegative $P \geq 0$.

We transform the right-hand side of equation (27). We perform integration over the volume V like integration by capillary length L and cross-section area S. We take into account that in stationary conditions fluxes do not depend on the z coordinate, but the intensity of the radiation varies along the z axis due to absorption by gaseous medium. The dependence of z coordinate radiation intensity can be accounted for by introduction of the linear absorption coefficient β. Then the amplitude of the wave radiation electric field and, as a consequence, a square of the Rabi frequency proportional to the intensity of the radiation, in accordance with Bouguer law, can be written as

$$E(z) = E(0)\exp(-\frac{\beta}{2} z), \qquad G_{mn}^2(z) = G_{mn}^2(0)\exp(-\beta z). \quad (28)$$

where $G_{mn}(0)$ is the Rabi frequency associated with the intensity of radiation at the capillary entrance.

The integration in the right-hand side of eqn. (27) along the capillary within appropriate pressure and temperature limits with regard to (28), allows us to write this equation as:

$$P = \sum_k \mathrm{J}_k X_k \tag{29}$$

where the generalized fluxes J_k and forces X_k are defined as follows

$$\mathrm{J}_\mathrm{p} = \sum_i \int dS \int v_z f_i d\mathbf{v} = \sum_i \int dS \int v_z f_{i0} h_i d\mathbf{v} \tag{30}$$

$$\mathrm{J}_\mathrm{T} = \sum_i \int dS \int V_z \left(\frac{m_0 V^2}{2} + E_i \right) f_i d\mathbf{v} = k_B T_0 \sum_i \int dS \int v_z \left(c^2 - \frac{5}{2} + \varepsilon_i - \overline{\varepsilon} \right) f_{i0} h_i d\mathbf{v} \tag{31}$$

$$\mathrm{J}_\mathrm{S} = k_B \sum_i \int dS \int \chi_S^- f_i \left(\frac{n_{j0}}{n_{i0}} - 1 \right) d\mathbf{v} = k_B \sum_i \int dS \int \chi_S^- f_{i0} h_i \left(\frac{n_{j0}}{n_{i0}} - 1 \right) d\mathbf{v} \tag{32}$$

$$X_P = -k_B \ln\left(\frac{p_2}{p_1} \right), \quad X_T = -\frac{1}{T_0} \ln\left(\frac{T_2}{T_1} \right), \quad X_S = \left(\frac{G_{mn}(0)}{k\overline{v}} \right)^2 \frac{1}{\beta} \left[1 - \exp(-\beta L) \right] . \tag{33}$$

$$\chi_S^- = \frac{\Gamma}{\left(\Gamma / k\overline{v} \right)^2 + \left(\Omega / k\overline{v} - c_z \right)^2} - \frac{\Gamma}{\left(\Gamma / k\overline{v} \right)^2 + \left(\Omega / k\overline{v} + c_z \right)^2},$$

$$\mathbf{V} = \mathbf{v} - \mathbf{U} , \quad k = \frac{2\pi}{\lambda} , \quad \overline{v} = \left(\frac{2k_B T_0}{m_0} \right)^{1/2} , \quad c_i = \frac{v_i}{\overline{v}} .$$

Here $\mathbf{V}$ is molecule own velocity with regard to the gas average macroscopic velocity $\mathbf{U}$, λ is the laser wavelength, S is the cross-section area of the capillary; p_1, T_1 and p_2, T_2 are pressure and temperature of the gas at the ends of the capillary.

The fluxes defined in eqn. (30) – (32) have an obvious physical meaning: J_p and J_T are respectively the mass and heat fluxes averaged over the cross-section of the capillary, J_S is the entropy flux calculated per unit length of the capillary.

Forces X_P and X_T for the case $(p_2 - p_1)/p_1 << 1$ and $(T_2 - T_1)/T_1 << 1$ were used in ref. [14] to describe non-isothermal gas motion in the channel. The force X_S is associated with the intensity of optical radiation, that is convenient because it is directly recorded in experiments [7].

4. Kinetic coefficients

Perturbation functions in linear approximation are a linear combination of generalized forces

$$h_i = h_i^P X_P + h_i^T X_T + h_i^S X_S \tag{34}$$

Substituting (34) into eqn. (30) – (32), we obtain to the fluxes

$$\mathrm{J}_l = \sum_k \Lambda_{lk} X_k \qquad (k,l) = p, T, S \tag{35}$$

where the kinetic coefficients Λ_{lk} are defined by the following equations

$$\Lambda_{pp} = \left(\varphi_P, h_i^P \right), \qquad \Lambda_{pT} = \left(\varphi_p, h_i^T \right), \qquad \Lambda_{pS} = \left(\varphi_p, h_i^S \right),$$

$$\Lambda_{Tp} = k_B T_0 \left(\varphi_T, h_i^p\right), \qquad \Lambda_{TT} = k_B T_0 \left(\varphi_T, h_i^T\right), \qquad \Lambda_{TS} = k_B T_0 \left(\varphi_T, h_i^S\right), \tag{36}$$

$$\Lambda_{Sp} = k_B \left(\varphi_S, h_i^p\right), \qquad \Lambda_{ST} = k_B \left(\varphi_S, h_i^T\right), \qquad \Lambda_{SS} = k_B \left(\varphi_S, h_i^S\right).$$

$$\varphi_p = v_z, \qquad \varphi_T = v_z\left(c^2 - \frac{5}{2} + \varepsilon_i - \bar{\varepsilon}\right), \qquad \varphi_S = \chi_S^-(v_z)\left(\frac{n_{j0}}{n_{i0}} - 1\right),$$

$$\left(\varphi_l, h_i\right) = \sum_{i=n,m} \int_S dS \int \varphi_l f_{i0} h_i d\mathbf{v}, \qquad l = p, T, S, \qquad i, j = n, m; \qquad i \neq j.$$

Here Λ_{lk} is the kinetic coefficient which determines contribution of generalized force X_k proportional component to the flux J_l. In so-called direct processes, force X_k defines the bound component of flux J_k, which is characterized by coefficient Λ_{kk}. The Poiseuille flux Λ_{pp} and the heat flux Λ_{TT} caused by the pressure and temperature gradients, respectively, are direct processes. The kinetic coefficient Λ_{SS} characterizes the entropy flux in the gas due to velocity selective radiation absorption by molecules. At the same time the forces X_k stimulate non-bound fluxes $J_i (i \neq k)$ characterized by kinetic cross-coefficients Λ_{ik}. In particular, temperature gradient stimulates gas motion along the capillary (thermal flux Λ_{pT}) and the entropy flux (thermo-entropic flux Λ_{ST}), and the pressure gradient causes heat flux (mechanocaloric effect Λ_{Tp}, [1]) and entropy flux (baro-entropic flux Λ_{Sp}). The action of resonant laser radiation leads to the drift of the gas Λ_{pS} and the heat flux Λ_{TS} which are directed across the baro- Λ_{Sp} and thermo-entropic Λ_{ST} fluxes respectively.

5. Onsager reciprocal relations

According to thermodynamics of nonequilibrium processes cross-coefficients in steady state must satisfy Onsager reciprocal relations:

$$\Lambda_{lk} = \Lambda_{kl} \tag{37}$$

To prove this we use the method proposed in ref. [12] and the modified in this study. Integrate the kinetic equation (11) for the odd perturbation functions h_i relatively to v_z along the length of the capillary. Substitute equation (34) for h_i in the obtained equations and then select summands under identical forces. As a result, we obtain the following system of equations for the functions h_i^l:

$$\mathbf{v}_\perp \frac{\partial h_i^P(\mathbf{v})}{\partial \mathbf{r}} - \sum_{j=n,m} L_{ij}\left(h_i^P(\mathbf{v})\right) = \frac{\varphi_P(v_z)}{k_B L}$$

$$\mathbf{v}_\perp \frac{\partial h_i^T(\mathbf{v})}{\partial \mathbf{r}} - \sum_{j=n,m} L_{ij}\left(h_i^T(\mathbf{v})\right) = \frac{T_0 \varphi_T(v_z)}{L} \tag{38}$$

$$\mathbf{v}_\perp \frac{\partial h_i^S(\mathbf{v})}{\partial \mathbf{r}} - \sum_{j=n,m} L_{ij}\left(h_i^S(\mathbf{v})\right) - b_i h_m^S(\mathbf{v}) = \frac{\varphi_S(v_z)}{L}$$

In the equations (38) for h_i^P and h_i^T the terms describing the radiative decay of the excited level are absent. This is due to the fact that the pressure and temperature gradients anywhere in the gas do not stimulate local perturbations of equilibrium populations of the particles energy levels.

By analogy with eqn. (10) we present the perturbation functions h_i^l as the sum of the even h_i^{l+} and odd h_i^{l-} parts with respect to the transverse component of the molecules velocity vector $\mathbf{v}_\perp$. Then the equations (38) can be divided into even and odd with respect to the vector $\mathbf{v}_\perp$

$$\mathbf{v}_\perp \frac{\partial h_i^{P-}(v_z,\mathbf{v}_\perp)}{\partial \mathbf{r}} - \sum_{j=n,m} L_{ij}\left(h_i^{P+}(v_z,\mathbf{v}_\perp)\right) = \frac{\varphi_P(v_z)}{k_B L} \tag{39}$$

$$\mathbf{v}_\perp \frac{\partial h_i^{T-}(v_z,\mathbf{v}_\perp)}{\partial \mathbf{r}} - \sum_{j=n,m} L_{ij}\left(h_i^{T+}(v_z,\mathbf{v}_\perp)\right) = \frac{T_0\varphi_T(v_z)}{L} \tag{40}$$

$$\mathbf{v}_\perp \frac{\partial h_i^{S-}(v_z,\mathbf{v}_\perp)}{\partial \mathbf{r}} - \sum_{j=n,m} L_{ij}\left(h_i^{S+}(v_z,\mathbf{v}_\perp)\right) - b_i h_m^{S+}(v_z,\mathbf{v}_\perp) = \frac{\varphi_S^-(v_z)}{L} \tag{41}$$

$$\mathbf{v}_\perp \frac{\partial h_i^{P,(T)+}(v_z,\mathbf{v}_\perp)}{\partial \mathbf{r}} = \sum_{j=n,m} L_{ij}\left(h_i^{P,(T)-}(v_z,\mathbf{v}_\perp)\right) \tag{42}$$

$$\mathbf{v}_\perp \frac{\partial h_i^{S+}(v_z,\mathbf{v}_\perp)}{\partial \mathbf{r}} - \sum_{j=n,m} L_{ij}\left(h_i^{S-}(v_z,\mathbf{v}_\perp)\right) - b_i h_m^{S-}(v_z,\mathbf{v}_\perp) = 0 \tag{43}$$

In (39) – (43) we take into account that the collision integrals for the odd (even) perturbation functions are the odd (even) functions of the velocity vector $\mathbf{v}_\perp$.

Reciprocity for thermal and mechanocaloric fluxes ($\Lambda_{pT} = \Lambda_{Tp}$) was proved in ref. [14]. Noteworthy that the proof of kinetic coefficients reciprocity that describe the transport processes is associated with exposure to radiation. Let us prove reciprocity relation for light-induced drift and baro -entropic flux $\Lambda_{pS} = \Lambda_{Sp}$. The equalities $\Lambda_{pT} = \Lambda_{Tp}$ and $\Lambda_{TS} = \Lambda_{ST}$ for cross-coefficients are proved similarly.

We multiply the equations (39) by h_i^S and the equations (41) by h_i^P. We subtract from the first obtained equation the second, summing over the states of the particles and integrating over velocity space and over the gas volume V. Taking into account the symmetry properties of the collision integral (commutativity of integral brackets) [13], we obtain

$$\Lambda_{pS} - \Lambda_{Sp} = k_B \sum_i \int_V dV \int f_{i0} \mathbf{v}_\perp \left(h_i^{S+}\nabla h_i^{P-} - h_i^{P+}\nabla h_i^{S-}\right) d\mathbf{v} + k_B \sum_i \int dV \int f_{i0} h_i^{P+} b_i h_m^{S+} d\mathbf{v} \tag{44}$$

The first term on the right side of equation (44) taking (42) and (43) is equal to

$$\begin{aligned} & k_B \sum_i \int_V dV \int f_{i0} \mathbf{v}_\perp \left(h_i^{S+}\nabla h_i^{P-} - h_i^{P+}\nabla h_i^{S-}\right) d\mathbf{v} = \\ & = k_B \sum_{i,j=m,n} \int dV \int f_{i0} \left[\mathbf{v}_\perp \nabla (h_i^{P-} h_i^{S+} - h_i^{P+} h_i^{S-}) - h_i^{P-} b_i h_m^{S-}\right] d\mathbf{v} \end{aligned} \tag{45}$$

We substitute the equation (45) into eqn. (44)

$$\Lambda_{pS}-\Lambda_{Sp}=k_B\sum_i\int_V dV\int f_{i0}\mathbf{v}_\perp\nabla\left(h_i^{S+}h_i^{P-}-h_i^{P+}h_i^{S-}\right)d\mathbf{v}+ \\ +k_B\sum_i\int dV\int f_{i0}b_i(h_i^{P+}h_m^{S+}-h_i^{P-}h_m^{S-})d\mathbf{v} \quad (46)$$

According to the theorem of the equivalence of the infinitesimal functions [15], the second term on the right-hand side of (46) is the infinitesimal magnitude of the higher (fourth) order of smallness and should be omitted. After applying Gauss transform to the first term in right part of eqn. (46), we get

$$\Lambda_{pS}-\Lambda_{Sp}=k_B\sum_i\int_{S_w} dS\int f_{i0}(\mathbf{v}\cdot\mathbf{n})(h_i^{S-}h_i^{P+}-h_i^{P-}h_i^{S+})d\mathbf{v} \quad (47)$$

Demonstrate that the integral in the right-hand side of (47) vanishes.

On the surface of the capillary the even h_i^{l+} and odd h_i^{l-} parts of perturbation functions have discontinuity and satisfy the following conditions:

$$h_i^{l+}(\mathbf{v})=\frac{h_i^l(-\mathbf{u})+h_{iw}^l(\mathbf{u})}{2} \quad (48)$$

$$h_i^{l-}(\mathbf{v})=\mp\frac{h_i^l(-\mathbf{u})-h_{iw}^l(\mathbf{u})}{2}=-\frac{|\mathbf{v}\cdot\mathbf{n}|}{(\mathbf{v}\cdot\mathbf{n})}\frac{h_i^l(-\mathbf{u})-h_{iw}^l(\mathbf{u})}{2} \quad (49)$$

$$\mathbf{u}=\mathbf{v}\,\mathrm{sgn}(\mathbf{v}\cdot\mathbf{n}),\qquad l=P,T,S\,.$$

Here $h_{iw}^l(\mathbf{u})$ are perturbation functions for the molecules reflected from the surface of capillary. The minus sign after the first equality of eqn. (49) is used at $(\mathbf{v}\cdot\mathbf{n})>0$, plus sign at $(\mathbf{v}\cdot\mathbf{n})<0$.

Eliminating from (48) and (49) $h_i^l(-\mathbf{u})$, we obtain the connection between the functions $h_i^{l-}(\mathbf{v})$ and $h_i^{l+}(\mathbf{v})$ at the boundary surface:

$$(\mathbf{v}\cdot\mathbf{n})h_i^{l-}(\mathbf{v})=-|\mathbf{v}\cdot\mathbf{n}|(h_i^{l+}(\mathbf{v})-h_{iw}^l(\mathbf{u})) \quad (50)$$

Substituting (48) and (50) into (47) for the case $l=P,S$ we obtain the equality

$$\Lambda_{pS}-\Lambda_{Sp}=\frac{k_B}{2}\sum_i\int_{S_w} dS\int f_{i0}|\mathbf{v}\cdot\mathbf{n}|\left[h_i^P(-\mathbf{u})h_{iw}^S(\mathbf{u})-h_i^S(-\mathbf{u})h_{iw}^P(\mathbf{u})\right]d\mathbf{v} \quad (51)$$

Consider the integral

$$Z=\frac{k_B}{2}\sum_i\int_{S_w} dS\int h_i^P(-\mathbf{u})f_{i0}|\mathbf{v}\cdot\mathbf{n}|h_{iw}^S(\mathbf{u})d\mathbf{v} \quad (52)$$

Replace $h_{iw}^S(\mathbf{u})$ with (9) and integrate over the half space of the velocities $\mathbf{v}\cdot\mathbf{n}>0$:

$$Z=\frac{k_B}{2}\sum_{i,j}\int_{S_w} dS\int_{\mathbf{v}\cdot\mathbf{n}>0} h_i^P(-\mathbf{v})d\mathbf{v}\int_{\mathbf{v}'\cdot\mathbf{n}<0}|\mathbf{v}'\cdot\mathbf{n}|f_{j0}(\mathbf{v}')h_j^S(\mathbf{v}')\cdot B(j,\mathbf{v}'\to i,\mathbf{v})d\mathbf{v}' \quad (53)$$

Reciprocity for the scattering kernel (5) gives the equation (53) the form:

$$Z = \frac{k_B}{2}\sum_{i,j}\int_{S_w} dS \int_{\mathbf{v}'\cdot\mathbf{n}<0} h_j^S(\mathbf{v}')d\mathbf{v}' \int_{\mathbf{v}\cdot\mathbf{n}>0} |\mathbf{v}\cdot\mathbf{n}| f_{i0}(\mathbf{v}) h_i^P(-\mathbf{v}) B(i,-\mathbf{v}\to j,-\mathbf{v}')d\mathbf{v} \tag{54}$$

We replace $\mathbf{v}'\to-\mathbf{v}$, $-\mathbf{v}'\to\mathbf{v}$, $i\to j,\ j\to i$. Then

$$Z = \frac{k_B}{2}\sum_{i,j}\int_{S_w} dS \int_{\mathbf{v}\cdot\mathbf{n}>0} h_i^S(-\mathbf{v})d\mathbf{v} \int_{\mathbf{v}'\cdot\mathbf{n}<0} |\mathbf{v}'\cdot\mathbf{n}| f_{j0}(\mathbf{v}') h_j^P(\mathbf{v}')\cdot B(j,\mathbf{v}'\to i,\mathbf{v})d\mathbf{v}' \tag{55}$$

When returning to complete space of velocities, with regard to (9) we get:

$$Z = \frac{k_B}{2}\sum_{i,j}\int_{S_w} dS \int h_i^S(-\mathbf{u})d\mathbf{v} \int_{\mathbf{v}'\cdot\mathbf{n}<0} |\mathbf{v}'\cdot\mathbf{n}| f_{j0}(\mathbf{v}') h_j^P(\mathbf{v}')\cdot B(j,\mathbf{v}'\to i,\mathbf{v})d\mathbf{v}' =$$
$$= \frac{k_B}{2}\sum_{i}\int_{S_w} dS \int f_{i0}|\mathbf{v}\cdot\mathbf{n}| h_i^S(-\mathbf{u}) h_{iw}^P(\mathbf{u})d\mathbf{v} \tag{56}$$

Substituting (56) into (51) we get $\Lambda_{pS}=\Lambda_{Sp}$, Q.E.D.

We note that the reciprocity relations are proved for arbitrary Knudsen (Kn) numbers and for any laws of interaction of the molecules with each other and with boundary surface.

6. Non-equilibrium stationary states

Discuss now the nonequilibrium stationary processes when the boundary conditions imposed on the system do not allow it to reach the equilibrium state. These processes have been considered in the linear approximation in ref. [16]. In stationary states the entropy production unlike the equilibrium states does not disappear. According to Prigogine's principle [1] in the stationary weakly non-equilibrium states the entropy production is minimal. If n of thermodynamic forces act on the system with fixed k forces then steady state of k-order is established in the system. Now the fluxes associated with other n-k forces are equal to zero.

Consider the possible stationary states. We start with the well-known effects.

6.1. *Thermomolecular pressure difference* (*TPD-effect*) [1, 14, 17].

Take the gas is in a closed system, which consists of two volumes connected by a capillary. If the radiation is absent, $X_S=0$, and the temperature difference at the ends of the capillary is fixed, $X_T=const$, then the pressure drop is established at the ends of the capillary. This phenomenon is known as thermomolecular pressure difference. In this case, the gas flow in the capillary is equal to zero, $J_p=0$, i.e. the thermal transport is compensated by the Poiseuille flux. With the use of (33), (35) and (36) we get [17]:

$$\frac{p_2}{p_1}=\left(\frac{T_2}{T_1}\right)^{\gamma_T}, \qquad \gamma_T=-\frac{1}{k_BT_0}\frac{\Lambda_{pT}}{\Lambda_{pp}} \tag{57}$$

where γ_T is the TPD index, defined as the ratio of the thermal creep of the Poiseuille flux. The value γ_T depends on the parameters of gas interaction with the surface of the capillary and Kn number. It is known [17] that in a slip-flow regime $\gamma_T \sim Kn^2$, and in free-molecular limit ($Kn >> 1$) in the case of specular-diffuse reflection of molecules from the capillary surface $\gamma_T=1/2$.

6.2. Mechanocaloric effect [1, 14].
This effect is responsible for the thermal energy transfer under isothermal movement of the gas. If the radiation is absent, $X_S = 0$ and the gas temperature is uniform, $X_T = 0$, then, according to (35), the heat flux is defined as

$$\mathrm{J}_T = -\Lambda_{Tp} k_B \ln\left(\frac{p_2}{p_1}\right). \tag{58}$$

The magnitude of q_p^* indicates the amount of heat that is transferred per unit time through the cross-section of the capillary under the action of pressure drop calculated per molecule:

$$q_p^* = \frac{\mathrm{J}_T}{\mathrm{J}_p} = \frac{\Lambda_{Tp}}{\Lambda_{pp}}. \tag{59}$$

Given the symmetry of cross kinetic coefficients, $\Lambda_{Tp} = \Lambda_{pT}$ from (57) and (59) we find connection between values $q_p^{\bullet}$ and γ_T

$$q_p^* = -k_B T_0 \gamma_T. \tag{60}$$

6.3. Light-induced pressure difference (LIPD-effect) [7, 18] is the emergence of the gas pressure drop at the ends of closed capillary under the action of resonant optical radiation.

Gas with uniform response to temperature ($X_T = 0$) is located in a closed capillary. The ends of the capillary can freely pass the laser radiation. The light-induced drift of the gas will stimulate the pressure redistribution along the capillary. The pressure inhomogeneity will cause the Poiseuille flux. In the steady state the light-induced drift will be compensated by the Poiseuille flux:

$$\mathrm{J}_p = \Lambda_{pp} X_p + \Lambda_{pS} X_S = 0. \tag{61}$$

From here taking into account (33) we obtain

$$\frac{p_2}{p_1} = \exp(\gamma_{ST}), \qquad \gamma_{ST} = \frac{1}{\beta k_B}\left(\frac{G_{mn}(0)}{k\overline{v}}\right)^2 \left(\frac{\Lambda_{pS}}{\Lambda_{pp}}\right)\left[1 - \exp(-\beta L)\right]. \tag{62}$$

In the case of an optically thin medium ($\beta L << 1$) expression for the parameter γ_{ST} simplifies and becomes

$$\gamma_{ST} = \frac{L}{k_B}\left(\frac{\Lambda_{pS}}{\Lambda_{pp}}\right)\left(\frac{G_{mn}(0)}{k\overline{v}}\right)^2. \tag{63}$$

The value of the index γ_{ST} depends on the parameters of the gas interaction with the capillary surface, detuning frequency radiation from the center of the absorption line, the intensity and wavelength of the radiation, as well as on *Kn* number. As shown in [19], in a slip-flow regime $\gamma_{ST} \sim Kn^2$, and in the free-molecular regime γ_{ST} does not depend on *Kn*.

6.4. Thermo-optical pressure difference (TOPD-effect) [19] is the phenomenon of the establishment of the gas pressure drop at the ends of closed capillary under the joint action of the resonant optical radiation and temperature inhomogeneity. This is a stationary state of second order in which two forces: X_T and X_S are fixed.

Longitudinal pressure gradient arises in such a system as a result of the light-induced drift of gas and thermal creep. In turn it will cause a Poiseuille flux. As a result of competition of non-equilibrium processes in the system is settles in some steady state of dynamic equilibrium. It will be the state with a minimum entropy production at which the gas flux J_p associated with the unfixed force X_p must be equal to zero:

$$J_p = \Lambda_{pp} X_p + \Lambda_{pT} X_T + \Lambda_{pS} X_S = 0 \quad . \tag{64}$$

From here follows:

$$\ln\left(\frac{p_2}{p_1}\right) = -\frac{1}{k_B T_0}\left(\frac{\Lambda_{pT}}{\Lambda_{pp}}\right)\ln\left(\frac{T_2}{T_1}\right) + \frac{1}{k_B}\left(\frac{\Lambda_{pS}}{\Lambda_{pp}}\right) X_S \quad . \tag{65}$$

Taking into account (33) we obtain the following expression for the pressure drop of gas at the ends of the capillary

$$\frac{p_2}{p_1} = \left(\frac{T_2}{T_1}\right)^{\gamma_T} \exp(\gamma_{ST}) \quad . \tag{66}$$

6.5. Optocaloric effect is the light–induced heat transfer.

Pressure and temperature of the gas are uniform, $X_p = 0$, $X_T = 0$, the radiation force is fixed $X_S = const$. Then according to (35) the heat flux and the drift gas are governed only by the radiation force X_S. Magnitude of q_S^* characterizes the quantity of heat which is transferred per unit time through a cross-section of the capillary under action the resonant optical radiation in calculation per molecule:

$$q_S^* = \frac{J_T}{J_p} = \frac{\Lambda_{TS}}{\Lambda_{PS}} \quad . \tag{67}$$

6.6. Light-induced baro- and thermo-effect [9] is phenomenon of simultaneous occurrence of pressure and the gas temperature difference at the ends of a closed thermally insulated capillary under exposure to resonant optical radiation. This is a stationary state of the first order when only the force of X_S is fixed. Light-induced gas drift and light-induced heat flux cause longitudinal pressure and temperature gradients in the system. These gradients, in turn, cause the own fluxes that will compete with light-induced fluxes. In the system some steady state of dynamic equilibrium is eventually established. According to Prigogine's principle [1], it will be state with a minimum entropy production at which the fluxes J_P and J_T, associated with unfixed forces X_p and X_T must be equal to zero:

$$J_p = \Lambda_{pp} X_p + \Lambda_{pT} X_T + \Lambda_{pS} X_S = 0, \tag{68}$$

$$J_T = \Lambda_{Tp} X_p + \Lambda_{TT} X_T + \Lambda_{TS} X_S = 0.$$

Equations for the light-induced differential pressure and temperature follow from here:

$$-k_B \ln\left(\frac{p_2}{p_1}\right) = X_S\,(\Lambda_{pT}\Lambda_{TS} - \Lambda_{pS}\Lambda_{TT})/\Delta_0$$

$$-\frac{1}{T_0}\ln\left(\frac{T_2}{T_1}\right) = X_S\,(\Lambda_{Tp}\Lambda_{pS} - \Lambda_{pp}\Lambda_{TS})/\Delta_0 \qquad (69)$$

$$\Delta_0 = \Lambda_{pp}\Lambda_{TT} - \Lambda_{Tp}\Lambda_{pT}$$

We see that longitudinal redistribution of gas pressure in the capillary is caused by light-induced drift (LID) and thermal transport, and also depends on heat fluxes. Thermal fluxes lead to the inhomogeneity of temperature that, in turn, is the cause of the thermal transport. The temperature difference is due to the light-induced heat transfer and the mechanocaloric flux, and depends also on Poiseuille flux and light-induced drift. Fluxes of the gas make the pressure gradient and hence the mechanocaloric flux.

Thus the resonant laser radiation can perform not only function of the piston compressing the gas, but also a sort of heat engine, capable to redistribute thermal energy in a gas.

7. References

[1] De Groot S R and Mazur P 1984 *Non-Equilibrium Thermodynamics* (Dover Publications)
[2] Gel'mukhanov F Kh and Shalagin A M 1979 *JETP Lett.* **29** 711
[3] Bennett W R Jr 1962 *Phys. Rev.* **126** 580
[4] Gel'mukhanov F Kh and Shalagin A M 1980 *JETP* **51** 839
[5] Ghiner A V, Stockman M I and Vaksman M A 1983 *Phys. Lett.* A **96** 79
[6] Chermyaninov I V and Chernyak V G 1988 *J. Eng. Phys. Thermophys.* **55** 1343
[7] Hoogeveen R W M, Van der Meer G J and Hermans L J F 1990 *Phys. Rev.* A **42** 6471
[8] Chermyaninov I V, Chernyak V G and Vilisova E A 2007 *JETP* **105** 511
[9] Chermyaninov I V and Chernyak V G 2013 *JETP* **117** 11
[10] Gel'mukhanov F Kh and. Il'ichov L V 1984 *Khim. Fiz.* **3** 1544
[11] Kucher Ivan 1971 *Surface Science* **25** 225
[12] Bishaev A M and Rykov V A 1983 *Zhurnal Vychislitel'noi Matematiki I Matematicheskoi Fiziki* **23** 954
[13] Cercignani C 1975 *Theory and Application of the Boltzmann equation* (Edinburgh and London: Scottish Academic Press)
[14] Loyalka S K 1971 *J. Chem. Phys.* **55** 4497
[15] Manturov O V and Matveev N M 1986 *Course of Higher Mathematics* (Moscow: Graduate School)
[16] Chermyaninov I V and Chernyak V G 2016 *Physics of Fluids* **28**, 092007
[17] Suetin P E, Skakun S G and Chernyak V G 1971 *Zh.Tekh.Fiz.* **41** 1738
[18] Chernyak V G, Vintovkina V A and Chermyaninov I V 1993 *JETP* **76** 768
[19] Chermyaninov I V and Chernyak V G 2014 *Physics of Fluids* **26** 092001

Acknowledgments
This work is supported by the Russian Federation Ministry of Education and Science (the research project number 3.6064.2017/БЧ).

IOP Conf. Series: Materials Science and Engineering **192** (2017) 012014 doi:10.1088/1757-899X/192/1/012014

Analysis of interface kinetics: solutions of the Gibbs-Thomson-type equation and of the kinetic rate theory

A Salhoumi[1,2] and P K Galenko[3,4]

[1] University of Hassan II Casablanca, Faculty of Sciences Ben M'Sik, Department of Physics, Laboratory of Condensed Matter Physics (LPMC), BP 7955 Casablanca, Morocco

[2] University of Hassan II Casablanca, Faculté des Sciences Juridiques, Economiques et Sociales Ain Sebaa, Département des Statistiques et Mathématiques Appliquées à l'Economie et à la Gestion (SMAEG), Laboratoire de la Modélisation Appliquée à l'Economie et à la Gestion (MAEG), BP 2634 Casablanca, Morocco

[3] Friedrich-Schiller-Universität Jena, Physikalisch-Astronomische Fakultät, D-07743 Jena, Germany

[4] Ural Federal University, Laboratory of Multi-Scale Mathematical Modeling, 620002 Ekaterinburg, Russia

E-mail: peter.galenko@uni-jena.de · ahmedsalhoumi@gmail.com

Abstract. Rapidly moving solid-liquid interface is treated analytically and numerically. Derivation and qualitative analysis of interface propagation kinetics is presented. Quantitative predictions of solutions, which follow from the Kinetic Rate Theory and the solution of Gibbs-Thomson-type equation, are compared with Molecular Dynamics simulation data (MD-data) on crystallization and melting of fcc-lattice of nickel. It is shown in the approximation of a linear behavior of the interface velocity versus undercooling that the Gibbs-Thomson-type equation and kinetic rate theory describe MD-data well enough, in the range of small growth velocity and within the range of relatively small undercooling, with a relative error for the obtained values of kinetic coefficient of the order 1.1%. Within the small-and long range of undercooling, in nonlinear behavior of the interface velocity versus undercooling, the kinetic rate theory disagrees sharply with MD-data, qualitatively and quantitatively, unlike to the Gibbs-Thomson-type equation which is in a good agreement with MD-data within the whole range of undercooling and crystal growth velocity.

1. Introduction

Methods of atomistic simulation represent now the robust quantitative techniques, which serve to obtain equilibrium and kinetic properties of phase interfaces. In equilibrium, these properties concern the surface energy and have physical relevance for the atomistic thickness of interfaces. In dynamics, these properties are connected to the interfacial self- and inter-diffusion as well as the mobility of interfaces. The latter property plays a crucial role in the description of both the interface propagation and pattern formation [1]. Indeed, special attention in atomistic simulations is paid to melting/crystallization kinetics and diffusion dynamics [2, 3]. More specifically, using MD-data of simulation [4, 5], as well as verification of various kinetic models and tests of phase field models have been provided [1, 6, 7].

IOP Conf. Series: Materials Science and Engineering **192** (2017) 012014 doi:10.1088/1757-899X/192/1/012014

In the present work, we use MD-data as a benchmark to verify outcomes from two theoretical approaches: the Kinetic Rate Theory (KRT) and Phase Field Model (PFM). We shall check both the qualitative behavior and the quantitative consistency of the kinetic equation for interface motion and acceleration dependent Gibbs-Thomson equation with respect to MD-data for solidification and melting of nickel crystal growing in the $< 100 >$–direction.

2. Kinetic rate theory

The Kinetic Rate Theory (or "the theory of rate kinetics") describes the interface motion due to the driving force which provides the overcome of an energetic barrier between the phases that have different energies [8]. In this section, we briefly describe the Kinetic Rate Theory due to Chernov [9]. This theory is applied to the growing crystal from an undercooled melt. When a particle (atom, molecule, or ion), moves from the crystal into the medium under the equilibrium conditions, it must change its energy by

$$\varepsilon_M - \varepsilon_S = T_m \left(s_M - s_S\right) \equiv \Delta H. \tag{1}$$

where ε_S and ε_M are the average energies of the particles, which occupy the equilibrium positions in the crystal and the medium, T_m is the melting temperature, s_S and s_M are the entropy in solid and in the medium, respectively, and ΔH is the enthalpy of the melting.

At the crystal-liquid interface, the particle must generally overcome a potential barrier E ("the activation energy" E) which depends on the configuration of the activated complex in the liquid. The real role is played by the value of E which is minimal with respect to the various configurations of the activation complex. These interfacial configurations depend on the structures of both the liquid and the solid phases, and the probability that the complex most suitable for the transition will appear is determined by the easiness with which the short-range order in the liquid changes to that in the solid. The atomic configurations also change in a viscous flow of the liquid. Therefore for the crystal-melt interface, E was expected to have the values of the order of the activation energy for a viscous flow.

The numbers j_+ of atoms that go from the melt over the crystal per unit time at a single kink (interface), and the opposite flow j_- from crystal to the melt can be written as

$$j_+ = \widetilde{\nu} \exp\left(-\frac{\Delta s}{k}\right) \exp\left(-\frac{E}{kT}\right), \tag{2}$$

$$j_- = \widetilde{\nu} \exp\left(-\frac{\Delta s}{k}\right) \exp\left(-\frac{E+\Delta H}{kT}\right). \tag{3}$$

where $\widetilde{\nu}$ is the frequency of thermal vibrations of an atom in the crystal and the liquid, Δs is the entropy of the melting, T is the absolute temperature, k is Boltzmann's constant, and $\exp\left(-\Delta s/k\right)$ is the probability of finding an atom of the liquid in the immediate vicinity of the kink in the most advantageous activation complex corresponding to barrier E.

If the average distance between the kinks is λ_0, the probability of finding a kink on the surface is equal to $(a/\lambda_0)^2$ with a the crystal lattice parameter. Then, the velocity of the phase boundary is given by

$$\begin{aligned} V &= \left(\frac{a}{\lambda_0}\right)^2 a\left(j_+ - j_-\right) \\ &= \left(\frac{a}{\lambda_0}\right)^2 a\widetilde{\nu} \exp\left(-\frac{\Delta s}{k}\right) \exp\left(-\frac{E}{kT}\right) \left\{1 - \exp\left[-\frac{\Delta H}{k}\left(\frac{1}{T} - \frac{1}{T_m}\right)\right]\right\}. \end{aligned} \tag{4}$$

which is reduced in the condition $\Delta s \Delta T/kT << 1$, i.e., the condition of low undercooling at growth front such as

$$V \simeq \beta\left(T\right) \Delta T \,. \tag{5}$$

where the kinetic coefficient of growth from the melt $\beta(T)$ and the undercooling at the solid-liquid interface ΔT are given respectively by

$$\beta(T) \simeq \left(\frac{a}{\lambda_0}\right)^2 a\tilde{\nu}\frac{\Delta s}{kT}\exp\left(-\frac{\Delta s}{k}\right)\exp\left(-\frac{E}{kT}\right), \tag{6}$$

$$\Delta T = T_m - T\,. \tag{7}$$

3. Phase field model

The evolution of a binary dilute system to equilibrium, which consist of A-atoms (solvent) together with a tiny amount of B-atoms (solute) with the concentration C under constant temperature T and constant pressure, is described by the following set of hyperbolic equations [10, 11]:

$$\tau_D\frac{\partial^2 C}{\partial t^2} + \frac{\partial C}{\partial t} = \nabla\cdot\left[M_C\left(\frac{\partial^2 f}{\partial C^2}\nabla C + \frac{\partial^2 f}{\partial C\partial\phi}\nabla\phi\right)\right], \tag{8}$$

$$\tau_\phi\frac{\partial^2\phi}{\partial t^2} + \frac{\partial\phi}{\partial t} = M_\phi\left(\varepsilon_\phi^2\nabla^2\phi - \frac{\partial f}{\partial\phi}\right), \tag{9}$$

where f is the free energy density, ϕ is the phase field variable, τ_D is the relaxation time for the diffusion flux, M_C is the mobility of B-atoms, τ_ϕ is the time scale for the relaxation of the rate of change of the phase field $\partial\phi/\partial t$, M_ϕ is the mobility of the phase field, and ε_ϕ^2 is the gradient energy factor.

According to Zhang et al. [12], the phase field dynamics at a given constant driving force ΔG with the average value, that is taken in the normal direction over the interface, is represented by

$$\Delta G = -\langle\Lambda^*(T, C, \phi)\rangle, \tag{10}$$

with

$$\Lambda^*(T, C, \phi) = \frac{RT}{v_m}\left[\frac{(1-k_e)C}{k_e + (1-k_e)p(\phi)} - \frac{1-k_e}{m_e}(T - T_A)\right]. \tag{11}$$

where $p(\phi) = \phi^2(3-2\phi)$ is the interpolation function, m_e is the slope of the liquidus line in the equilibrium phase diagram, T_A is the solidification temperature of the solvent, R is the gas constant, v_m is the molar volume (assumed equal for A- and B-atoms), and k_e is the equilibrium solute partition coefficient.

Under the assumption of constant averaged driving force, Eqs. (10) and (11), we have obtained from the system of governing equations (8) and (9) the following sole hyperbolic phase field equation,

$$\tau_\phi\frac{\partial^2\phi}{\partial t^2} + \frac{\partial\phi}{\partial t} = \nu\left[\nabla^2\phi - \frac{9}{2\delta^2}\frac{dg(\phi)}{d\phi} - \frac{1}{2\sigma\delta}\Delta G\frac{dp(\phi)}{d\phi}\right]. \tag{12}$$

where $g(\phi) = \phi^2(1-\phi)^2$ is the double-well function, ν is the phase field diffusion parameter, δ is the interfacial width, and σ is the interface energy. Note that the interface energy and the interfacial width are connected via the gradient energie factor such as $\varepsilon_\phi^2 = 2\sigma\delta$.

Using the averaging method widely known in the mechanics of continuous media [13, 14], in solidification processes [15, 16], and, particularly, in the derivation of conservation equations for multiphase systems [14, 17, 18, 19], we have evaluated the spatial and the temporal derivatives *via* interface thickness, velocity and acceleration which are driven by the average force ΔG and the interface curvature, κ [20]. These spatial and the temporal derivatives are found by choosing the phase-field profile ϕ in the dynamical regimes consistently with the traveling waves, Eq. (13), [11, 12] which are propagating in the non-stationary regime,

$$\phi(n,t) = \frac{1}{2}\left[1 + \tanh\left(\frac{n}{\ell(t)}\right)\right]. \tag{13}$$

Note that the unit normal vector exterior to the solid phase $\mathbf{n}_s$ and the exterior unit normal vector to the liquid phase $\mathbf{n}_l$ are defined such as $\mathbf{n}_s = \mathbf{n}$, and $\mathbf{n}_l = -\mathbf{n}_s = -\mathbf{n}$, respectively. Also, according to Ref. [17] and references therein, the module of a phase field gradient $|\nabla\phi|$, Eq. (14), the average unit normal vector exterior to the solid phase $\mathbf{n}$, Eq. (15), and the curvature of the solid-liquid interface κ, Eq. (16), are respectively given by

$$|\nabla\phi| = \frac{\partial\phi}{\partial n}, \tag{14}$$

$$\mathbf{n} = -\frac{\nabla\phi}{|\nabla\phi|}, \tag{15}$$

$$\kappa = \nabla\cdot\mathbf{n} = -\frac{1}{|\nabla\phi|}\left[\nabla^2\phi - \frac{(\nabla\phi\cdot\nabla)|\nabla\phi|}{|\nabla\phi|}\right], \tag{16}$$

where the curvature takes the negative sign for the convex interfaces, $\kappa < 0$.

Therefore we have found the following spatial and temporal derivatives, Eqs. (17)-(19),

$$\nabla^2\phi = -\kappa|\nabla\phi| + \frac{(\nabla\phi\cdot\nabla)|\nabla\phi|}{|\nabla\phi|} = -\kappa\frac{\partial\phi}{\partial n} + \frac{\partial^2\phi}{\partial n^2}, \tag{17}$$

$$\frac{\partial\phi}{\partial t} = -V|\nabla\phi| = -V\frac{\partial\phi}{\partial n}, \tag{18}$$

$$\frac{\partial^2\phi}{\partial t^2} = -\gamma|\nabla\phi| - V\frac{\partial(|\nabla\phi|)}{\partial t} = -\gamma|\nabla\phi| + \frac{V}{\ell}\frac{\partial\ell}{\partial t}\frac{\partial\phi}{\partial n} + V^2\frac{\partial^2\phi}{\partial n^2}, \tag{19}$$

where V is the velocity of the interface directed by the normal vector $\mathbf{n}$, $\gamma = \partial V/\partial t$ is the interface acceleration, and $\ell = \ell(t)$ is the time-dependent effective thickness of the interface which plays the role of the correlation length.

By incorporating the above spatial and temporal derivatives (17)-(19) in Eq. (12), we have obtained a nonlinear equation yielding the expression for the effective thickness, ℓ, Eq. (20)

$$\ell(t) = \frac{2\delta}{3}\left[1 - V^2(t)/(V_\phi)^2\right]^{1/2}, \tag{20}$$

and the generalized relationship between acceleration, γ, velocity, V, curvature, κ, of the interface and the driving force, ΔG, for the interface motion as [20]:

$$\tau_\phi\frac{\gamma}{[1 - V^2/(V_\phi)^2]^{3/2}} + \frac{V}{[1 - V^2/(V_\phi)^2]^{1/2}} = \frac{\nu}{\sigma}\Delta G + \frac{\nu\kappa}{[1 - V^2/(V_\phi)^2]^{1/2}}. \tag{21}$$

where the maximum speed V_ϕ of the phase field propagation is defined by the relaxation time τ_ϕ, $V_\phi = (\nu/\tau_\phi)^{1/2}$.

Note that with the increase of the interface velocity, ℓ becomes smaller and, in the limit $V \to V_\phi$, the correlation length tends to zero as predicted by Eq. (20). Since the correlation length ℓ is the distance at which the essential spatial variation of the phase field occurs, one can mark that the limit $V \to V_\phi$ corresponds to a sharp change in the phase field variation. Obviously, in the steady state regime of interface propagation, i.e., in a motion with the constant normal velocity, $V \equiv const$, the correlation length (20) does not depend on time. Due to inclusion of acceleration and high interface velocity, Eq. (21) represents a general case of a

Gibbs-Thomson interfacial condition. This generalized condition can be used at the steady state regimes with constant interface velocity ($\gamma = 0$ and $V = const$) as well as at the non-stationary regimes of interface propagation ($\gamma \neq 0$ and $V \neq const$).

The limited numbers of cases and existing theories, which are connected with Eq. (21) were discussed from a theoretical point of view in [20]. Among the other important results of the dynamic boundary condition Eq. (21) for the rapidly moving curved interface which includes both velocity and acceleration of the interface and can move by mean curvature and due to the driving force given by a deviation of temperature and concentration from their equilibrium values, the following should be mentioned: (*i*) a generalization for the well-known velocity dependent Gibbs-Thomson relation, and (*ii*) a generalized Born-Infeld equation for the hyperbolic motion by mean curvature and under driving force in the traditional Cartesian coordinates. These (*i*) and (*ii*) generalizations are reduced to the classical equation for the motion by mean curvature (which follows from the Cahn-Allen equation [21, 22]) and to the standard Born-Infeld equation [23, 24] (which follows from the undamped Klein-Gordon equation).

Otherwise if the interface velocity is comparable to the maximum speed of disturbances, which have been propagating in the phase field, i.e., with $V \lesssim V_\phi$, the non-stationary propagation of the curved interface is described by the full equation (21). It is clear that the normal velocity to the interface, V, cannot overcome and be larger than the maximum speed of disturbance propagation in the phase field because the phase field itself determines the interface, its velocity and acceleration. In the case of the fast propagating planar interface, the presence of the constant velocity ($V = const$ and $\kappa = 0$), Eqs. (20) and (21) denote the velocity V as

$$V = \frac{\nu(\Delta G/\sigma)}{\sqrt{1 + (\nu/V_\phi)^2\,(\Delta G/\sigma)^2}}\,, \tag{22}$$

where the driven force (Gibbs free energy difference on transformation) ΔG for a pure (chemically one component, elemental) system is given by

$$\Delta G = -\left\langle \Lambda^*\left(T, C, \phi\right)\right\rangle \equiv \frac{\Delta H}{T_m}\Delta T\ . \tag{23}$$

Table 1. Material parameters for Ni used in calculations.

Material parameter	Dimension	Value	Reference
Melting temperature, T_m	K	1706	[4]
Enthalpy of melting, ΔH	$J{\cdot}m^{-3}$	2.66×10^9	
	$J{\cdot}mol^{-1}$	17200	[25]
Entropy of melting, $\Delta s = \frac{\Delta H}{T_m}$	$J{\cdot}m^{-3}{\cdot}K^{-1}$	1.56×10^6	
	$J{\cdot}mol^{-1}{\cdot}K^{-1}$	10.07	[25]
Interface energy, σ	$J{\cdot}m^{-2}$	0.25	present work
Crystal lattice parameter, a (at T_m)	m	3.565×10^{-10}	[26]
Mean distance between steps at the crystal surface, λ_0	m	$\approx 3a$	[9]
Frequency of phonon vibration, $\widetilde{\nu}$	s^{-1}	10^{14}	[9]
Activation energy, E	$J{\cdot}m^{-3}$	2.61×10^8	
	$J{\cdot}mol^{-1}$	1720	[25]
Diffusion coefficient for phase field, ν	$m^2{\cdot}s^{-1}$	1.15×10^{-7}	present work
Maximum speed for phase field propagation, V_ϕ	$m{\cdot}s^{-1}$	185	present work

4. Comparison with MD simulation in the case of nickel

Using material parameters for the nickel (see Table 1) we have tested the outcomes of the Kinetic Rate Theory and Phase Field Model in comparison with MD-data obtained in works [4, 5]. We have especially checked these approaches in a small range of undercooling (where the solid-liquid interface moves by the linear "velocity-undercooling" relationship) and for a large range of undercooling (where a clear non-linearity in the "velocity-undercooling" relationship appears in crystal growth [1]).

4.1. Small range of undercooling and overheating

Now, we compare the predictions of the equations from kinetic rate theory [9] and the nonlinear solution of the phase field model [20] with MD-data of computer simulation of Mendelev et al. [4]. These data was obtained for relatively small values of the undercooling $0 < \Delta T(\mathrm{K}) < 50$ and overheating $-55 < \Delta T(\mathrm{K}) < 0$ at the solid-liquid interface of the nickel crystal. Figure 1 shows that MD-data exhibits linear behavior for the interface velocity, both in melting and in crystallization, for the undercooling interval $-55 < \Delta T(\mathrm{K}) < 50$. It is seen clearly that Eq. (5) and Eq. (22) describe these well enough and even if we take into account the exponential equation of the kinetic rate theory, namely Eq. (4), only a very slight difference in comparison with the linearized equation; Eqs. (5) appears.

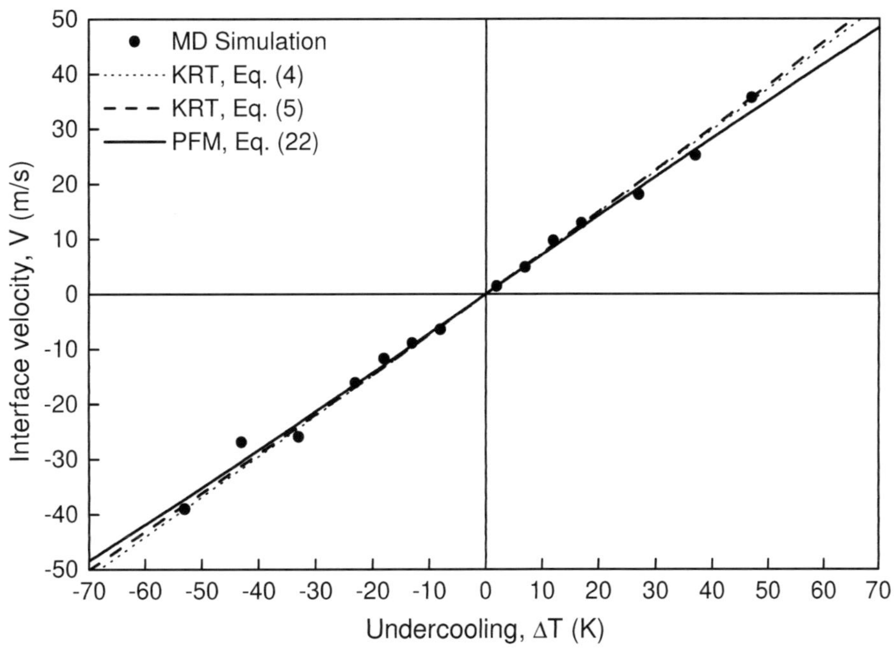

Figure 1. Comparison of the predicted interface velocity V by Eq. (4) ······ and Eq. (5) - - - - of Kinetic Rate Theory (KRT) [9], and by Eq. (22) —— of Phase Field Model (PFM) [20] with MD-data of computer simulation ● by Mendelev et al. [4] for Ni in the case of small range of undercooling $0 < \Delta T(\mathrm{K}) < 50$ and overheating $-55 < \Delta T(\mathrm{K}) < 0$.

Furthermore we have found that, for the Kinetic Rate Theory, the better correspondence to the simulated MD-data is obtained for the kinetic coefficient $\beta(T) = 0.711$ m/(s·K). This latter is a very close value, with a relative error of the order 1.1%, in comparison with the kinetic coefficient value $\beta = 0.719$ m/(s·K) given by Mendelev et al. [4], which was computed from the results of MD-simulations. Therefore, both approaches [solutions of the phase field model Eq. (22) and linearized-and exponential equations of rate kinetics Eq. (5) and Eq. (4), respectively] confirm the ability to describe MD-data in the range of linear behavior $V \sim \Delta T$ and, relatively close to equilibrium point $\Delta T = 0$ K.

4.2. Large range of undercooling

In the present subsection, we compare the equations from kinetic rate theory [9] and the nonlinear solution of the phase field model [20] with MD-data of computer simulation obtained by Hoyt et al. [5] for large values of the undercooling $0 < \Delta T(\mathrm{K}) < 710$ at the solid-liquid interface of nickel crystal. Figure 2 illustrates the comparison of the exponential-and linearized equations of the kinetic rate theory, Eq. (4) and Eq. (5) respectively, and of the nonlinear solution of the phase field equation, Eq. (22), with the MD-data of Hoyt et al.

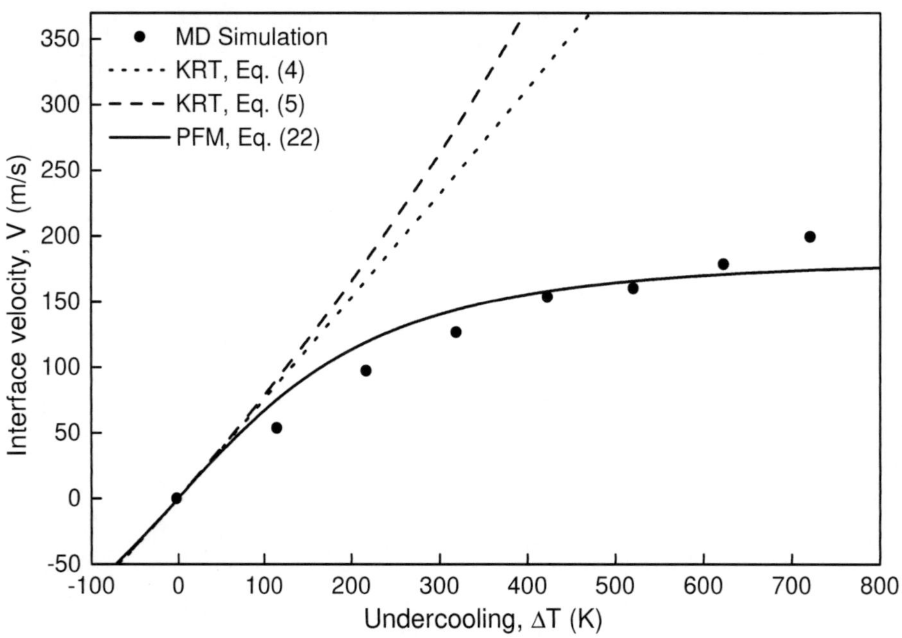

Figure 2. Comparison of the predicted interface velocity V by Eq. (4) ······ and Eq. (5) - - - - of Kinetic Rate Theory (KRT) [9], and by Eq. (22) —— of Phase Field Model (PFM) [20] with MD-data of computer simulation ● due to Hoyt et al. [5] for Ni in the case of large range of undercooling $0 < \Delta T(\mathrm{K}) < 710$.

It can be seen that both equations of the kinetic rate theory behave inconsistently in comparison with MD-data qualitatively and, as a consequence, quantitatively. By contrast, non-linear solution of the phase field model predicts "velocity-undercooling" relationship in agreement with MD-data. Namely, the phase field equation, Eq. (22), shows that the interface velocity V gradually deviates from the linear law to smaller values as the undercooling increases. Such a behavior in crystal growth kinetics has been found by MD-simulations of elemental systems [1], qualitatively confirmed in our work [20] and presently confirmed in quantitative comparison of Eq. (22) with the MD-data of Hoyt et al. [5].

5. Conclusions

Solidification kinetics of a pure elemental substance has been analyzed. A qualitative and quantitative comparison of solid-liquid interface motion has been made. Predictions from the Kinetic Rate Theory and from the phase field model have been discussed. As a consequence of the Kinetic Rate Theory, we have considered the kinetic equation for the solid-liquid interface. From the analysis of the phase field model, we have considered the solution of Gibbs-Thomson-type equation. The kinetic equation of the Kinetic Rate Theory and the solution of Gibbs-Thomson-type equation are compared with the data of atomistic simulations obtained by method of Molecular Dynamics.

It has been shown that the Molecular-Dynamics-data of simulation for solidification or melting of a pure nickel [4] are described well by both approaches in the range of small growth velocity and within the range of relatively small undercooling (up to 50K) or small overheating (up to −55K). Within the whole (small and large) range of undercooling (up to 700K), the Kinetic Rate Theory disagrees with MD-data [5] qualitatively and quantitatively. Instead, the Gibbs-Thomson-type equation is in a good agreement with MD-data [5] within the whole range of undercooling and nickel crystal growth velocity.

6. References

[1] Hoyt J J, Asta M and Karma A 2003 Atomistic and continuum modeling of dendritic solidification *Mater. Sci. Eng.* **41** 121-64
[2] Hoyt J J, Asta M, Haxhimali T, Karma A, Napolitano R E, Trivedi R, Laird B B and Morris J R 2004 Crystal–mlt interfaces and solidification morphologies in metals and alloys *MRS Bulletin* **29** (12) 935-39
[3] Kerrache A, Horbach J and Binder K 2008 Molecular-dynamics computer simulations of crystal growth and melting in $Al_{50}Ni_{50}$ *Euro. Phys. Lett.* **81** 58001-1-6
[4] Mendelev M I, M I, Rahman M J, Hoyt J J and Asta M 2010 Molecular-dynamics study of solid–liquid interface migration in fcc metals *Model. Simul. Mater. Sci. Eng.* **18** 074002
[5] Hoyt J J, Sadigh B, Asta M and Foiles S M 1999 Kinetic phase field parameters for the Cu-Ni system derived from atomistic computations *Acta Mater.* **47** 3181–87
[6] Bragard J, Karma A, Lee Y H and Plapp M 2002 Linking phase-field and atomistic simulations to model dendritic solidification in highly undercooled melts *Inter. Sci.* **10** 121–36
[7] Berghoff M, Selzer M, and Nestler B 2013 Phase-field simulations at the atomic scale in comparison to molecular dynamics *The Scientific World Journal* **2013**, Article ID 564272 8 pages
[8] Christian J W 1975 *The Theory of Transformations in Metals and Alloys* (Oxford: Pergamon)
[9] Chernov A A 1984 *Modern Crystallography III – Crystal Growth* (Berlin: Springer)
[10] Galenko P and Jou D 2005 Diffuse-interface model for rapid phase transformations in nonequilibrium systems *Phys. Rev.* E. **71** 046125
[11] Galenko P K, Abramova E V, Jou D, Danilov D A, Lebedev V G and Herlach D M 2011 Solute trapping in rapid solidification of a binary dilute system: A phase-field study *Phys. Rev.* E. **84** 041143-1-17
[12] Zhang L, Danilova E V, Steinbach I, Medvedev D, Galenko P K 2013 Diffuse-interface modeling of solute trapping in rapid solidification: Predictions of the hyperbolic phase-field model and parabolic model with finite interface dissipation *Acta Mater.* **61** 4155–68
[13] Sou S 1971 *Hydrodynamics of Multiphase Systems* (Moscow: Mir)
[14] Drew D A 1983 Mathematical modeling of two-phase flow *Ann. Rev. Fluid Mech.* **15** 261–91
[15] Zhuravlev V A and Kitaev E M 1974 *Thermophysics of Continuous Ingot Formation* (Moscow: Metallurgia)
[16] Vorobiov I L 1980 *Proceedings of Bauman-MVTU* **30** 31
[17] Beckermann C, Diepers H J, Steinbach I, Karma A and Tong X 1999 Modeling melt convection in phase-field simulations of solidification *J. Comput. Phys.* **154** 468–96
[18] Hassanizadeh M. and Gray W.G. 1979 General conservation equations for multi-phase systems: 1. Averaging procedure *Adv. Water Resour.* **2** 131–44
[19] Ni J and Beckermann C A 1991 Volume-averaged two-phase model for transport phenomena during solidification *Metall. Trans.* B. **22** 349–61
[20] Salhoumi A and Galenko P K 2016 Gibbs-Thomson condition for the rapidly moving interface in a binary system *Physica* A **447** 161–71
[21] Cahn J W and Allen S M 1977 microscopic theory of domain wall motion and its experimental verification in Fe-Al alloy domain growth kinetics *J. Phys. Colloq.* **38** C7–51–54

[22] Allen S M and Cahn J W 1979 A microscopic theory for antiphase boundary motion and its application to antiphase domain coarsening *Acta Metall.* **27** 1085–95
[23] Born M and Infeld L 1934 Foundations of the new field theory *Proc. R. Soc. A Math. Phys. Eng. Sci.* **144** 425–51
[24] Whitham G B 1974 *Linear and Nonlinear Waves* (New York: Wiley-Interscience)
[25] Galenko P K and Danilov D A 2000 Hyperbolic self-consistent problem of heat transfer in rapid solidification of supercooled liquid *Phys. Lett.* A. **278** 129–38
[26] Davey W P 1925 Precision measurements of the lattice constants of twelve common metals *Phys. Rev.* **25** 753–61

Acknowledgments
The authors are grateful to Mark Asta for useful discussions and exchanges. Supports from the RSF, grant number 16-11-10095 and from the project "MULTIPHAS", contract Nr. 50WM1541 by the European Space Agency, German Aerospace Center and Friedrich-Schiller-Universität Jena are acknowledged. One of us, A.S., thanks Prof. M. Bennai and Prof. R. Guerbaz for incorporating the present work within the research activities of their respective laboratories LPMC and MAEG.

IOP Conf. Series: Materials Science and Engineering **192** (2017) 012015 doi:10.1088/1757-899X/192/1/012015

The influence of melt flow on grain structure of tin and its alloys produced by ultrafast quenching from the melt

Olga Gusakova[1] **and Vasiliy Shepelevich**[2]

[1]Department of Nuclear and Radiation Safety, International Sakharov Environmental Institute of Belarussian State University, Dolgobrodskaya st. 23, 220070 Minsk, Belarus
[2]Department of Solid-State Physics, Belarussian State University, Nezavisimosti pr.4, 220050 Minsk, Belarus

E-mail: ol.gusakova@gmail.com

Abstract. The results on the grain structure and texture of tin and tin-based foils produced by ultrafast quenching from the melt are presented in this work. The formation of fine equiaxed and large grains elongated along the spreading direction of the foil was observed. Detailed studies of subgrain structure showed that the crystallographic orientation of elongated grains changes monotonically within the grain. It can be assumed that the reason for the change of crystallographic orientation is the deformation of solidified material. The deformation is caused by the viscous force of undercooled melt which moves above the crystallization front in the direction of the foil spreading.

1. Introduction

The classical methods to improve mechanical properties of material have already been well studied. Currently, intense research work on the materials production in non-equilibrium solidification conditions is conducted. Strongly nonequilibrium crystallization conditions are implemented in such techniques as: melt spinning technique, ultrafast quenching from the melt, laser, ion-plasma, compression-plasma surface treatment. These techniques allow reaching the purposeful influence on the functional material properties. In these techniques, the cooling rate of the melt reaches 10^4–10^7 K/s [1–2]. Solidification begins at deep undercooling and proceeds with a high liquid – solid interface velocity in the range 1–10 m/s [3–4]. In this case, the melt which is above the solidification front is not stationary but moves in the tangential direction relative to the interface. Thus, during ultra-fast quenching from the melt, velocity is comparable to the linear crystallizer velocity and is a few meters per second. It is known that the viscosity of the melt decreases when its temperature increases, rising by fifteen orders from the melting point to the glass transition temperature [5]. The directional motion of the viscous supercooled melt causes the shear stress on the surface of the solidified layer, leads to the material deformation and the microstrain appearance in the grain volume, i.e. strain hardening. In the present study grain structure of foils of tin and its alloys with Bi, Pb, Zn and In obtained by ultrafast quenching from the melt is investigated.

2. Experimental

Foils were prepared by splashing of melt drops of mass 0.2–0.3 g on a polished inner surface of a rotating hollow copper cylinder (mold). The diameter of the mold was 25 cm and the linear velocity of rotation was 15 m/s. Melt temperature was controlled by a thermocouple placed in a furnace in the vicinity of the melt, and was set in the range 550–570 K. To homogenize the melt, it was kept in the furnace for 60 seconds. Foils were prepared from tin, with purity better than 99.99%.

Studies of the grain structure and texture of the foils were performed by electron backscatter diffraction technique (EBSD), which was implemented using the phase analysis attachment «HKL CHANNEL 5» (Oxford Instruments, UK) to the scanning electron microscope LEO 1455 VP (Karl Zeiss, Germany). Grain structure was analyzed along the foil by scanning of the side adjacent to the mold, the inner layer, freely solidified side, as well as longitudinal and cross sections of the foil. Slices were prepared by grinding and polishing machine TegraPol by Struers technique.

3. Results

The foils obtained by ultrafast quenching from the melt were characterized by width ranging from 5 to 10 mm, thickness ranging from 40 to 50 μm and by the length up to 10 cm. Further, it will be assumed that the direction of the foils spreading is OX, direction perpendicular to it in plane parallel to the foil surface is OY and the direction perpendicular to the foil surface is OZ (figure 1).

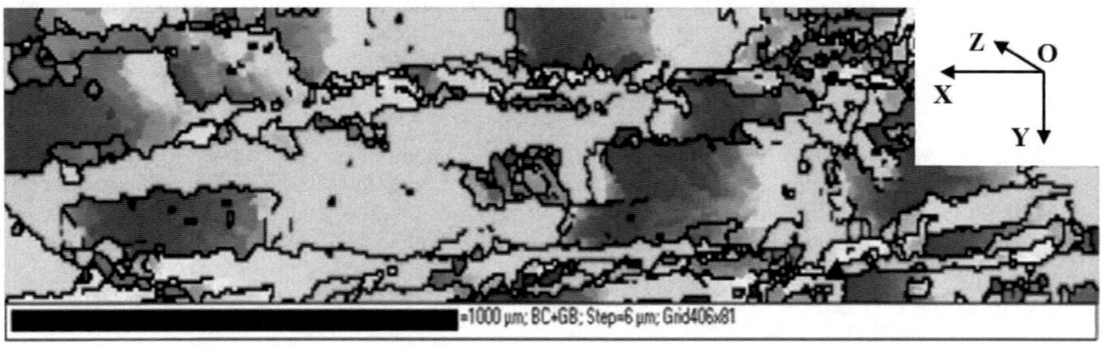

Figure 1. A typical view of the grain structure of the middle foil part.

As it was shown in [6], two types of grains are formed in the major part of the foil: large elongated along the spreading direction grains, and fine equiaxed grains located at the boundaries of large ones. The size of small grains is in the range of 10 - 80 microns, and their average diameter is 35 μm. The volume fraction of small equiaxed grains is 20 - 25%. Large elongated grains make up about 80% of the foil's volume. Both large and small grains have a columnar structure. The orientation of the elongated grains have similar character on the side adjacent to the mold, in the middle of the foil and at freely solidified side because of the columnar structure [7]. Thus, the foils have a bimodal grain structure consisting of small grains and large elongated grains.

It was revealed by EBSD technique that grains oriented so that the deviation of axis C_4 of β-Sn from foil spreading direction (OX) is less than 20 degrees occupy up to 80% of the total area. The grains of such orientation are shown in the figure 2a. Figure 2d shows the unit cell of β-Sn with it's C_4 axis of symmetry and figure 2b shows scheme of grains colouring according to the crystal orientation relative to the surface of the foil. According to the crystal orientation colouring such grains are colored in green or blue colors.

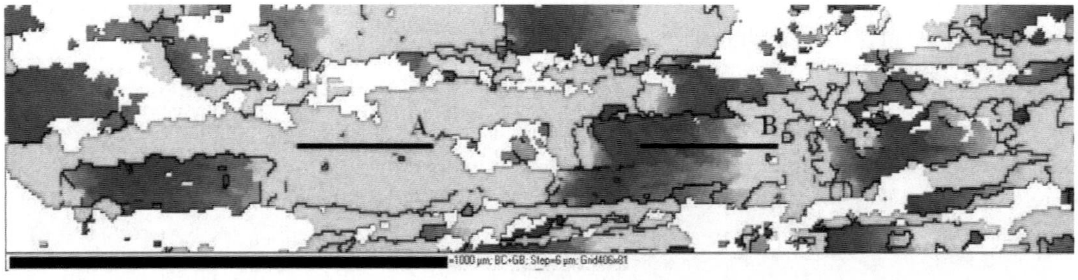

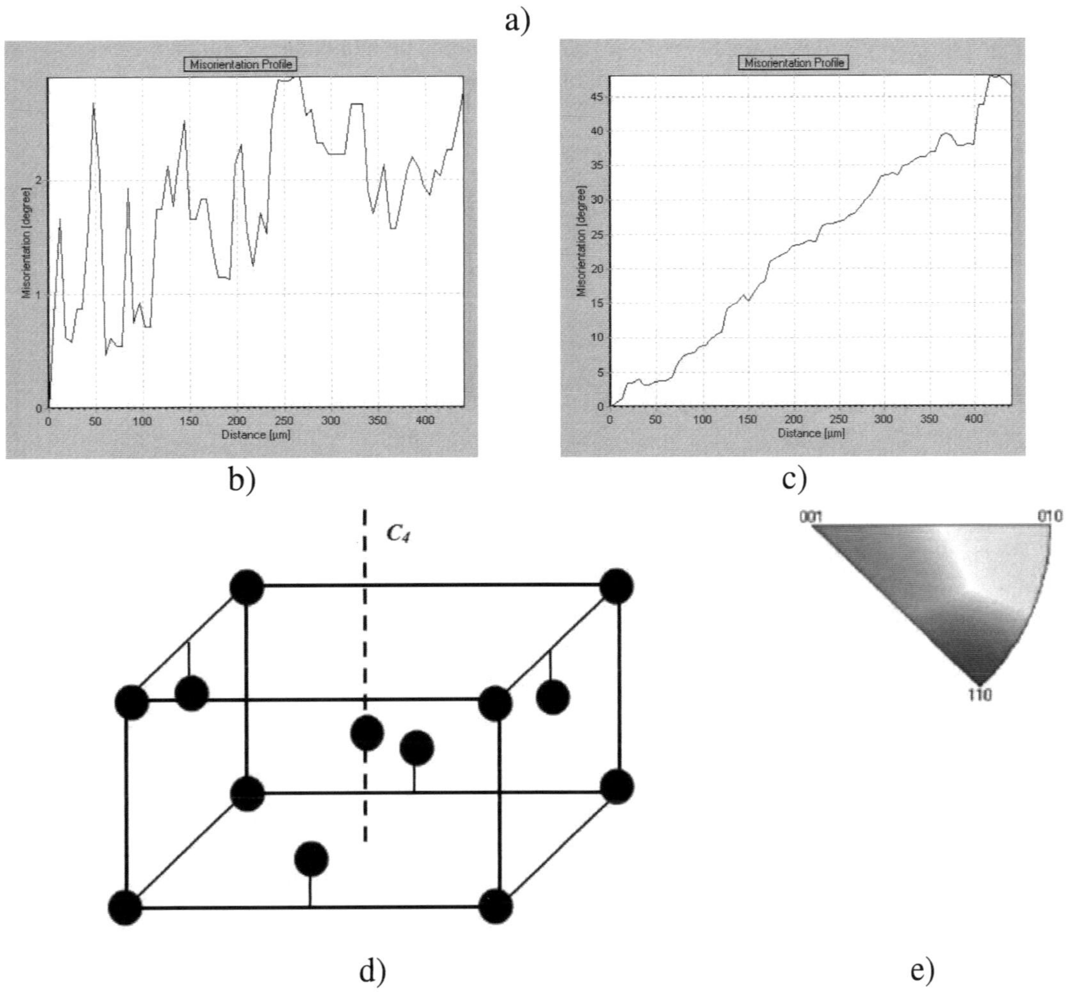

Figure 2. The elongated grains (a), changes of the angle between the identical crystallographic planes along the line at A-type grain (b) and B-type grain (c), unit cell of β-Sn (d), colouring scheme (e).

At figure 2a high-angle grain boundaries (the angle between the grains orientation of more than 10 degrees) are outlined by black lines. As it seen from figure 2a elongated grains are of two types of colouring: uniform (grain of A-type) and non-uniform (grain of B-type). Non-uniform coloration of B-type grains means that the crystal orientation is changed within the grain.

Figures 2 d and e shows changes of the angle between the identical crystallographic planes along the line which is parallel the spreading direction of the foil. Scanning grains of A- type shows that the adjacent parts of the grain are randomly disoriented by angle which is not higher than error bar of the EBSD-measurements. That means that there are now high-angle boundaries in such grains. A different kind of behavior of the angle between the same crystallographic planes along the line is observed for grains of B-type (figure 2c). In this case, the angle between the same crystallographic planes changes monotonically within a grain. In the given example the final angle change is 45 degrees (figure 2c).

It was found that the subgrain structure of large elongated grains depends on the crystallographic orientation of nucleation center. Uniformly coloured grains (A- type) grow in the case if (100) plane parallel to the foil surface or deviated by a small angle and the C_4 crystallographic axis is parallel to the spreading direction. B-type grains grow if the nucleation center is oriented so that the (100) plane parallel to the foil surface or deviated by a small angle but the C_4 crystallographic axis is not parallel to the spreading direction. In this case, the lattice orientation monotonically changes during the growth process. A sequence of low-angle boundaries (outlined by white lines) is revealed in the grains of type B by the EBSD technique. The lattice orientation changes at the points 1, 2, 3, 4 and 5 of the B-type grain showed at the figure 3. It was found that the unit cell orientation changes in a way that it is "rotated" around the C_4 axis. The unit cell "rotates" from the initial position until (110) plane becomes parallel to the foil surface.

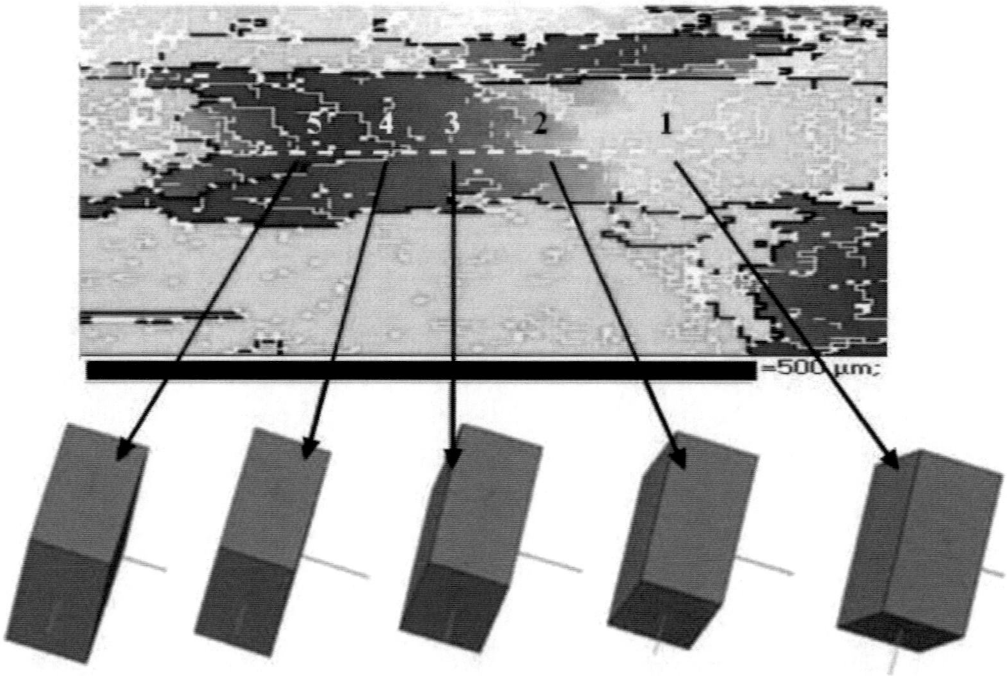

Figure 3. Changes of the lattice orientation in the grain of B-type.

Studies of foils grain structure of binary alloys of tin with lead and bismuth have shown that the addition of the alloying element with amount not higher than its solubility at ultrarapid quenching from the melt (up to 8 wt. % Bi, and 5 wt. % Pb) has no significant effect on the characteristics of the grain structure. Figure 4a and 4b shows the grain structure of foils of alloys Sn - 4 wt. % Bi, Sn- 2 wt. % Pb. It was also found that the elongated grains observed in the foils of ternary eutectic Sn - 8 wt. % Zn -3 wt. % Bi alloyed additionally by 5 wt.% In (figure 4c).

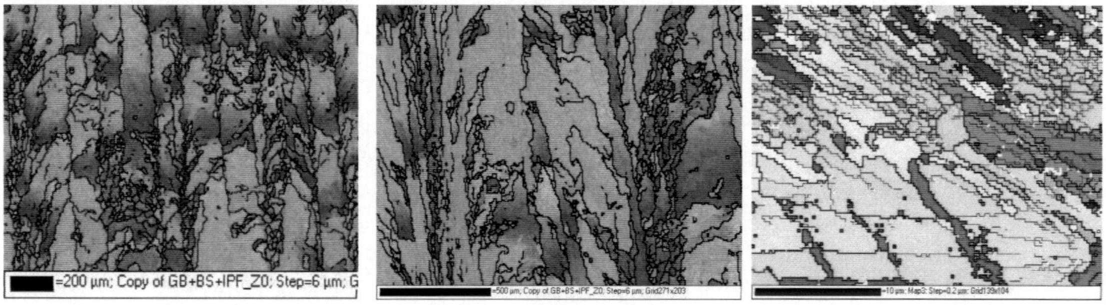

a) b) c)

Figure 4. Grain structure of foils of alloys Sn - 4 wt. % Bi (a), Sn- 2 wt. % Pb (b) and Sn - 8 wt. % Zn -3 wt. % Bi -5 wt.% In.

4. Discussion

The above-described elongated grains shape is formed because the grain grows both in the direction of the heat extraction (perpendicular to the mold) and in the direction of melt flow. This is possible in the case of relatively slow rotation of the mold and the high velocity of the liquid - solid interface motion [7]. According to our previous studies, an increase of the mold rotating velocity leads to the equiaxed shape of grains [6 - 7].

It was found that during ultra-fast quenching of the Sn from the melt the nucleation of the grain oriented by the (100) plane parallel to the surface of the mold is the most likely one. Since the (100) plane of β-Sn has the highest reticular density, the formation of nuclei in which this plane is parallel to the substrate provides the lowest surface energy.

The observed "rotation" of grain orientation is caused by the features of crystallographic structure of the tin and its mechanical properties. The distinctive feature of the tin is large number of sliding systems and high plasticity. According to the Ref. [8], the following sliding systems are realized in tin at room temperature (110) $[\bar{1}11]$, (110) [001], (100) [010], (100) [001], (101) $[\bar{1}01]$ и (121) $[\bar{1}01]$. The sliding systems (110) $[\bar{1}11]$ and (110) [001] are the most probable.

It can be assumed that the reason for the change of crystallographic orientation of the B-type grains is the deformation of just crystallized material by the viscosity force of the supercooled melt which moves above the crystallization front in the direction of foil spreading.

According to the Schmid-Boas equation a shear stress at sliding plane in the direction of the sliding can be calculated by following equation:

$$\tau = \sigma \sin(\chi) \cos(\lambda), \tag{1}$$

where σ is applied stress, χ is angle between applied stress direction (OX direction of foil spreading) and sliding plane, λ is angle between stress direction and sliding axis.

If the grain is positioned so that the projection of the shear stress on the plane (110) has a component along the [111] and [001] directions, the crystal «rotates» until the sliding direction is parallel the axis of stretching. Shear strain persists as long as the stress τ is zero, that is, until the (110) plane is not parallel to the direction of flow of the melt.

The fact that the above-described binary and multicomponent alloys based on tin have a similar grain structure with the grain structure of pure tin is caused by the formation of a supersaturated solid tin-based solution at the crystallization

5. Conclusions

It was found that the grains of two types are formed at the main foil body: large grains elongated in the direction of foil spreading and fine equiaxed grains localized on the border of the large ones. Subgrain structure of large elongated grains depends on the crystallographic orientation of the nucleus. If the nucleation center is oriented so that the (100) plane parallel to the foil surface or deviated by a small angle and the C_4 axis deviated from to the spreading direction by any angle, the lattice orientation monotonically changes during the growth process. It can be assumed that the reason for the change of the crystallographic orientation of type B grains is the deformation of solidified material by the viscosity force of undercooled melt, which moves above the crystallization front in the direction of the foil spreading. The easy slip systems of tin are (110) [111] and (110) [001]. Therefore, if a grain is oriented so that the projection of the shear stress on the plane (110) has a component along the [111] and [001] directions, the crystal rotates until the sliding direction becomes parallel to the foil spreading direction. Shear deformation persist until the shear stress is not zero, that is, until the (110) plane is parallel to the melt flow direction.

6. References

[1] Willnecker R, Herlach D M and Feuerbacher B 1989 *Physical Review Letters* **62** 2707-2710
[2] Ochoa F, Williams J J and Chawla N 2003 *JOM* **55** 56-60
[3] Herlach D M, Galenko P K and Holland-Moriz D 2006 *Metastable Solids from Undercooled Melts* (Amsterdam: Elsevier)
[4] Galenko P, Abramova E V, Jou D, Danilov D A, Lebedev V G and Herlach D M, *Physical Review E* 2011 **84** 041143-1– 041143-17
[5] Galenko P and Lebedev V 2008 *Int. J. Thermodynamics* **11** 21-29
[6] Gusakova O V and Shepelevich V G 2010 *Materials, technologies, instruments* **15** 54-57
[7] Gusakova O V and Shepelevich V G 2011 *Journal of Surface Investigation. X-ray, Synchrotron and Neutron Techniques* **5** 1011-1015.
[8] Yang F and Li J C M 2007 *J. Mater. Sci.: Mater. Electron.* **18** 191-210

Remote control system for high-perfomance computer simulation of crystal growth by the PFC method

Evgeny Pavlyuk [1], Ilya Starodumov[2], Sergei Osipov[3]

[1] Department of computational mathematics and computer science, Ural Federal University, 620000 Ekaterinburg, Russia
[2] Laboratory of Multi-Scale Mathematical Modelling, Ural Federal University, 620000 Ekaterinburg, Russia
[3] School of Natural Sciences and Mathematics, Ural Federal University, 620000 Ekaterinburg, Russia

E-mail: evgeny.pavluk@urfu.ru

Abstract. Modeling of crystallization process by the phase field crystal method (PFC) - one of the important directions of modern computational materials science. In this paper, the practical side of the computer simulation of the crystallization process by the PFC method is investigated. To solve problems using this method, it is necessary to use high-performance computing clusters, data storage systems and other often expensive complex computer systems. Access to such resources is often limited, unstable and accompanied by various administrative problems. In addition, the variety of software and settings of different computing clusters sometimes does not allow researchers to use unified program code. There is a need to adapt the program code for each configuration of the computer complex. The practical experience of the authors has shown that the creation of a special control system for computing with the possibility of remote use can greatly simplify the implementation of simulations and increase the performance of scientific research. In current paper we show the principal idea of such a system and justify its efficiency.

1. Introduction

Numerical modeling of physical processes is often associated with the solution of complex differential systems of equations. Such tasks can be solved only with the use of computer technology. In the case of complex three-dimensional simulations, a typical personal computer may not suffice, which requires modeling on high-performance computing systems. The phase field crystal method is one of such algorithms. The PFC model has been formulated in a parabolic form [1–3]. Then it has been extended to the form described by partial differental equation of a hyperbolic type (MPFC) [4–6, 6–10]. The MPFC model can be expressed using the free-energy functional

$$\mathcal{F}[\phi, \nabla\phi, \nabla^2\phi] = \int_\Omega \left[f(\phi) - |\nabla\phi|^2 + \frac{1}{2}(\nabla^2\phi)^2 \right] \mathrm{d}\Omega, \tag{1}$$

where Ω is the computational domain, in which it's possible to define the chemical potential as

$$\mu(\phi) = \frac{\delta \mathcal{F}}{\delta \phi} = \frac{df}{d\phi} + 2\nabla^2\phi + \nabla^4\phi, \tag{2}$$

where ϕ describes conserved order parameter. The free energy density $f(\phi)$ represents the homogeneous part of energy and in the present report it is written in the form of Landau-de Ginnes:

$$f(\phi) = \frac{1-\epsilon}{2}\phi^2 + \frac{\alpha}{3}\phi^3 + \frac{1}{4}\phi^4. \tag{3}$$

Here, ϵ is the dimensionless overcooling and α is the measure of metastability. Finally, the modified (hyperbolic) phase field crystal model can be expressed by the following equation:

$$\tau\frac{\partial^2\phi}{\partial t^2} + \frac{\partial\phi}{\partial t} = \nabla^2\mu, \tag{4}$$

where t is the time, τ is the relaxation time of the atomic flux to its stationary state.

2. HPC researches of the MPFC task

The research of the MPFC equation (4) is associated with a complex computational task. The necessity to solve differential equations of the 6th order in space and 2nd order in time requires the use of special numerical algorithms [10]. Analysis of the effectiveness of this method presented [11, 12]. that, despite the high rates of parallelization of computing processes, the method requires significant computer resources. This means that the implementation of simulations is necessary to use high-performance computing cluster. Access to such clusters, as a rule, has a number of significant limitations, which makes it difficult to carry out simulations, especially in a continuous mode. Among these limitations are the following:

1) The differences in the configuration of software and hardware do not allow for the calculation without the preliminary settings of computing programs, postprocessors and additional user's applications.
2) In the case of a large amount of data with results (in the calculations by the MPCF method this volume can reach hundreds of gigabytes), there is a need for a large data storage. Practice experience presents that usually HPC centers are not ready to provide the storage quota of significant capacity.
3) Regular transfer of large amounts of data from a cluster to a user's computer requires a stable high-speed network connection.
4) The need for special software to process the results of calculations limits the user in using only a preconfigured computer.

Additionally, the following system requirements are relevant:

1) View the task status and the system components statuses (for example, nodes load).
2) Task results postprocessing (with a choice of tools).
3) Ability to view the task results without the need to use special soft.
4) The ability to exchange data between clusters to distribute one task between different clusters.
5) Monitoring the task execution and restarting, if necessary, to ensure fault tolerance.
6) Ability to duplicate the critical elements of the system to ensure fault tolerance.
7) Performance of component interactions.
8) Scalability to support large numbers of components and interactions among components: users, clusters, tasks.
9) Modifiability of components to meet changing needs (even while the application is running).
10) Simplicity of a uniform interface.

3. Remote control system of computations for simulation

To solve the described problems the following special system was proposed in Figure 1:

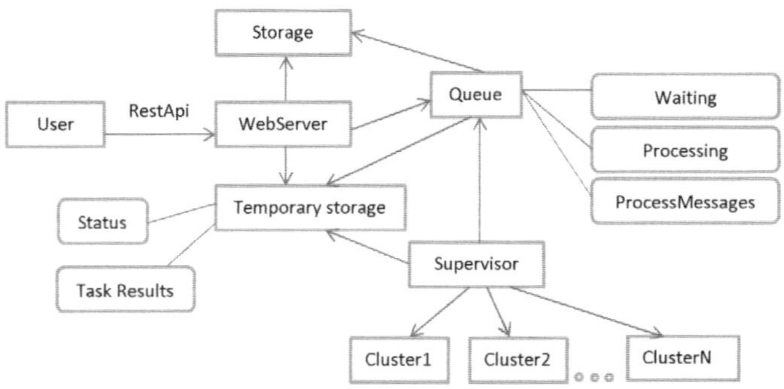

Figure 1. The remote control system of computations for simulation.

Web-server process user commands. Commands are passed through http requests. For example,

- Get system status : `GET /status`
- Get task status: `GET /status?tag="taskname_20170130"`
- Run task (precisely put in task queue): `PUT /task?tag="taskname_20170130"& clusters="clustername"&parameters='{ "node_count"=3, "time_limit"=10000}' d @"Problem_Conditions.csv"`
- Get task result: `GET /result?tag="taskname_20170130"`
- Add cluster to cluster set: `POST /add_cluster?parameters='{"ip":"192.168.99.100", "name":"Cluster1"}'`

This format allows to create console and web clients, to use the system as a server for calculations in some other software system. In addition, the web server can process a command to postprocess task results (the handler can be on a separate computer), save the results in the storage, view processed results via the web interface with ParaviewWeb. This allows to view the results from any computer that has just a browser.

Task queue is a system of robust storing and transmitting messages between system components based on AMQP (Advanced Message Queuing Protocol). The queue performs routing, guarantees delivery, distributes data flows, subscribes to the required message types. In addition, the queue must support horizontal scaling. The system have three separate queues:

- Waiting – for tasks that are waiting to be launched;
- Processing – for already running tasks. You may need to restart such a task;
- ProcessMessages – there will be data to exchange between clusters.

Using such a queue will eliminate data loss cases due to the impossibility of executing the request. Even if the whole system or its individual components are stopped, the data of user requests and intermediate calculations will not be lost and the calculations can be continued from the point of stop.

Temporary Storage is a key-value data base. It is for storing queue data, and ancillary data. Requirements for this storage are following:

- The ability to run copies on several independent nodes to ensure fault tolerance;
- Access speed is primarily important for tasks that exchange data between clusters.

Supervisor is a system that controls the set of clusters. Its main tasks are:

1) Start / Stop tasks. Each cluster can have specific control scripts. The supervisor calls these scripts through a single interface;
2) Collect the statuses of each cluster, server components and individual tasks, and save them to the Temporary Storage;
3) Collecting the task results and logs and saving in Temporary Storage for debugging and starting the task from the checkpoint load balancing;
4) Control over the task execution time and emerging problems.

The system has a modular architecture, each module can be replaced if necessary, without having to stop and restart the entire system.

4. Conclusions

In this paper, the problem of modeling crystal growth by the highly effective MPFC method on high-performance computing systems is considered. The authors describe the difficulties encountered by researchers in carrying out serial computer simulations aimed at solving complex problems. As a solution to these problems, the authors propose a scalable and fault-tolerant system for controlling computations, the concept of which is described in this paper. The system makes it possible to run tasks on a variety of clusters with a variety of configurations, allowing a variety of postprocessing of the results of calculations and having a convenient API (application programming interface) that allows to create console and web clients, use the system as a server for calculations in some other software system.

References

[1] Elder K R, Katakowski M, Haataja M and Grant M 2002 *Physical Review Letters* **88** 245701
[2] Elder K R and Grant M 2004 *Physical Review E* **70** 051605
[3] Elder K R, Provatas N, Berry J, Stefanovich P and Grant M 2007 *Physical Review B* **75** 064107
[4] Stefanovic P, Haataja M and Provatas N 2006 *Physical Review Letters* **96** 225504
[5] Stefanovic P, Haataja M and Provatas N 2009 *Physical Review E* **80** 046107
[6] Galenko P K, Danilov D A and Lebedev V G 2009 *Physical Review E* **79** 051110
[7] Galenko P K and Elder K R 2011 *Physical Review B* **83** 064113
[8] Wang C and Wise S M 2011 *SIAM Journal on Numerical Analysis* **49** 945–969
[9] Baskaran A, Hu Z, Lowengrub J S, Wang C, Wise S M and Zhou P 2013 *J. Comput. Phys.* **250** 270–292 ISSN 0021-9991
[10] Galenko P K, Gomez H, Kropotin N V and Elder K R 2013 *Phys. Rev. E* **88**(1) 013310
[11] Starodumov I, Pavlyuk E, Klyuev L, Kovalenko M and Medyankin A 2015 *Ceur-WS Proceedings* **1513** 109–122
[12] Starodumov I, Pavlyuk E, Abramov S, Klyuev L, Galenko P and Alexandrov D 2016 *Bulletin of Udmurt University. Mathematics, Mechanics* **26** 445–450 URL `http://dx.doi.org/10.1063/1.4959750`

5. Acknowledgments

Authors achnowledge support from the Russian Science Foundation under project number 16-11-10095.

IOP Conf. Series: Materials Science and Engineering **192** (2017) 012017 doi:10.1088/1757-899X/192/1/012017

On the theory of self-similar phase transitions with a mushy layer

D V Alexandrov, I V Alexandrova and A A Ivanov

Department of Theoretical and Mathematical Physics, Laboratory of Multi-Scale Mathematical Modeling, Ural Federal University, Ekaterinburg, 620000, Russian Federation

E-mail: dmitri.alexandrov@urfu.ru

Abstract. It is well-known that directional crystallization processes frequently occur in unsteady-state manner. One of such time-dependent crystallization regimes is the self-similar process characterizing by the scaled combination $\xi/\sqrt{t}$ between the spatial coordinate ξ (directed along the solidification axis) and time t. Such scaling relation usually establishes when the crystallization process is far from its initial stage $t = 0$. In this paper, we consider some approximate heuristic relations describing the self-similar solidification mode.

1. Introduction

A mathematical description of directional crystallization plays a very important role in different applications such as the crystal growth, engineering, geophysical and metallurgical process [1–9]. So, for example, the mathematical models allow to predict many properties of solids produced by solidification. When the liquid is a melt (a mixture of two or more components) its solidification process completely differs from solidification of pure liquid. In particular, various distributions of impurity in the solid phase leads to different mechanical and physical properties of ingots. This phenomenon arises due to the impurity displacement into the liquid phase by the moving solidification front. When the impurity displacement is rather large, the constitutional supercooling originates ahead of the planar solid-liquid interface [10] and, generally speaking, the two-phase zone (mushy layer) appears. Moreover, the solid nuclei in the form of newly born crystals may evolve in this layer. The authors of references [11–13] developed a full set of thermodynamic equations for a mushy zone and solved a mush-reduced set of them approximately for the constrained growth of a binary alloy. A more complete solution has since been given in references [14–18] for the transient and steady-state solidification conditions.

Note that the constitutional supercooling origination (this occurs if the concentration gradient $G_C = m\partial C_L/\partial\xi$ exceeds the temperature one $G = \partial\theta_L/\partial\xi$ at the planar front; here m is the liquidus slope determined from the phase diagram, C_L and θ_L are the concentration and temperature fields in the liquid phase) leads under certain circumstances to the mushy zone solidification and the classical description of crystallization by means of the planar front model becomes inapplicable. However, if the process is far from a cooled wall and the constitutional supercooling and/or the mushy layer are absent, the solidification process with a planar front (after a lapse of time) approaches to its self-similarity. This solidification scenario was experimentally studied by Huppert and Worster [19] on the basis of aqueous solution of sodium nitrate. At first, the solidification process occurs within the framework of the Stefan

IOP Conf. Series: Materials Science and Engineering **192** (2017) 012017 doi:10.1088/1757-899X/192/1/012017

thermodiffusion problem with a planar front. Then the planar shape of the solid-liquid interface destroys and the process takes place with a mushy layer (see, for example, Figure 3 in [19]).

2. Self-similar model

Let us consider a unidirectional solidification process of a binary melt or solution directed along the ξ axis. The cooled wall is placed at $\xi = 0$ and regions $0 < \xi < \Sigma(t,\zeta)$ and $\Sigma(t,\zeta) < \xi < \infty$ are filled with the solid and liquid phases, respectively. Before a point of time when the constitutional supercooling originates, the process is described within the framework of the classical thermodiffusion model with a planar front $\Sigma(t)$. Namely, the temperature and concentration fields in both the phases are governed by the following parabolic-type equations

$$\frac{\partial \theta_L}{\partial t} = a_L \Delta \theta_L, \ \frac{\partial C_L}{\partial t} = D_L \Delta C_L, \ \Sigma(t) < \xi < \infty, \tag{1}$$

$$\frac{\partial \theta_S}{\partial t} = a_S \Delta \theta_S, \ \frac{\partial C_S}{\partial t} = D_S \Delta C_S, \ 0 < \xi < \Sigma(t). \tag{2}$$

Here θ_S and C_S are the temperature and concentration in the solid, a_L and a_S are the thermal diffusivities in the liquid and solid phases, D_L and D_S are the diffusion coefficients in these phases, and $\Delta = \partial^2/\partial\xi^2 + \partial^2/\partial\zeta^2$ is the Laplacian. In the case of solidification with a planar front nothing depends on the spatial coordinate ζ directed perpendicular to the crystallization direction. The thermal and mass balance conditions as well as the equality of both temperatures to the phase transition temperature hold true at the front

$$K_S \frac{\partial \theta_S}{\partial \xi} - K_L \frac{\partial \theta_L}{\partial \xi} = L_V \frac{\partial \Sigma}{\partial t}, \ (1-k) C_L \frac{\partial \Sigma}{\partial t} + D_L \frac{\partial C_L}{\partial \xi} - D_S \frac{\partial C_S}{\partial \xi} = 0, \ \xi = \Sigma(t,\zeta), \tag{3}$$

$$\theta_L = \theta_S = \theta_0 + m C_L + \Gamma \theta_0 K, \ \xi = \Sigma(t,\zeta), \tag{4}$$

where K_S and K_L are the thermal conductivities in the solid and liquid phases, L_V is the latent heat parameter, k is the equilibrium partition coefficient, θ_0 is the phase transition temperature of pure system, Γ is the surface tension, and K is the front curvature. Moreover, we assume that the concentration jump at the front is also given, that is,

$$C_S = k C_L, \ \xi = \Sigma(t,\zeta). \tag{5}$$

The temperatures at the cooled wall and far from the front in the liquid phase are fixed

$$\theta_S = \theta_W, \ C_S = C_W = k C_{L\infty}, \ \xi = 0, \tag{6}$$

$$\theta_L \to \theta_{L\infty}, \ C_L \to C_{L\infty}, \ \xi \to \infty. \tag{7}$$

The model (1)-(7) is valid before the constitutional supercooling origination, that is, before the mushy zone initiation [5, 10]. If the mushy zone exists, the process must be described by means of a mushy layer model. Let us discuss below the mushy layer model when the nucleation within the mush is not taken into account. In this case, equations (1) and (2) are valid in regions $\Sigma_L < \xi < \infty$ and $0 < \xi < \Sigma_S$, where Σ_L and Σ_S stand for the mushy layer - liquid phase and solid phase - mushy layer interfaces while the region $\Sigma_S < \xi < \Sigma_L$ is filled with the mushy layer. The heat and mass transfer processes in the mushy region are governed by the following equations

$$\begin{gathered} \rho_m c_m \frac{\partial \theta_m}{\partial t} = \frac{\partial}{\partial \xi}\left(K_m \frac{\partial \theta_m}{\partial \xi}\right) + L_V \frac{\partial \varphi}{\partial t}, \\ \frac{\partial}{\partial t}\left((1-\varphi) C_m\right) = \frac{\partial}{\partial \xi}\left(D_m \frac{\partial C_m}{\partial \xi}\right) - k C_m \frac{\partial \varphi}{\partial t}, \ \Sigma_S < \xi < \Sigma_L, \end{gathered} \tag{8}$$

where θ_m and C_m are the temperature and concentration fields in the mush, φ is the bulk fraction of the solid phase, and $\rho_m c_m(\varphi) = \rho_L c_L(1-\varphi) + \rho_S c_S \varphi$, $K_m(\varphi) = K_L(1-\varphi) + K_S\varphi$, $D_m(\varphi) = D_L(1-\varphi) + D_S\varphi$. Here ρ_S and ρ_L are the densities in the solid and liquid phases, c_S and c_L are the thermal capacities in these phases. The two-phase layer is assumed to be in a state of thermodynamic equilibrium. This means that the temperature in both phases is equal to the phase transition temperature connected with the solute concentration. Let us write down the liquidus line equation in a linear form

$$\theta_m = \theta_0 + mC_m, \ \Sigma_S < \xi < \Sigma_L. \tag{9}$$

The set of equations (8) and (9) must be supplemented by the boundary conditions imposed at both interfaces, i.e.

$$\begin{aligned} &\varphi = \varphi_*, \ K_S\frac{\partial \theta_S}{\partial \xi} - K_m\frac{\partial \theta_m}{\partial \xi} = L_V(1-\varphi_*)\frac{\partial \Sigma_S}{\partial t}, \ C_S = kC_m, \\ &(1-k)(1-\varphi_*)C_m\frac{\partial \Sigma_S}{\partial t} + D_m\frac{\partial C_m}{\partial \xi} = 0, \ \theta_m = \theta_S, \ \xi = \Sigma_S, \end{aligned} \tag{10}$$

$$\varphi = 0, \ C_m = C_L, \ D_m\frac{\partial C_m}{\partial \xi} = D_L\frac{\partial C_L}{\partial \xi}, \ \theta_m = \theta_L, \ K_m\frac{\partial \theta_m}{\partial \xi} = K_L\frac{\partial \theta_L}{\partial \xi}, \ \xi = \Sigma_L. \tag{11}$$

In addition, we assume that the boundary conditions (6), (7) hold true far from the mushy layer.

In the case of self-similar crystallization, the spatial coordinate ξ and time t are connected by the relation $\xi/\sqrt{t}$ while the planar front position Σ as well as the solid phase - mushy layer Σ_S and the mushy layer - liquid phase Σ_L coordinates in accordance with two models above are expressed as $\Sigma = 2\lambda\sqrt{D_L t}$, $\Sigma_S = 2\lambda_S\sqrt{D_L t}$ and $\Sigma_L = 2\lambda_L\sqrt{D_L t}$, where the parabolic growth rate constants λ, λ_S and λ_L must be found from the model solution. Computations carried out by Worster [20] and instability analysis given by Alexandrov [5] show that the mushy zone originates only after the constitutional supercooling initiation whereas the planar front of solidification is always stable to small morphological perturbations. In other words, if the self-similarity establishes, the planar front shape cannot be destroyed by small temperature or mechanical perturbations, that is, the self-similar Stefan thermodiffusion problem is stable. However, when the constitutional supercooling arises (in this case the process must be described by a mushy layer model; not by the planar front model without a mush), the metastable supercooling zone exists ahead of the solid phase interface. Therefore, the solidification process with a planar front without a mushy layer must be described by the model (1)-(7) whereas the solidification process in the presence of constitutional supercooling and, as a consequence, in the presence of a mushy layer, must be described within the framework of the model (6), (7) and (8)-(11).

Figures 1 and 2 demonstrate the parabolic growth rate constants as functions of the temperature $T_1 \equiv mC_{L\infty} - T_W$ for two sets of parameters given in table 1 ($T_{L\infty} = 15^{\circ}$C and $C_{L\infty} = 14$). The solid curves illustrate the numerical solutions found in [20] and the crosses in Figure 2 represent the experimental values found in [19]. As is seen from the figures, there is a point of bifurcation (intersection of the curves) where $\lambda = \lambda_b \approx 0.15$, $T_1 = T_{1b} \approx 2^{\circ}$C (Figure 1) and $\lambda_b \approx 0.15$, $T_{1b} \approx 1.72^{\circ}$C (Figure 2). This point is responsible for the mushy zone incipience (this happens at $\lambda > \lambda_b$ and $T_1 > T_{1b}$). In other words, the solid phase - mushy layer (λ_S) and mushy layer - liquid phase (λ_L) boundaries appear. It is easy to see that the function $\lambda(T_1)$ has a root-like behavior. Therefore, let us use the following approximate heuristic expression

$$\lambda = \lambda_b + \sqrt{A\left(T_1 - T_{1b}\right)}, \tag{12}$$

where A is a constant. The dashed curves in Figures 1 and 2 are plotted in accordance with expression (12). Our calculations show that A is of the order of 0.75 (Figure 1) and 2 (Figure

Table 1. The physical parameters used for calculations [19, 20].

Property	Set a	Set b $NaNO_3 + H_2O$	Units
Thermal conductivity in liquid, K_L	$1.3 \cdot 10^{-3}$	$1.3 \cdot 10^{-3}$	cal /(cm s °C)
Thermal conductivity in solid, K_S	$1.3 \cdot 10^{-3}$	$5.3 \cdot 10^{-3}$	cal /(cm s °C)
Thermal diffusivity in liquid, a_L	$1.3 \cdot 10^{-3}$	$1.3 \cdot 10^{-3}$	cm^2/s
Thermal diffusivity in solid, a_S	$1.3 \cdot 10^{-3}$	$1.2 \cdot 10^{-2}$	cm^2/s
Diffusion coefficient in liquid, D_L	10^{-5}	10^{-5}	cm^2/s
Diffusion coefficient in solid, D_S	10^{-9}	10^{-9}	cm^2/s
Latent heat per unit volume, L_V	80	73.6	cal/sm^3
Liquidus slope, m	-0.4	-0.4	°C
Segregation coefficient, k	0	0	-
Density of liquid, ρ_L	1	1	gr/cm^3
Density of solid, ρ_S	1	0.48	gr/cm^3

2). Some small deviations of the dashed curves from the solid ones occur near the bifurcation point. This is due to the fact that the self-similar mushy layer is not established in this region. In the case of large values of temperature T_1, the expression above is in good agreement with experimental data.

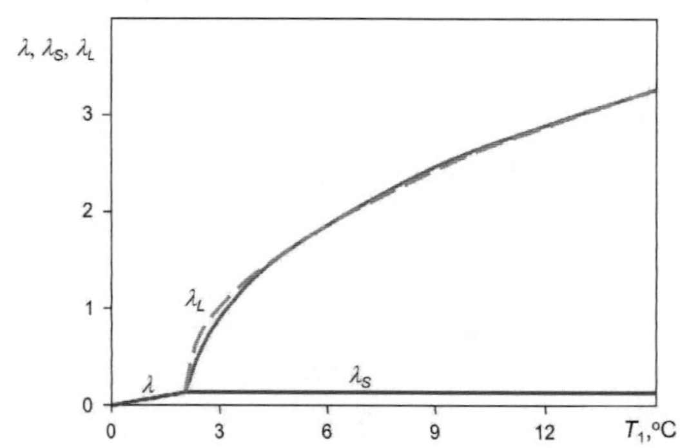

Figure 1. The parabolic growth rate constants for the Set a (table 1).

3. Conclusion

Thus, the main peculiarities of self-similar solidification process seem to be clear. At first, the solidification process occurs in accordance with the classical thermodiffusion model with a planar front. Let us especially emphasize that the velocity of solidification within the framework of this model is a function inversely proportional to the process time, i.e. $d\Sigma/dt = \lambda\sqrt{D_L/t}$. In other words, the front motion is decelerated. This means that the solidification process may occur in accordance with two scenarios. First of them happens if the impurity displacement is slow. If this is really the case, the concentration gradient (G_C) does not exceed the temperature one (G) at the front, and the front shape remains planar. The second possible scenario appears due to the constitutional supercooling origination and, as a consequence, due to the metastable mushy layer initiation. In this case, the solidification scenario must be described by a mushy layer

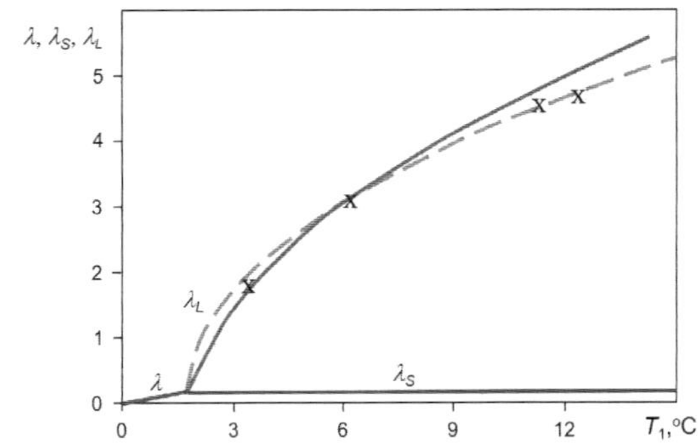

Figure 2. The parabolic growth rate constants for the Set b (table 1).

model. What is more, the mushy layer evolution occurs due to the constitutional supercooling so that if it is absent, the self-similar solidification occurs with a planar front.

The theory of unsteady-state solidification with a mushy layer should be extended to take into account the processes of nucleation and growth of newly born crystals in a metastable layer. This can be done on the basis of recently developed theories of transient nucleation for single-component and binary systems [21–26].

4. References

[1] Chalmers B 1961 *Physical Metallurgy* (New York: John Wiley)
[2] Laudise R A 1972 *The Growth of Single Crystals* (New York: Prentice Hall)
[3] Brice J S 1973 *The Growth of Crystals from Liquid* (Amsterdam: North Holland)
[4] Hobbs P V 1974 *Ice Physics* (Oxford: Clarendon)
[5] Alexandrov D V 2004 *Int. J. Heat Mass Trans.* **47** 1383–1389
[6] Herlach D M 2008 *Phase Transformations in Multicomponent Melts* (Weinheim: Wiley-VCH)
[7] Alexandrov D V and Malygin A P 2011 *Int. J. Heat Mass Trans.* **54** 1144–1149
[8] Alexandrov D V and Malygin A P 2012 *Int. J. Heat Mass Trans.* **55** 3196–3204
[9] Herlach D M 2015 *Crystals* **5** 355–375
[10] Alexandrova I V, Alexandrov D V, Aseev D L and Bulitcheva S V 2009 *Acta Physica Polonica A* **115** 791–794
[11] Hills R N, Loper D E and Roberts P H 1983 *Q. J. Appl. Math.* **36** 505–539
[12] Fowler A C 1985 *IMA J. Appl. Math.* **35** 159–174
[13] Borisov V T 1987 *Theory of Two-Phase Zone of Metallic Ingot* (Moscow: Metallurgia)
[14] Alexandrov D V 2001 *J. Crystal Growth* **222** 816–821
[15] Alexandrov D V, Nizovtseva I G, Malygin A P, Huang H-N and Lee D 2008 *J. Phys.: Condens. Matter* **20** 114105
[16] Alexandrov D V and Nizovtseva I G 2008 *Dokl. Earth Sci.* **419** 359–362
[17] Alexandrov D V, Ivanov A A and Malygin A P 2009 *Acta Physica Polonica A* **115** 795–799
[18] Alexandrov D V and Ivanov A A 2009 *Int. J. Heat Mass Trans.* **52** 4807–4811
[19] Huppert H E and Worster M G 1985 *Nature* **314** 703–707
[20] Worster M G 1986 *J. Fluid Mech.* **167** 481–501
[21] Alexandrov D V and Malygin A P 2013 *J. Phys. A: Math. Theor.* **46** 455101
[22] Alexandrov D V and Nizovtseva I G 2014 *Proc. R. Soc. A* **470** 20130647
[23] Alexandrov D V and Malygin A P 2014 *Modelling Simul. Mater. Sci. Eng.* **22** 015003
[24] Alexandrov D V 2014 *Phys. Lett. A* **378** 1501–1504
[25] Alexandrov D V 2014 *J. Phys. A: Math. Theor.* **47** 125102
[26] Alexandrov D V 2014 *Phil. Mag. Lett.* **94** 786–793

Acknowledgments

This work was supported by the Russian Science Foundation (grant no. 16-11-10095).

IOP Conf. Series: Materials Science and Engineering **192** (2017) 012018 doi:10.1088/1757-899X/192/1/012018

Shear elastic modulus of magnetic gels with random distribution of magnetizable particles

L Yu Iskakova and A Yu Zubarev

Ural Federal University, Ekaterinburg, 620000, Russian Federation

E-mail: larisa.iskakova@urfu.ru, andrey.zubarev@urfu.ru

Abstract. Magnetic gels present new type of composite materials with rich set of uniquie physical properties, which find active applications in many industrial and bio-medical technologies. We present results of mathematically strict theoretical study of elastic modulus of these systems with randomly distributed magnetizable particles in an elastic medium. The results show that an external magnetic field can pronouncedly increase the shear modulus of these composites.

1. Introduction

Magnetic gels and elastomers are composites of fine magnetic particles in soft polymer matrixes. Coupling of rich set of physical properties of polymer and magnetic materials is very promising for many modern and perspective technologies. Discussions of technical and biomedical application of these systems can be found, for example, in [1-12]. A short overview of works on mechanical properties and behavior of magnetic polymers is given in [13].

Uniaxial elongation and magnetostriction effects in magnetic gels have been studied in many works (see, for example [13-19]). The shear deformations of these systems also present significant interest both from scientific and practical points of view. Theoretical studies of the shear effects in the composites with the particles, united in linear chain-like aggregates, have been done in [20-22]. The general conclusion of these works is that an external magnetic field can significantly increase the shear modulus of these composites.

As a rule, the chain-like aggregates appear in magnetic polymers on the stage preceding the composite curing due to the action of an external magnetic field (field of polymerization). On the other hand, very often magnetic gels are prepared without this field. The spatial distribution of particles in these systems is rather random and isotropic (see, for example, [15, 17, 23]). The aim of this work is theoretical study of effect of an external magnetic field on the shear elastic modulus of magnetic gels with homogeneous and isotropic distribution of non Brownian particles in a continuous matrix. It should be noted that usually the Brownian effects are negligible for the magnetic particles with the diameter 100nm and more. Composites with the particles of these sizes present the main interest from the point of view of the magnetomechanic effects, since these effects in the systems with the smaller particles, as a rule, are very weak.

The matrix is supposed elastic with the linear law of deformation and incompressible. It should be noted that the last condition is fulfilled not for all gels; however, it allows us to restrict calculations and to get the final results in transparent forms. Analysis of effects of the composite compressibility can be considered as a natural generalization of this model.

The principal and not overcome problem of the theory of composite materials is account of multiparticle interactions, both the direct ones and interactions through the perturbations of the current matrix. Usually these effects are taken into account by using various empirical and semi empirical approaches, which accuracy a priori is unknown [24].

In order to achieve mathematically rigorous results, here we will consider the systems with low concentration of the particles and neglect any interactions between them. One needs to admit that the low-concentrated systems are not very interesting from the practical point of view. However this limiting model allows us to avoid intuitive and heuristic constructions. That is why the strict results can be considered as a robust asymptotic background for the analysis of the concentrated system with the interacting particles.

The structure of the paper is the following. In the part 2 we study the composite with identical spherical magnetically hard particles; each particle has a permanent magnetic moment bounded with the particle body. The part 3 deals with the systems of magnetically soft ellipsoidal particles with random orientations of the ellipsoids axes.

2. Magnetically hard particles

We consider a system of identical spherical non Brownian particles embedded in an elastic continuous medium. All particles have the permanent magnetic moment m "frozen" in the particle body. This means that the moment can turn round only with the particle. We suppose that the volume concentration φ of the particles is low and will neglect any interactions between them.

Let us suppose that the composite is placed in a uniform magnetic field H and experiences small shear deformation in the plane perpendicular to the field. Since the concentration of the particles in the composite is supposed small, we will not take into account the difference between the external field H and the field inside the sample.

It is convenient to introduce a Cartesian coordinate system with the axis Oz in the field direction and the axis Ox in the direction of the shear. By using the mathematical similarity between the Navier– Stokes equation of Newtonian incompressible fluid flow and the Lame equation of deformation of an elastic incompressible medium [24], as well as the results of theory of dilute magnetic fluids (see, for example, [25]), one can present the needed component of the macroscopic (measurable) stress σ in the composite as:

$$\sigma = G_0(1 + 2.5\varphi)\gamma + \mu_0 \frac{m}{2v} < \mathrm{e}_x > H, \qquad \sigma = \sigma_{xz} \,. \tag{1}$$

Here G_0 is the shear modulus of the pure polymer matrix, $\gamma = \frac{\partial u_x}{\partial z}$, u_x is the component of the macroscopic (measurable) vector $\boldsymbol{u}$ of the composite displacement, μ_0 is the magnetic permeability of vacuum, v is volume of the particle, e_x is the component of the unit vector $\boldsymbol{e}$ directed along the magnetic moment $\boldsymbol{m}$ of a particle, the angle brackets <...> mean the averaging over the orientations of all particles. Our aim now is to determine the mean component $< e_x >$.

To this end we will consider an arbitrary particle situated in the field $\boldsymbol{H}$ and denote by $\boldsymbol{e}_0$ its initial (before the composite deformation) vector $\boldsymbol{e}$. Taking into account the mathematical identity of the Navier–Stokes and Lame equations and using the equations [25-27] of dynamics of a spherical non-Brownian magnetic particle in a viscous fluid, after simple transformations we come to the following equations for the components e_x and e_z of the unit vector $\boldsymbol{e}$ of orientation of an arbitrary particle:

$$\begin{aligned} \delta e_x &= \frac{1}{2}\gamma e_z - \kappa e_x e_z \\ \delta e_z &= -\frac{1}{2}\gamma e_x + \kappa(1 - e_z^2) \\ \kappa &= \frac{\mu_0 m H}{6 G_0 v} \end{aligned} \tag{2}$$

Here $\delta e_i = e_i - e_{i0}$, $i = x, z$, e_{i0} is the initial (before macroscopic shear and the field application) component of the particle vector $\boldsymbol{e}$. It should be noted that the equations (2) are derived in the linear approximation with respect to the shear strain γ.

After simple transformations in the linear approximation in γ we get:

$$\delta e_x = \frac{1}{2}\gamma\left(\frac{e_{z0}}{1+\kappa e_{z0}} + \frac{1-e_{z0}^2+e_{x0}^2}{(1+\kappa e_{z0})(1+2\kappa e_{z0})}\right) - \kappa^2 \frac{e_{x0}(1-e_{z0}^2)}{(1+\kappa e_{z0})(1+2\kappa e_{z0})} \ . \quad (3)$$

We suppose that initially the particles had random orientation of their magnetic moments, i.e. $< e_{x0} > = 0$. Therefore

$$< e_x > = < e_{x0} > + < \delta e_{x0} > = < \delta e_{x0} > \ . \quad (4)$$

It is convenient to introduce the spherical coordinate system with the polar θ and azimuthal ϕ angles, so that:

$$e_{z0} = \cos\theta, \quad e_{x0} = \sin\theta\cos\phi \ . \quad (5)$$

By using (5), one can get:

$$< \delta e_x > = \frac{1}{4\pi}\int_0^{\pi} \sin\theta d\theta \int_0^{2\pi} \delta e_x d\phi. \quad (6)$$

Substituting (3) and (5) into (6), we obtain:

$$< \delta e_x > = \frac{1}{4}\gamma\left(\int_{-1}^{1}\frac{x}{1+\kappa x}dx + 1.5\int_{-1}^{1}\frac{1-x^2}{(1+\kappa x)(1+2\kappa x)}dx\right). \quad (7)$$

The integrals (7) can be calculated analytically; however, they have cumbersome forms, that is why we omit these forms here.

Parameter κ presents the ratio of the magnetic and elastic torques, acting on the particle. The Lame equations of the elastic deformation of a continuum are valid only in the case of small deformations of this medium. In part, this means that the angle of the particle turn, under the action of the magnetic and elastic torques, must be small for these equations applicability. This leads to the condition $\kappa < 1$ of restriction of the linear approximation.

Combining equations (7) and (1), we come to the following relations:

$$\sigma = G_1\gamma \ ,$$

$$G_1 = G_0\left(1 + \frac{5}{2}\varphi + q(\kappa)\varphi\right), \quad (8)$$

$$q(\kappa) = \frac{3}{4}\kappa\left(\int_{-1}^{1}\frac{x}{1+\kappa x}dx + 1.5\int_{-1}^{1}\frac{1-x^2}{(1+\kappa x)(1+2\kappa x)}dx\right).$$

The parameter G_1 is the effective shear modulus of the composite with the magnetically hard spheres, q reflects the addition to G_1 due to the magnetic field effect. The results of calculation of this parameter are shown in the figure 1.

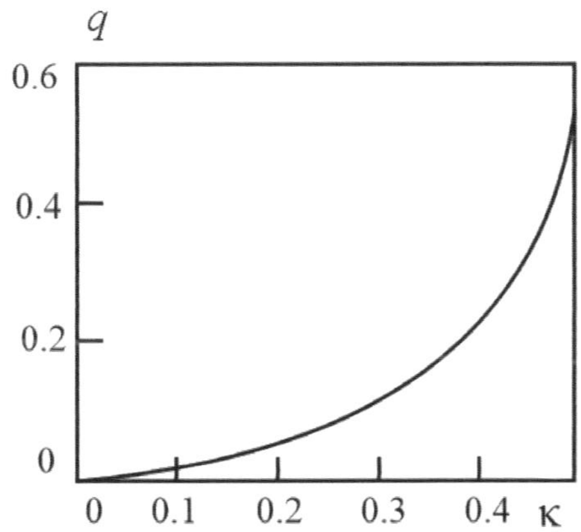

Figure 1. Parameter q in (8) versus the dimensionless magnetic field κ.

Within the framework of the used approximation, the effect of the magnetic field on the effective shear modulus of the composite can achieve about one fourth of the effect of the solid inclusions described by the Einstein term 2.5 φ. It should be noted that the situation when $\kappa \sim 1$ or more is beyond the approximation of small deformation inside the elastic matrix. The linear Lame equations cannot be used for the description of the particle rotation under the magnetic and mechanic torques for the large values of κ. The analysis of this case requires numerical solution of non-linear equations of the polymer matrix deformation.

3. Magnetically soft ellipsoidal particles

In this part we consider a system of ellipsoidal magnetically soft particles randomly distributed in an elastic matrix. For the maximal simplification of calculations and to get transparent physical results, we will restrict ourselves by the approximation of linearly magnetizable particles. The generalization to the nonlinear magnetization is not difficult, but leads to cumbersome calculations and final results.

We again suppose that the composite experiences the deformation of simple shear with the mean displacement $\boldsymbol{u}$ in the direction Ox and the gradient of the displacement along the axis Oz. The magnetic field $\boldsymbol{H}$ is aligned along the axis Oz.

By using the results of [26-28], we can present the component of the macroscopic shear stress as follows:

$$\sigma = G_0\Big\{1 + \varphi\Big[\alpha + \tfrac{1}{2}[(\xi + \beta\lambda)(< e_x^2 > + < e_z^2 >) + \beta(< e_z^2 > - < e_x^2 >) + \\ +2(\chi - 2\lambda\beta) - < e_x^2 e_z^2 >]\gamma + \varphi g h^2 < e_x e_z >\Big]\Big\}, \quad (9)$$

$$\sigma = \sigma_{xz}, \quad \gamma = \frac{\partial u_x}{\partial z}, \quad g(r) = \frac{(\mu_p - 1)^2 (N_\perp - N_\parallel)}{(1 + (\mu_p - 1)N_\perp)(1 + (\mu_p - 1)N_\parallel)}, \quad h = \sqrt{\frac{\mu_0 H^2}{2G_0}}.$$

Here $\boldsymbol{e}$ is the unit vector aligned along the particle axis of symmetry; $\alpha, \beta, \lambda, \xi$ and χ are the functions on the aspect ratio r of the ellipsoidal particle (the ratio of the particle axis of symmetry to its diameter), μ_p is the particle relative magnetic permeability, $N_\parallel$ and $N_\perp$ are the demagnetizing factors of the particle along and perpendicular to its axis of symmetry respectively. The explicit forms of the shape-functions $\alpha, \beta, \lambda, \xi, \chi$ as well as of the factors $N_\parallel$ and $N_\perp$ are given in the appendix.

We present again the components of the vector $\boldsymbol{e}$ of an arbitrary particle as $e_i = e_{i0} + \delta e_i$, $i = x, z$. Because of the initial random orientation of the particles we get $< e_{i0}^2 > = \frac{1}{3}$; $< e_{x0} e_{z0} > = 0$; $< e_{z0}^2 > = \frac{1}{15}$. By using these relations in (9), in the linear approximation with respect to γ one can obtain:

$$\sigma = G_0\left\{1+\varphi[\alpha+\frac{1}{3}(\xi+\beta\lambda)+\frac{1}{15}(\chi-2\lambda\beta)]\gamma+\varphi g h^2(< e_{x0}\delta e_z > + < e_{z0}\delta e_x >)\right\}. \quad (10)$$

Following [27], one can present the equations for the vector $\boldsymbol{e}$ of an arbitrary particle as:

$$\delta e_x = \frac{1}{2}\gamma[\lambda(1-2e_x^2)+1]e_z - \psi e_x e_z^2 \quad (11)$$

$$\delta e_z = \frac{1}{2}\gamma[\lambda(1-2e_z^2)-1]e_x + \psi(e_z - e_z^3)$$

$$\psi(r) = h^2\frac{g(r)}{3\delta(r)}\#(12)$$

Here $\delta(r)$ is a function of the particle aspect ratio r. Its explicit form is given in the appendix. The parameter h presents the ratio of the magnetic and elastic torques acting on the particles. Similar to the previous case of the magnetically hard particles, the linear Lame equations of the small deformations of the elastic matrix are applicable only when the inequality $h < 1$ holds true.

Substituting the form $e_i = e_{i0} + \delta e_i$ into equation (11), after simple transformations, in the linear approximation we come to the relations:

$$< e_{z0}\delta e_x > = \frac{1}{2}\gamma\left[< \frac{Ae_{z0}}{1-\psi e_{z0}^3} > - \psi < \frac{2Be_{x0}e_{z0}}{(1-\psi e_{z0}^3)(1-\psi(1-3e_{z0}^2))} >\right], \quad (13)$$

$$< e_{x0}\delta e_z > = \frac{1}{2}\gamma < \frac{e_{x0}B}{1-\psi(1-3e_{z0}^2)} >.$$

Here

$$A = [\lambda(1-2e_{x0}^2)+1]e_{z0}, \quad B = [\lambda(1-2e_{z0}^2)-1]e_{x0}.$$

Combining (10) and (13), we get:

$$\sigma = G_2\gamma G_2 = G_0\{1+\varphi f(r,h)\}, \quad (14)$$

$$f(r,h) = [f_1(r) + f_2(r,h)],$$

$$f_1 = \alpha + \frac{1}{3}(\xi+\beta\lambda) + \frac{1}{15}(\chi - 2\lambda\beta),$$

$$f_2 = \frac{3}{2}\psi\delta(r)\left[< \frac{Ae_{z0}}{1-\psi e_{z0}^3} > - h^2 < \frac{2Be_{x0}e_{z0}}{(1-\psi e_{z0}^3)(1-\psi(1-3e_{z0}^2))} > + < \frac{e_{x0}B}{1-\psi(1-3e_{z0}^2)} >\right].$$

Here G_2 is the effective shear modulus for the composite with the magnetically soft ellipsoidal particles. The term f_1 describes the effect of the rigid randomly oriented particles on this modulus; the term f_2 reflects the influence of the magnetic field H on G_2, the term f indicates the total effect of the particles on the elastic modulus G_2. For the spherical particles ($r = 1$) the relations, given in the Appendix, read: $\alpha = \frac{5}{2}$; $\beta, \lambda, \xi, \chi, \psi$ =0. Therefore, the Einstein formula $G_2 = G_0\left(1+\frac{5}{2}\varphi\right)$ for these particles is fulfilled.

Some results of calculations of the terms f, f_1 and f_2 versus the dimensionless magnetic field h as well as versus the particle aspect ratio r, are shown in the figures 2 and 3 respectively.

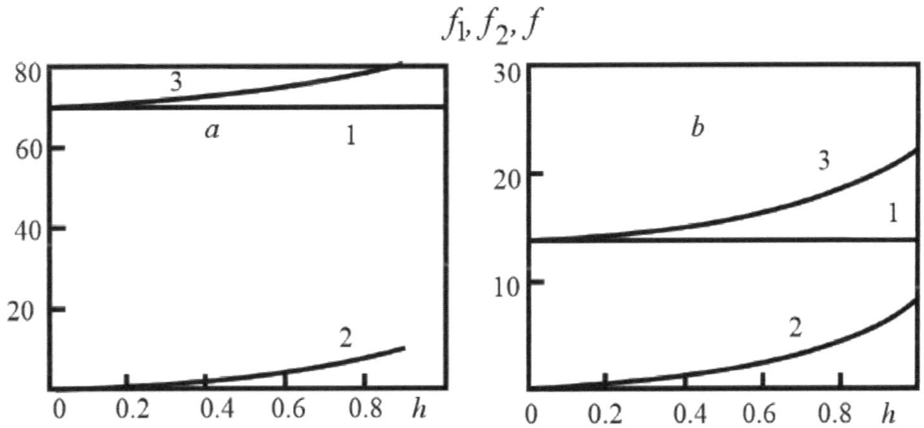

Figure 2. Parameters which determine the effective shear modulus G_s of the composite with the magnetically soft spheres vs. the dimensionless magnetic field h. (*a*) and (*b*) - the particle aspect ratio $r = 0.01$ and 10 respectively. Figures near the curves: 1 - f_1; 2 - f_2; 3 - $f = f_1 + f_2$.

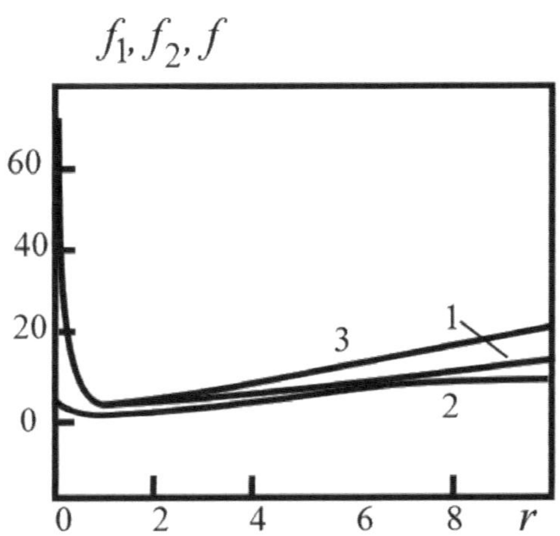

Figure 3. Parameters f_1, f_2, f vs the particle aspect ratio r. The dimensionless magnetic field $h = 1$.

These results demonstrate that the "magnetic" term f_2 can give the contribution to G_s very close to the "rigid particles" term f_1. This contribution is especially significant for the highly elongated particles ($r \gg 1$). For the oblate particles ($r < 1$) and the relatively weak fields ($h < 1$) the term f_2 is much less than f_1.

4. Conclusion

We present the results of theoretical study of effect of uniform magnetic field on the shear modulus of a ferrogels, consisting of magnetic particles randomly distributed in a polymer matrix. In order to achieve mathematically strict results, we have restricted ourselves by the analysis of the dilute systems and neglected any interactions between the particles. The results show that magnetic field increases the modulus and this effect can be quite comparable with the effect, provided by the particles as rigid inclusions in the composite. We believe that the results, obtained in the limiting case of the low concentrated systems, can be a robust background for the development of a theory of the moderately and highly concentrated soft magnetic composites. It should be noted that the we restricted ourselves by the spherical shape of the magnetically hard particles just for maximal simplification of mathematical part of the work. Combining the approaches, considered in the parts 2 and 3, one can easily generalize this analysis for the magnetically hard ellipsoids.

5. Appendix

The shape-coefficients $\alpha, \dots \chi$ as functions of the particle aspect ratio r have the following form [25]:

$$\alpha(r) = \frac{1}{r\alpha_0'}; \quad \beta(r) = \frac{2(r^2-1)}{r(r^2\alpha_0 + \beta_0)}; \quad \xi(r) = \frac{4}{r(r^2+1)\beta_0'} - \frac{2}{r\alpha_0'};$$

$$\chi(r) = \frac{2\alpha_0''}{r\alpha_0'\beta_0''} - \frac{8}{r(r^2+1)\beta_0''} + \frac{2}{r\alpha_0'}; \qquad \lambda(r) = \frac{r^2-1}{r^2+1}; \qquad \delta(r) = \frac{\beta(r)}{3\lambda(r)}.$$

Here

$$\alpha_0 = -\frac{1}{r^2-1}\left[\frac{2}{r} + \frac{1}{\sqrt{r^2-1}}\ln\left(2r^2 - 1 - 2r\sqrt{r^2-1}\right)\right];$$

$$\beta_0 = \frac{1}{r^2-1}\left[r - \frac{1}{2\sqrt{r^2-1}}\ln\left(2r^2 - 1 + 2r\sqrt{r^2-1}\right)\right];$$

$$\alpha_0' = \frac{1}{4(r^2-1)^2}\left[r(2r^2-5) - \frac{3}{2\sqrt{r^2-1}}\ln\left(2r^2 - 1 - 2r\sqrt{r^2-1}\right)\right];$$

$$\beta_0' = \frac{1}{(r^2-1)^2}\left[\frac{r^2+2}{r} - \frac{3}{2\sqrt{r^2-1}}\ln\left(2r^2 - 1 + 2r\sqrt{r^2-1}\right)\right];$$

$$\alpha_0'' = \frac{1}{4(r^2-1)^2}\left[r(2r^2+1) - \frac{4r^2-1}{2\sqrt{r^2-1}}\ln\left(2r^2 - 1 + 2r\sqrt{r^2-1}\right)\right];$$

$$\beta_0'' = -\frac{1}{(r^2-1)^2}\left[3r + \frac{2r^2+1}{2\sqrt{r^2-1}}\ln\left(2r^2 - 1 - 2r\sqrt{r^2-1}\right)\right];$$

The demagnetizing shape-factors $N_\parallel$ and $N_\perp$ are [28]:

$$N_\parallel = \begin{cases} \dfrac{r}{2(r^2-1)^{3/2}}\left[\ln\dfrac{r+\sqrt{r^2-1}}{r-\sqrt{r^2-1}} - 2\dfrac{\sqrt{r^2-1}}{r}\right], & r > 1 \\ r\dfrac{r+\sqrt{1-r^2}}{2(1-r^2)^{3/2}}\left[\dfrac{\sqrt{1-r^2}}{r} - a\tan\dfrac{\sqrt{1-r^2}}{r}\right], & r < 1 \end{cases}$$

$$N_\perp = \frac{1-N_\parallel}{2}.$$

6. References

[1] Filipcsei G, Csetneki I, Szilagyi A and Zrınyi M 2007 *Advanced Polymer Science* **206** 137-189
[2] Boczkowska A and Awietjan S F 2010 *Tuning Active Magnetorheological Elastomers for Damping Applications, Materials Science Forum* **636-637** 766-771
[3] van Bruggen M and van Zon J 2010 *Sensors and Actuators, A Physics* **158** 240–248
[4] van Berkum S, Dee J T, Philipse A P and Erné B H 2013 *International Journal of Molecular Science* **14** 10162-10177
[5] Bañobre-López M, Piñeiro-Redondo Y, de Santis R, Gloria A, Ambrosio L, Tampieri A. et al. 2011 *Journal of Applied Physic* **109** 07B313-1–07B313-3
[6] Das B, Mandal M, Upadhyay A, Chattopadhyay P and Karak N 2013 *Biomedical Materiarls* **8** 035003
[7] Gloria A, Russo R, d'Amora U, Zeppetelli S, d'Alessandro T, Sandri M et al. 2013 *Journal*

Royal Society Interface **10** 20120833
[8] Li Y, Huang G, Zhang X, Li B, Chen Y, Lu T et al. 2013 *Advanced Functional Materials* **23** 660-672
[9] Panseri S, Cunha C, 'Alessandro T, Sandri M, Giavaresi G, Marcacci M, et al. 2012 *Journal Nanobiotechnology* **10** 32
[10] Zeng X B, Hu H, Xie L Q, Lan F, Jiang W, Wu Y et al. 2012 *International Journal of Nanomedicine* **7** 3365
[11] Singh R K, Patel K D, Lee J H, Lee E J, Kim J H, Kim T H et al. 2014 *PLoS One* **9** e91584
[12] Lopez-Lopez M T, Scionti G, Oliveira A C, Duran J D G, Campos A, Alaminos M and Rodriges I A 2015 *PLoS One* DOI:10.1371/journal.pone.0133878
[13] Lopez-Lopez M T, Durán Juan D G, Iskakova L Yu and Zubarev A Yu 2016 *Journal of Nanofluids* **5** 479-495
[14] Kchit N, Lancon P and Bossis G 2009 *Journal of Physics D, Applied Physics* **42** 105506-105514
[15] Stepanov G V, Abramchuk S S, Grishin D A, Nikitin L V, Kramarenko E Y and Khokhlov A R 2007 *Polymer* **48** 488–495
[16] Shen Y, Golnaraghi M F and Heppler G R 2004 *Journal of Intellectual Materials and Structures* **15** 27–35
[17] Stepanov G, Borin D, Raikher Yu, Melenev P and Perov N S 2008 *Journal of Physics: Condensed Matters* **20** 204121-204126
[18] Diguet G, Beaugnon E and Cavaillé J Y 2010 *Journal of Magnetism and Magnetic Materials* **322** 3337-3341
[19] Romeis D, Toshchevikov V and Saphiannikova M 2016 *Soft Matter* **12** 9364-9376
[20] Jolly M R, Carlson J D, Muñoz B C and Bullions T A 1996 *Journal of Intelligent Material Systems and Structures* **6** 613-622
[21] Danas K, Kankanala S V and Triantafyllidis N 2012 *Journal of Mechanics and Physics of Solids* **60** 120-138
[22] Han Y, Hong W and Faidley L E 2013 *International Journal of Solids and Structures* **50** 2281–2288
[23] Khimi S R and Pickering K L 2015 *Composites Part B: Engineering* **83** 175-183
[24] Cristensen R 1979 *Mechanics of Composite Materials* (New York: Wiley)
[25] Rosensweig R 1985 *Ferrohydrodynamics* (Cambridge: Cambridge University Press)
[26] Jeffery J B 1922 *Proceeding of Royal Society of London, Series A* **102** 161-179
[27] Pokrovskiy V N 1978 *Statistical Mechanics of Dilute Suspensions* (Moscow: Nauka)
[28] Landau L D and Lifshitz E M 1960 *Electrodynamics of Continuum Media* (London: Pergamon Press)

Acknowledgments
This work has been done under the financial support of the Russian Fund of Fundamental Investigations, project 16-58-12003 and the Program of Russian Federation Ministry of Science and Education, project 3.1438.2017/PCh.

IOP Conf. Series: Materials Science and Engineering **192** (2017) 012019 doi:10.1088/1757-899X/192/1/012019

The diagram of phase-field crystal structures: an influence of model parameters in a two-mode approximation

V Ankudinov[1] and P K Galenko[2,3]

[1] Udmurt State University, Universitetskaya str. 1, building 4, 426034, Izhevsk, Russia
[2] Friedrich-Schiller-Universität Jena, Physikalisch-Astronomische Fakultät, D-07743 Jena, Germany
[3] Ural Federal University, Laboratory of Multi-Scale Mathematical Modeling, 620000 Ekaterinburg, Russia

E-mail: vladimir@ankudinov.org

Abstract. Effect of phase-field crystal model (PFC-model) parameters on the structure diagram is analyzed. The PFC-model is taken in a two-mode approximation and the construction of structure diagram follows from the free energy minimization and Maxwell thermodynamic rule. The diagram of structure's coexistence for three dimensional crystal structures [Body-Centered-Cubic (BCC), Face-Centered-Cubic (FCC) and homogeneous structures] are constructed. An influence of the model parameters, including the stability parameters, are discussed. A question about the structure diagram construction using the two-mode PFC-model with the application to real materials is established.

1. Introduction

The phase-field crystal model (PFC-model) is suitable for simulations of the dynamics of atomic densities on the diffusion time scales. This model can be used to calculate the coexistence of structures having various crystal lattices and to model a wide spectrum of processes [1].

The amplitude's approximation for the PFC-model is based on the limitation of the periodic atomic density field by the amplitude envelope as a non-conserved order parameter. The PFC-model [2–5] was formulated to describe continual transitions from the homogeneous to the periodic crystal state (similarly to the Landau–Brazovskii transition [6–8]) and between different periodic states [9] over diffusion times. The model can be derived as a reformulation of the Swift-Hohenberg equation for the thermal fluctuation fields [10] or it follows as a first approximation of the Density Functional Theory [11–13]. The model uses the free energy as functional of the atomic density field which is n-periodic in the solid (crystal) phase and homogeneous in the liquid state. Periodicity of the field n naturally takes into account the elastic energy and symmetry of crystals that is described as the motion equation for the conservative order parameter. In such a way, it is possible to simulate a wide class of phenomena including, e.g., epitaxial growth and ordering of nanostructures on micrometer scales [9], crystallization and high-speed regimes of front propagation [14, 15], the motion of dislocations and plastic flow, the formation of a disordered amorphous state, premelting of grain boundaries, crack spreading, rearrangement of microscopic structure of interfaces, and the dynamics of colloidal systems and polymers [16].

The determination of equilibrium structures and their coexistence in "average atomic density–transition driving force" diagrams takes a special place in the phase-field model. Such diagrams determine the structures to which unstable or metastable states of a material must evolve. Diagrams of two dimensional crystalline structures for transitions from an unstable state were constructed using the atomic density functional, amplitude expansion and the Maxwell equal-area rule [3, 17]. The structural diagram for transitions for two-mode functionals from the metastable state was also constructed earlier for different control parameters [1, 18–22]. In the present study, for the structure diagram construction, we use a numerical method for minimizing of free energy functionals. This method makes it possible to determine the parameter range for the existence of three dimensional structures in materials. Using the free energy functional in two-mode approximation (that extends our previous study [23]), the construction for the coexistence of homogeneous (liquid) phase, Body-Centered-Cubic (BCC), Face-Centered-Cubic (FCC) crystal lattices of materials is made. The structure diagram is constructed in coordinates "temperature – atomic density".

2. Free energy for crystalline states

To evaluate the various properties of the different crystal states it is useful to use two-mode approximation (accounting a second atomic coordination sphere) instead of the original one-mode PFC approximation. Using this approximation we can construct the set of the free energy functionals for the given phases in the equilibrium state. This approach lead to the more accurate and exact calculations for complex crystal structures.

The free energy functional for the first-order phase transformations is given by [1, 18, 20]

$$F(n,T) = \int_V dV\left(\Delta B_0\frac{n^2}{2} + B_0^x\frac{n}{2}\mathfrak{D}_i n - 2a\frac{n^3}{3} + 15v\frac{n^4}{4}\right), \tag{1}$$

where the $n = (\rho - \rho_{liquid})/\rho_{liquid}$ is dimensionless atomic density, ρ is the local atomic density, ρ_{liquid} is the reference atomic density of the homogeneous state (liquid), $\Delta B_0 = B_0^\ell - B_0^x$ is the driving force taken as a difference of compressibility and elastic modulus of system, a and v are phenomenological parameters which can be selected for certain type of material. To describe the energy of crystal we introduce the nonlinear operator $\mathfrak{D}_i$:

$$\mathfrak{D}_i = \begin{cases} r_0 + (q_0^2 + \nabla^2)^2, & \text{one-mode approximation, i=1,} \\ [r_0 + (q_0^2 + \nabla^2)^2][r_1 + (q_1^2 + \nabla^2)^2], & \text{two-mode approximation, i=2,} \end{cases} \tag{2}$$

where r_0 and r_1 are shifts of the first two wave vectors, responsible for the relative stability of structures, and q_0 and q_1 are the modules of first two sublattice wave vectors. This extension of the previously introduced two-mode PFC model (Wu et al. [20] with $r_0 = 0$) has been proposed by Asadi and Zaeem [21]. The role of the parameter r_0 will be revealed below, here we just can mention that r_0 can be used for the exact minimization of the free energy by the parameter q to find the wave number of the equilibrium state for the crystal or homogeneous (liquid) phase.

To get a compact and dimensionless form of the free energy (1), it is convenient to use the following substitutions:

$$\tilde{n} = \sqrt{\frac{15v}{B_0^x}}\left(n - \frac{2a}{45v}\right), \quad \varepsilon = \frac{1}{B_0^x}\left(\frac{4a^2}{45v} - \Delta B_0\right), \quad F(n, \Delta B_0) \to \mathfrak{F}(\tilde{n}, \varepsilon). \tag{3}$$

Hence the free energy (1) transforms to

$$\mathfrak{F} = \int dV\left(\frac{\tilde{n}}{2}[-\varepsilon + \mathfrak{D}_i]\tilde{n} + \frac{\tilde{n}^4}{4}\right), \tag{4}$$

where we neglect the high order terms for the simplicity. Then the dimensionless operators $\mathfrak{D}_i$ (2) are:

$$\mathfrak{D}_i \begin{cases} (1+\nabla^2)^2, & \text{one-mode approximation, i=1,} \\ [R_0+(1+\nabla^2)^2][R_1+(Q_1^2+\nabla^2)^2], & \text{two-mode approximation, i=2,} \end{cases} \tag{5}$$

with $R_0 = r_0/q_0^4$, $R_1 = r_1/q_0^4$, $Q_1 = q_1/q_0$.

As a result, the free energy (4) with $\mathfrak{D}_2$ from eq. (5) has three parameters: R_0, R_1 and ε. The first two parameters control the stability of structures. The third parameter, as the driving force parameter ε, is the relative temperature $\varepsilon = (T_c - T)/T$ which determines the excess over the transition temperature $T = T_c$ and the relation between elastic properties of the system in terms of quantities ΔB_0 and B_0^x. In the next section we will use the two-mode form of the functional which implies the form of the amplitude's expansion coefficients and number of the parameters.

3. Crystalline structure determination and selection of parameters

Using the summation of reciprocal lattice vectors [1,19,21] we can make the amplitude expansion and get the dimensionless density profiles indicating the symmetry and properties of a given phase (homogeneous, BCC, FCC):

$$n_{homogeneous} = \tilde{n}, \tag{6}$$

$$\begin{aligned} n_{BCC} = \tilde{n} + 4\eta_1 \left[\cos(qx)\cos(qy) + \cos(qx)\cos(qz) + \cos(qy)\cos(qz)\right] \\ +2\eta_2\left[\cos(2qx) + \cos(2qy) + \cos(2qz)\right], \end{aligned} \tag{7}$$

$$n_{FCC} = \tilde{n} + 4\eta_1 \cos(qx)\cos(qy)\cos(qz) + \eta_2\left[\cos(2qx) + \cos(2qy) + \cos(2qz)\right], \tag{8}$$

where η_1 and η_2 are the density amplitudes for the first and second reciprocal lattice vectors and q is the unit cell size. Substituting Eqs. (6-8) into the free energy (4) with $\mathfrak{D}_2$ from Eq. (5) and integrating over the elementary cell with size $2\pi/q_{eq}$ one gets the free energy for every phase which depends of the equilibrium lattice number q_{eq}.

In one-mode approximation, the wave number q_{eq} can be found by minimization of free energy analytically. In the two-mode approximation, obtaining of q_{eq} can be made only numerically. The work [21] proposes the way to minimize the functional by accepting the parameters $q_{eq,BCC} = 1/\sqrt{2}$ and $q_{eq,FCC} = 1/\sqrt{3}$ [18–20] and to correct the minimization error $\partial\mathfrak{F}/\partial q_{eq}|_{q_{eq}} \neq 0$. Asadi and Zaeem [19,21] suggest to introduce the condition for R_0 as:

$$\left.\frac{\partial\mathfrak{F}}{\partial q_{eq}}\right|_{q_{eq}} = 0 \quad\Rightarrow\quad R_0 = R_1\frac{\eta_2^2}{\eta_1^2}. \tag{9}$$

In this way it is possible to prevent the minimization error and incorporate the condition on the R_0 parameter value at the same time. The model of Mkhonta et al. [24] for the multi-mode PFC approximation states parameters b_1 and b_2 as independent. These parameters are similar to the R_0 and R_1 in our model for number of modes $N = 2$. Therefore, we are calculating structure diagrams using the approaches of Asadi-Zaeem and Mkhonta et al., i.e., we shall use the dependent R_0 by (9) and the independent case for positive and negative values of R_0.

Because interplanar distances are different for different crystal structures we utilize the relevant $Q_{1,BCC} = \sqrt{2}$, $Q_{1,FCC} = 2/\sqrt{3}$ for each phase. For the homogeneous phase, the parameter Q_1 is given by the coexistent crystalline phase. The minimization procedure of $\mathfrak{F}$ with regard to the density amplitudes η_1 and η_2 is given by the gradient descent algorithm. This

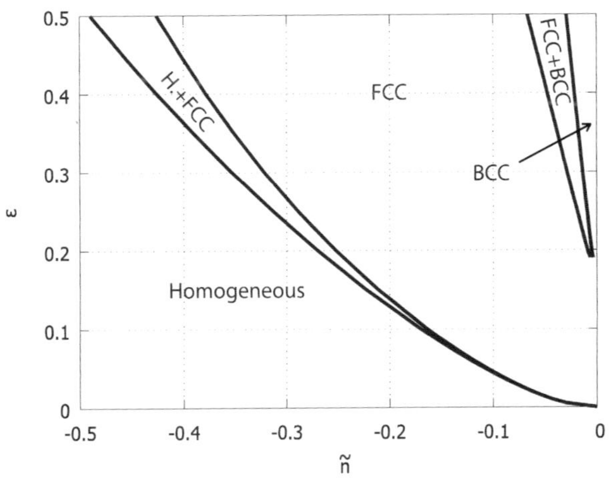

Figure 1. The structure phase diagram for BCC, homogeneous (liquid), FCC-structures at $R_1 = 0.05$ with the approximation for the R_0 to dispose the wave number minimization error. The parameter Q_1 is taken as different for corresponding phases. In high-temperature region (small $\varepsilon = 0..0.15$), the stable BCC-structure is absent (due to $\mathfrak{F}_{FCC} < \mathfrak{F}_{BCC}$). For this set of parameters and approximations, the high temperature transformation of BCC-FCC is absent. The phase transition at much lower temperature T, namely at $\varepsilon = 0.2$, between BCC- and FCC-structures appears in the solid state similarly to the low temperature FCC-BCC phase transitions in a pure iron. Label "H." on the diagram means "homogeneous".

predict the stable structure's coexistence for obtained amplitudes even for small values of ε and negative values of R_0 and R_1. Using the obtained amplitudes we find the free energy functionals for each phase $\mathfrak{F}_{homogeneous}(\varepsilon, \tilde{n}, R_0, R_1)$, $\mathfrak{F}_{BCC}(\varepsilon, \tilde{n}, R_0, R_1)$, $\mathfrak{F}_{FCC}(\varepsilon, \tilde{n}, R_0, R_1)$. With these free energies, we construct the diagram of crystal structures with their coexistence at values of $R_0 = -0.1, 0, 0.1$ and $R_1 = -0.1, 0, 0.1$ for independent R_0 case; and $R_1 = 0, 0.05, 0.1$ for the dependent parameter R_0.

4. Construction of structure diagram and discussion

4.1. Structure diagram for three dimensional crystals

In this section the structure diagrams are calculated using the solution of the Maxwell area rule, using the chemical potentials and free energy functionals for each phase. The selection mechanism is based on the thermodynamic rule of the minimal energy for a virtually existed structure.

Figure 1 shows the domains of existence of the homogeneous (liquid) phase, the crystalline BCC structure and FCC structure in coordinates "normalized average atomic density $\tilde{n}$ vs. dimensionless temperature ε" for the dependent parameter R_0 and $R_1 = 0.05$. We calculate the coexistence separately for the each pair of the structure and then select the structure by its minimal free energy. For the positive R_1, the BCC region is located in the region of large ε. This region decreases with the decrease of R_1 to the zero. The condition on the parameter R_0 prevents shifting of the BCC structure region under the FCC area. With large negative values of R_0, the BCC-structure becomes unstable.

4.2. Structure phase diagram for the independent control parameter R_0

Figure 2 presents the structure diagram for the phases with independent values of $R_0 = -0.1$ and $R_1 = -0.1$. We calculated a set of diagrams to find the influence of parameters R_0 and R_1 on the position of lines for coexisted structures. Note that the parameter Q_1 has to be chosen exactly for a given structure to take in account the symmetry of the density profile and the correct free energy of a structure. For the presently constructed diagrams, the overlapping of

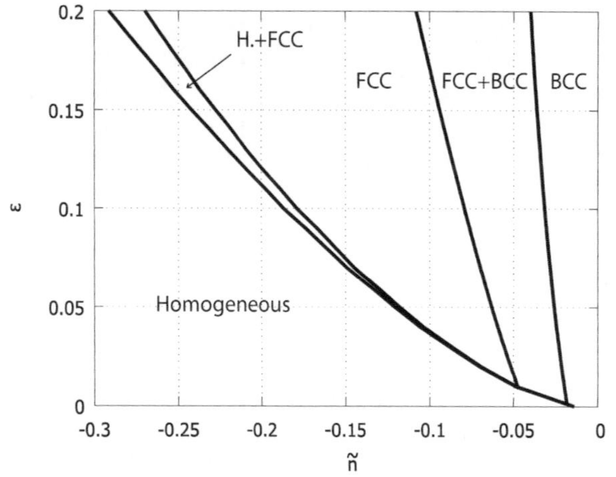

Figure 2. The structure diagram for BCC, homogeneous (liquid), FCC-structures at $R_0 = -0.1, R_1 = -0.1$. The parameter Q_1 is taken different for corresponding phases. The region of existence of the BCC structure is enlarged relatively to the Fig. 1 and it shows the possibilities of the coexistence of BCC and homogeneous structure. Nevertheless there is no alternation in structures with the increase of ε at the fixed $\tilde{n}$. For the complicated sequence of phase transformations in the region of small undercoolings $\varepsilon = 0 \div 0.15$, one can expect that FCC-BCC-FCC can appear when the parameter R_0 is free. Label "H." on the diagram means "homogeneous".

regions BCC-FCC-BCC is absent. At small undercooling, i.e. at high temperature the sequence of structures should be BCC-FCC (i.e., high temperature ferrite – austenite in the pure iron). The possible opportunity to obtain such sequence is to introduce in the model an additional atomic interaction (which may lead to appearing of additional structural transformation).

For the positive values of R_0 and R_1, the BCC region is located in the region of large ε, the lines of coexisted structures do not changed dramatically compared to the Fig. 1. The same situation is observed for the positive R_1 and fixed R_0. The independent parameter R_0 successfully controls the width of BCC zone and location of the lower part of the region. Simultaneous converging of the parameters to zero leads to widening and getting down of the permitted BCC region. With the positive values of the R_1 there is no strong influence of the $R_0 = 0$ on the form of the coexisted curves. The negative values of the parameter R_1 allows us to broad the BCC region and to enhance the incline of the "FCC+Homogeneous" line. For independent values of R_0, there is no limitation for the width of the BCC zone. This leads to appearance of the "Homogeneous(Liquid)-BCC" transition (small $\varepsilon < 0.02$ in Fig. 2). The "FCC+Homogeneous(Liquid)" structure detaches the BCC region following the thermodynamic phase selection rule.

5. Conclusions

Structure diagrams were calculated for the two-mode PFC model with different control parameters. These parameters are linked with stability of the three dimensional crystal structures.

The parameter R_0 can be dependent that limits the ability to form complex crystal coexistence with small values of the driving force ε. On the other hand, we can reduce the degrees of freedom of the entire system. In such procedure, the reduction of the minimization error is very important.

We have shown the possibilities to find the equilibrium structure diagrams for independent parameter R_0. With this aim we have used the numerical minimization of the density amplitudes in the dimensionless atomic density expansion. The resulting diagram showed the stable solution for the coexisted curves in the wide region of driving force.

The pair of parameters R_0 and R_1 gives the capability to vary the properties of the material. Despite of the minimization error, it seems to be necessary to separate R_0 and R_1 making

R_0 as independent parameter. Minimization error could be corrected during the numerical minimization of amplitudes. The proposed method makes it possible to extend the diagrams description of the complex three dimensional crystalline structures using the two-mode PFC-model.

6. References

[1] Provatas N and Elder K 2010 *Phase-Field Methods in Materials Science and Engineering* (Weinheim: Wiley)
[2] Elder K R, Katakowski M, Haataja M et al. 2002 *Phys. Rev. Lett.* **88** 245701
[3] Elder K R and Grant M 2004 *Phys. Rev. E* **70** 051605
[4] Elder K R, Provatas N and Berry J et al. 2007 *Phys. Rev. B* **75** 064107
[5] Brazovskii S 1995 *World Sicentific Series in 20th Century Physics* **11** 109-13
[6] Landau L D 1937 *Phys. Z. Sowjet.* **11** 26
[7] Kats E I, Lebedev V V and Muratov A R 1993 *Physics Reports* **228** (1-2) 1
[8] Zhang H and Jiang K 2014 *Discrete and Continuous Dynamical Systems B* **19** (2) 607
[9] Elder K R, Rossi G and Kanerva P et al. 2012 *Phys. Rev. Lett.* **108** 226102
[10] Swift J and Hohenberg P C 1977 *Phys. Rev. A* **15** (1) 319
[11] Ramakrishnan T V and Yussouff Y 1979 *Phys. Rev. B* **19** 2775
[12] Evans R 1979 *Adv. Phys.* **28** 143
[13] Singh Y 1991 *Phys. Rep.* **207** 351
[14] Galenko P, Danilov D and Lebedev V 2009 *Phys. Rev. E* **79** 051110
[15] Galenko P K and Elder K R 2011 *Phys. Rev. B* **83** 064113
[16] Emmerich H, Löwen H and Wittkowski R et al. 2012 *Advances in Physics* **61** 665
[17] Galenko P K, Gomez H, Kropotin N V et al. 2013 *Phys. Rev. E* **88** 013310
[18] Jaatinen A and Ala-Nissila T 2010 *J. Phys.: Condens. Matter* **22** 205402
[19] Asadi E and Zaeem M A 2015 *Comput. Mat. Sci.* **105** 101-9
[20] Wu K A, Adland A and Karma A 2010 *Phys. Rev. E* **81** (6) 061601.
[21] Asadi E, Zaeem M A 2015 *Comput. Mat. Sci.* **105** 110-113
[22] Emdadi A, Asadi E and Zaeem M A 2016 *Comput. Mat. Sci.* **123** 139-47
[23] Ankudinov V E, Galenko P K, Kropotin N V and Krivilyov M D 2016 *JETP* **122** (3) 298–309
[24] Mkhonta S K, Elder K R and Huang Z-F 2013 *Phys. Rev. Lett.* **111** 035501

Acknowledgments

We thank Kenneth Elder for permanent and helpful discussions on the topic of this study. This work was supported financially by the Russian Foundation for Basic Research project No. 16-38-00839 mol_a. P.K.G. acknowledges supports from the RSF, grant number 16-11-10095 and from the project "MULTIPHAS", contract Nr. 50WM1541 by the European Space Agency, German Aerospace Center and Friedrich-Schiller-Universität Jena.

IOP Conf. Series: Materials Science and Engineering **192** (2017) 012020 doi:10.1088/1757-899X/192/1/012020

Two-phase nanofluid-based thermal management systems for LED cooling

V Kiseev, D Aminev and O Sazhin

Ural Federal University, Lenin av. 51, Ekaterinburg, 620000 Russia

E-mail: oleg.sazhin@urfu.ru

Abstract. This research focuses on two-phase thermal control systems, namely loop thermosyphons (LTS) filled with nanofluids, and their use as LED cooling devices. The behavior of the fluid in the thermosyphons and the mechanisms explaining the possible impact of nanoparticles on thermal properties of the working fluid as well as the processes in the LTS are addressed. Nanoparticle distribution in the nanofluid, methods of preparation of nanofluids and nanofluid degradation processes (aging) are studied. The results are obtained from a set of experiments on thermosyphon characteristics depending on the thermophysical properties of the working fluid, filling volume, geometry and materials of radiators. The impact of nanofluids on heat-transfer process occurring inside thermosyphon is also studied. Results indicate strong influence of nanoparticles on the thermal properties of the thermosyphons, with up to 20% increase of the heat transfer coefficient. Additionally, a method of calculating the hydrodynamic limit of the LTS is proposed, which allows for estimation of the maximum heat flux that can be transferred by means of the LTS. Possible ways for further improvement of the model are proposed. The nanofluids are shown to be effective means of enhancing two-phase systems of thermal management.

1. Introduction

Two-phase heat transfer systems are widely used in different technical devices, especially in spacecraft thermal control systems, as well as in electronic devices such as microprocessors, semiconductor power electronics, modern powerful laser sources, optoelectronic devices, light-emitting diode (LED) lighting [1-3], etc.

Recent findings indicate that a new type of liquids with addition of metal or metal oxide nanoparticles may enhance thermal characteristics of two-phase systems. Over the last 12 years, extensive studies of nanofluids have been conducted containing both positive and negative results of the use of nanoparticles in heat transfer systems. Effect of nanoparticles on the system parameters is characterized by change of thermal resistance, critical heat flux, heat transfer coefficient, thermal conductivity, intensification of boiling and changes in the operating mode of the system. Some studies demonstrate viscosity increase, reduction of flow resistance, and change of wetting angle along with changes in flow mode between laminar and turbulent [4-9].

However, the conducted studies primarily concern conventional thermosyphons or focus on defining thermal properties of nanofluids, thus leaving LTS outside the scope of research.

Considering the above, we have set the following objectives for our study:

1. Examine nanofluids as a working fluid for thermal-management devices; compare the obtained results with the performance of clear working fluids;

2. Study nanoparticle distributions;
3. Develop a thermal control system based on loop thermosyphons for LEDs with a flat cylindrical evaporator and front heat input;
4. Optimize working fluid volume, come up with recommendations for working fluid choice and devise a filling procedure for nanofluids (omitted from this short version);
5. Compute the maximum heat flux based on the hydrodynamic limit using MathCad.

2. Experimental apparatus

In order to determine the change in thermal properties of the system, in this work we utilize temperature measurement by thermocouples at specific locations throughout the thermosyphon. Using thin, lightweight copper-constantan thermocouples allows for placing them at otherwise hard-to-reach locations of the thermosyphon. Temperature recording has been conducted as follows: After applying voltage and measuring power, the data acquisition system began to take temperature over a period of time. Then, when the temperature stabilized, the experimental run continued for one hour, with the temperature still being recorded every second by the DAC (Owen TRM-148). Finally, after the experiment was finished, the measured temperatures were averaged and standard deviation was calculated. For all experimental runs, the total error was less than ±0.2 °C.

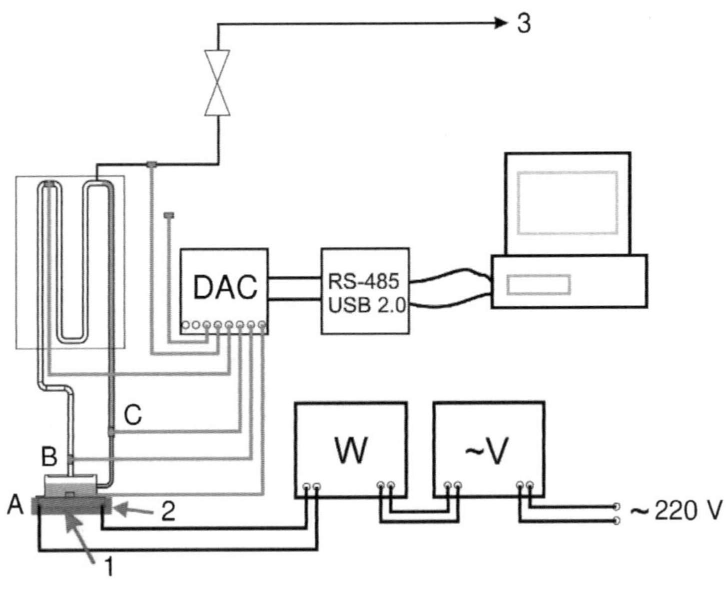

Figure 1. Conceptual scheme of experimental setup for temperature measurements at specific locations of the thermosyphon. 1 – heater, 2 – thermal isolation, 3 – filling/evacuation; A, B, C – thermocouple locations; DAC – data-acquisition system.

Figure 1 illustrates the schematic experimental setup and location of thermocouples. Experimental study assumes obtaining temperatures at specific points on a loop thermosyphon depending on the nominal load of LEDs or heat load, both measured in Watts. The heat load is intended to simulate the actual LED heat flux. The following locations on the thermosyphon have been selected: A – temperature of the heater or diodes, located between the evaporator and the heater T_h; B – vapor temperature at the top point of the evaporator T_v; C – liquid temperature by the condenser pipes at the evaporator's entrance T_l.

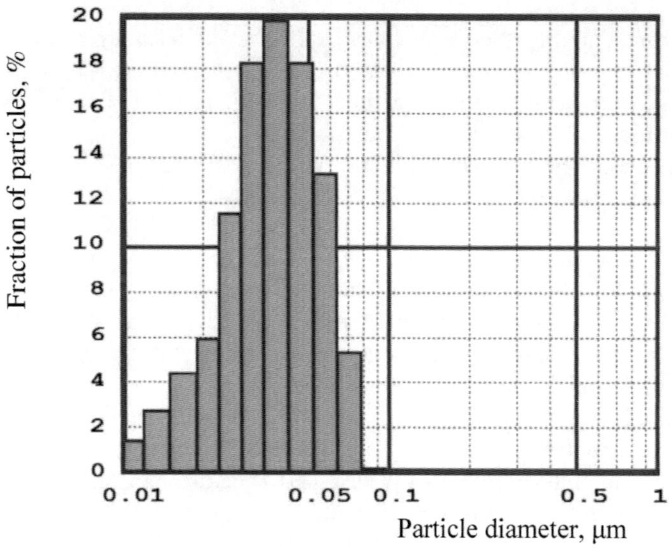

Figure 2. Particle size distribution in nanofluid. Mean size is 36 nm.

Preparation of nanofluid included the following 3-step procedure. First, laser-ablated highly dispersed powder of Fe_2O_3 was added to bi-distilled water in a cylinder. The resulting mixture was placed into an ultrasonic dispergator for 10-15 minutes. During dispergation, in order to cool down the mixture and prevent boiling, the cylinder was wrapped with pipes flowing cold water, since boiling causes nanoparticles to exit the liquid and condense elsewhere in the environment. Total ultrasonic power was 5kW. Finally, in order to assess the quality of the prepared nanofluid, the mixture was placed into the laser-based particle size analyzer SHIMADZU SALD-7101, which measured the particle size distribution. The measurement results are presented in figure 2. The Fe_2O_3 nanoparticle powder has been selected for its non-reactive nature and affinity to the materials used in the system. This procedure was repeated for each solution. Prepared solutions had the following mass concentrations: w = 2.0%, 1.5%, 1.0%, 0.8, 0.3%.

3. Results and discussion

3.1. Calculation of maximum heat flux of the thermosyphon

The calculation of the hydrodynamic limit was based on the necessary working condition that causes the circulation of working fluid in thermosyphon. This condition can be written as the inequality $\Delta P_g = (\rho_l - \rho_v)gH \geq \Delta P_v + P_l$, where ρ_l – density of liquid phase; ρ_v – density of vapor phase; H – the height difference between the evaporator and the condenser (gravity force that actually returns the liquid to the evaporator); ΔP_v and P_l are the pressure drop in liquid phase and vapor phase, respectfully.

Using Hagen-Poiseuille equation for laminar flow of vapor and Fanning equation for turbulent flow and considering formulas for local and integral pressure drop coefficients (left out), the final equation will be:

$$Q(H) = \begin{cases} \dfrac{-(E+D) + \sqrt{(E+D)^2 + 4(O+P)\cdot[(\rho_l - \rho_v)\cdot g \cdot H \cdot 0{,}75]}}{2\cdot(O+P)} & \text{at } Re \leq 2100 \\ \dfrac{-(C+D) + \sqrt{(C+D)^2 + 4(O+P)\cdot[(\rho_l - \rho_v)\cdot g \cdot H \cdot 0{,}75]}}{2\cdot(O+P)} & \text{at } Re > 2100, \end{cases}$$

where

$$D = 128\frac{\eta_l(T_l)}{\rho_l(T_l)}\cdot\frac{L_v}{\pi d_l^4}\cdot\frac{1}{H_{ev}(T_v)}, P = 0{,}0175\lambda_k\cdot\delta\cdot\frac{R_0}{d_v}\cdot\frac{1}{2\cdot H_{ev}^2\cdot\rho\cdot S^2}, O = \zeta_{loc}\cdot\frac{1}{2\cdot H_{ev}^2\cdot\rho\cdot S^2},$$

$$E = 128\frac{\eta_v(T_v)}{\rho_v(T_v)}\cdot\frac{L_v}{\pi d_v^4}\cdot\frac{1}{H_{ev}(T_v)} \text{ and } C = 0{,}6328\cdot \mathrm{Re}^{0.75}\cdot\frac{\eta_v(T_v)}{\rho_v(T_v)}\cdot\frac{L_v}{\pi d_v^4}\cdot\frac{1}{H_{ev}(T_v)}.$$

The equation above includes physical properties of liquid and vapor phases, namely viscosity, density, heat capacity, and enthalpies of condensation and vaporization. Therefore, knowing the listed properties, it's possible to obtain functional dependency Q from H for laminar and turbulent flow respectively. Figure 3 presents the graph plotted for a set of commonly used working fluids as a function $Q(H)$. As apparent from the graph, the selected model adequately describes the processes of heat and mass transfer in thermosyphon even though the model doesn't take into account boiling process and leaves out certain details of heat transfer mechanisms that occur at liquid-vapor phase boundaries.

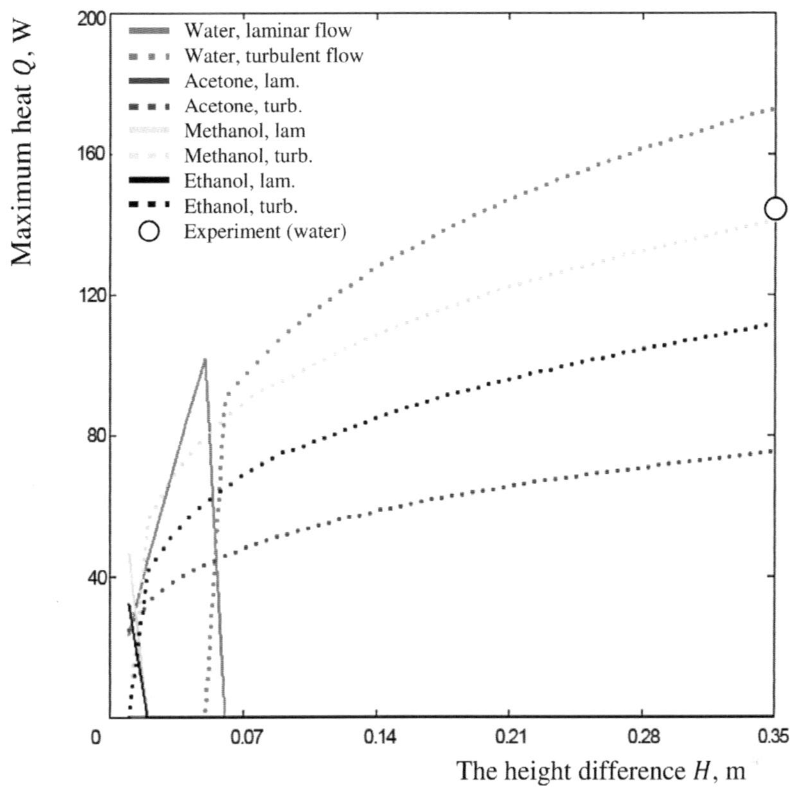

Figure 3. Maximum heat Q that can be transferred by LTS as a function of the height difference H between the evaporator and the condenser for a set of commonly used working fluids.

Possibly, employing the finite-element (FEM) method alongside the finite-volume method (FVM), depending on modeling situation, can significantly improve the consistency of the model with the experimental results.

3.2. Two-phase thermosyphons with nanofluids

To determine the influence of nanofluid on the process occurring in the thermosyphon, we have calculated the heat transfer coefficient using the equation $\alpha = Q \ / \ S \ / \ (T_h - T_v)$, where Q – the applied heat; S – the surface area of heat input $19.2 \cdot 10^{-4}$ m^2; T_h and T_v are the heater and vapor temperatures at the locations A and B (Fig.1), respectively.

The experimental results show that there's a substantial increase of the heat transfer coefficient (see Figure 4). For concentration w=2.0% the increase is as high as 20%, taking into account that thermosyphon itself is a very effective system. A possible explanation of how the nanoparticles may cause the heat transfer coefficient increase is presented in Figure 5.

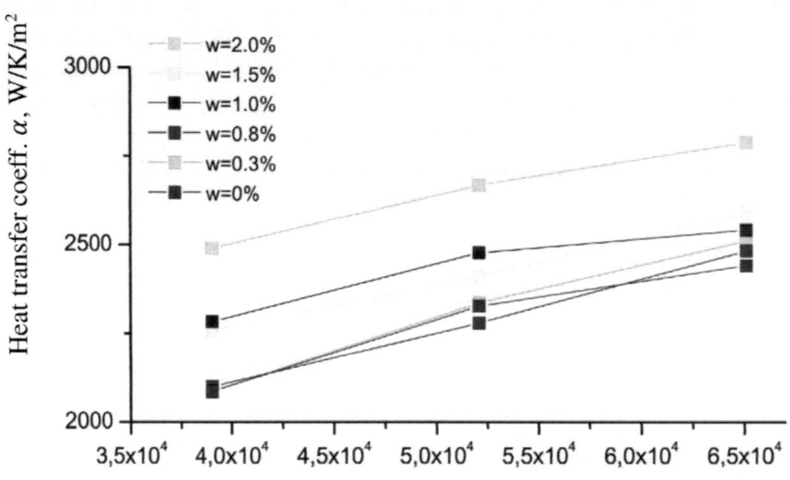

Figure 4. Nanofluid's influence on the heat transfer coefficient α for different mass concentration w of nanoparticles.

Nanosystems are subject to processes of degradation due to high values of free energy, which can be reduced via an agglomeration process. Some of the studies have shown a degradation over time (aging) process caused by aggregation of nanoparticles in the solution and their settling. To estimate the possible negative impact of these processes, the experiments were repeated the next day and in one week. No difference between the one-week and one-day results has been observed, so the final results include only the one-day degradation dependency.

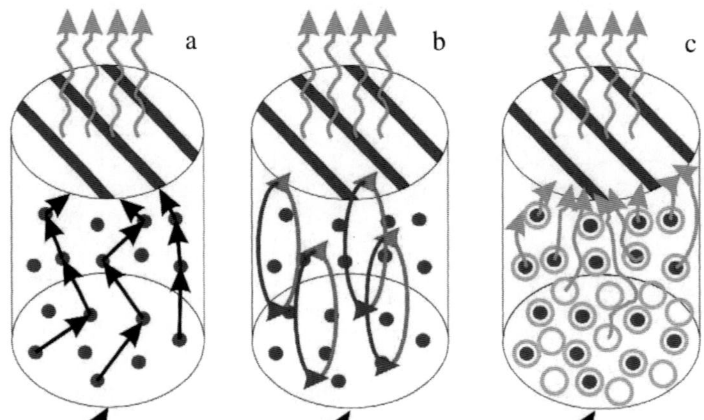

Figure 5. Possible explanation of the heat transfer increase. a – nanoparticles have greater heat conductivity than liquid, and heat is conducted via Brownian motion of the particles; b – nanoparticles' influence on Rayleigh–Benard cells formation; c – nucleate boiling (particles act as nucleation centers); Q is the applied heat.

Figure 6 presents the process of the nanofluid degradation over time. Indeed, obtained results show that nanofluids are subject to steady decrease in thermal conductivity, most probably due the process of aggregation and settling of nanoparticles.

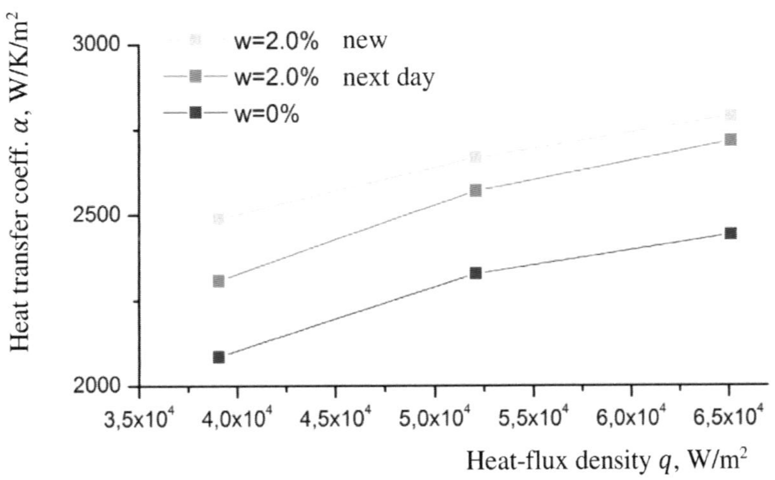

Figure 6. The process of the nanofluid degradation (aging of nanofluid) over time; w is mass concentration of nanoparticles.

To address this problem, an introduction of surface-active agents is proposed. Alternatively, a non-stop operation mode is recommended for such systems. These measures can effectively prevent nanoparticles from segmentation, thereby stabilizing the thermal system.

4. Conclusion

1. New thermal management devices have been proposed, tested and proved to be effective for LED cooling.
2. In this study, we have proposed a model for calculating maximal heat flux based on the thermal properties of working fluids and height difference between the evaporator and the condenser. This model has shown good consistency with experimental data. Possible ways of improvement of the model have also been discussed.
3. The experimental study of the nanoparticles' effect on the processes occurring in the thermosyphon has also been conducted. The obtained results indicate that nanofluids show great promise as a working fluid in two-phase systems, effectively increasing the heat transfer coefficient for up to 20%. Particularly the study of stability of nanofluids has revealed that in time nanoparticles tend to agglomerate and settle down, which can be prevented by introducing surface-active substances. However, further studies of these process need to be conducted. Additionally, the mechanism of nanoparticles improving the thermal properties of the base liquid is yet to be explained. The study of nanoparticles of different substances also merits attention. Metallic particles promise greater thermal conductivity than oxides, though the former are prone to oxidation, which may result in a substantially lowered thermal properties. These studies have not yet been conducted.

5. References

[1] Kiseev V M, Pogorelov N P and Menkin L I 1992 *Proc. of 8th Int. Heat Pipes Conf. (Beijing)* p 673
[2] Kiseev V M, Aminev D S, Cherkashin V G and Murzin R M 2011 *Solid-State Lighting* **3** 27
[3] Kiseev V M, Aminev D S, Cherkashin V G and Murzin R M 2011 *Thermal Processes in Engineering* **11** 523
[4] Do K H and Jang S P 2010 *Int. J. Heat Mass Tran.* **53** 2183
[5] Khandekar S, Joshi Y M and Mehta B 2008 *Int. J. Therm. Sci.* **47** 659

[6] Huminic G and Huminic A 2011 *Exp. Therm. Fluid Sci.* **35** 550
[7] Lun-Chun L and Zhen-Hua L 2008 *Heat Mass Transfer* **45** 1
[8] Noie S H, Heris S Z, Kahani M and Nowee S M 2009 *Int. J. Heat and Fluid Fl.* **30** 700
[9] Parametthanuwat T, Rittidech S and Pattiya A 2010 *Int. J. Heat Mass Tran.* **53** 4960

Acknowledgements
The support by the Ministry of Education and Science of the Russian Federation through the base part of the State task for high educational institutions (the research project # 3.6064.2017/БЧ) is gratefully acknowledged.

IOP Conf. Series: Materials Science and Engineering **192** (2017) 012021 doi:10.1088/1757-899X/192/1/012021

The local–nonequlibrium kinetics of growth of stoichiometric compounds under isothermal condition: phase–field approach

V G Lebedev[1], **A A Lebedeva**[1] **and M G Vasin**[2]

[1]Udmurt State University, Department of Matematics, Informatics and Physics, 426034 Izhevsk, Russia

[2]Physical-Technical Institute of UrB RAS, 426000 Izhevsk, Russia

E-mail: lvg@udsu.ru

Abstract. The presence of line (stoichiometric) compounds in alloys of light metals greatly improves their mechanical and strength properties. The formation of Al_2Y and Al_3Y line compounds in a highly supercooled Al–Y melt is investigated using the phase field method and numerical modeling of the directional solidification of the melt. The main problem in the phase-field description of phase transitions in line compounds is finding the driving forces of these transitions. One of the possible ways to handle this problem is development of a thermodynamically consistent phase-field approach based on the minimization of the full Gibbs energy of the system, nonequilibrium thermodynamics and separation of diffusion processes in different phases. The resulting equations are analyzed in the equilibrium case. The model dynamics is examined numerically for the case of the one–dimensional problem of directional solidification of the Al–Y melt under isothermal condition. The analysis of the numerical solutions shows that initially the transformation of the melt from liquid to solid happens due to the driving forces of the phase transition, while further formation of the spatially–nonhomogeneous laminated structure is caused by the interfacial barriers.

1. Introduction

Many insoluble intermetallic precipitates, or precipitates with extremely small solubility, are known as "line compounds". The free energy of these compounds, by definition, is represented by a single value at a given temperature. The phase-field description of stoichiometric compound growth, similarly to the description of growth of phases with variable composition, poses serious challenges, since the derivative of the free energy with respect to concentration does not exist. One of the possible solutions is the approximation of the line compound free energy by the parabolic function of its composition [1, 2]. Hu with co-authors [3] pointed out the inconsistency of this approach and noted the difficulties encountered in processing the logarithmic contributions to the Gibbs potentials [4]. Hu presented [3] a thermodynamically more consistent view on the solidification kinetics of stoichiometric phases. His assumption about equality of chemical potentials of the compound and variable composition phase leads to the tangent construction for the driving forces of the phase transition. However, in paper [3] the diffusion equation is written for average concentrations of different phases, without consideration of dynamic peculiarities in each phase. It is easy to show that if diffusion in different phases is separated, then we get the correct equation of interaction between the compounds and the phases with variable composition.

As an example let us consider the phase-field description and modeling of peritectic reaction which is frequently encountered in technologically important materials (e.g., steels, brass, bronze, intermetallic compounds, magnetic materials and $YBa_2Cu_3O_x$ superconductors). Many interesting microstructures have been found during directional solidification of peritectic alloys, which have attracted much attention since the last decade. In the present work, in order to investigate the growth behavior of the Al_3Y phase as a peritectic phase modeling of the directional solidification haves been performed for the Al–Y peritectic alloy previously examined experimentally [5].

2. Formulation of the problem

We consider an isothermal problem of phase transformations in the melt of Al-Y with the average molar concentration x_1 of yttrium and suggest that this concentration is limited to some small interval the $x_1 \approx 0.25$ atomic fraction. Then we assume that there is a small temperature gradient of G=20 K/mm and the initial temperature is equal to $T_0 = 980°C$, Fig. 1. Under these conditions except the liquid phase only two solid stoichiometric phases are relevant. Dependences of the densities of Gibbs potentials [6] of liquid and stoichiometric compounds on the concentration of solutes are shown in Fig. 2 in relative values.

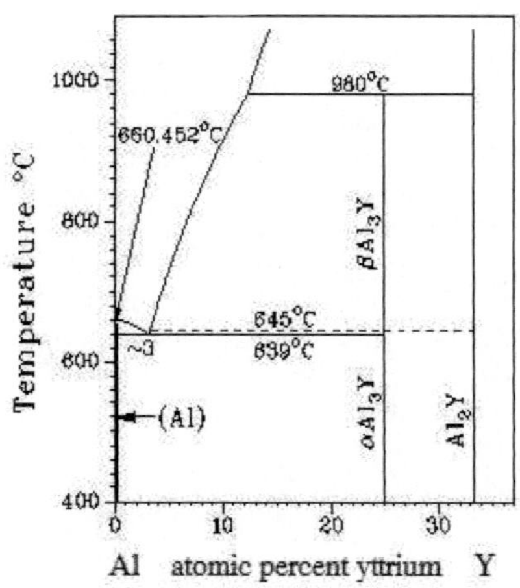

Figure 1. A part of the Al–Y phase diagramm.

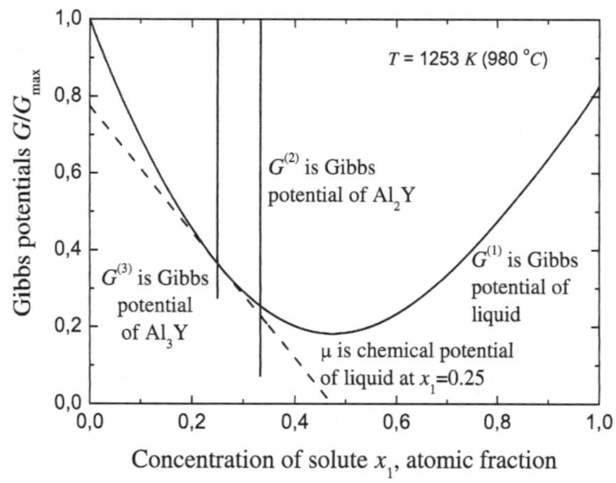

Figure 2. Gibbs potentials of Al_2Y, Al_3Y stoichiometric compounds and liquid at $T = 1253\,K$ (980° C).

In order to describe the state of the solution phase we confine ourselves to the ternary phase field $(\varphi_1, \varphi_2, \varphi_3)$, such that at each point of space the decomposition of unity should is satisfied:

$$\sum_{\alpha=1}^{3} \varphi_\alpha = 1. \tag{1}$$

The phase fields indices, 1, 2, 3, denote, respectively, the liquid, Al_2Y, and Al_3Y phases. Each of the phase field is equal to unity in the appropriate phase and zero in other phases. The unambiguity of the local phase state description is violated only near the interface with a small but finite thickness. Bulk properties of the phases (for example, the proportion of phases per unit volume) within the diffuse interfaces are interpolated by functions of the form:

$$\xi_\alpha = \varphi_\alpha^2(3 - 2\varphi_\alpha) + 2\varphi_1\varphi_2\varphi_3, \qquad \alpha \in \{1, 2, 3\}, \tag{2}$$

choice of which is determined by the requirement of the phases stability [7] and by the condition: $\sum_{\alpha=1}^{3} \xi_\alpha = 1$ following from $\left(\sum_{\alpha=1}^{3} \varphi_\alpha\right)^3 = 1$.

3. Thermodynamics of multiphase systems

The thermodynamic potential determining the relaxation of a non-equilibrium isothermal system under the fixed pressure conditions is the Gibbs potential. The full Gibbs potential can be represented as: $G = \int \Big(G_{eq} + G_{ne} + \lambda(\sum_\alpha \varphi_\alpha - 1)\Big)dV$, where λ is the Lagrange multiplier, which ensures the condition (1). Neglecting the change of the molar volume, v_m, in the process of phase transition (and, consequently, internal stresses in the solidifying system), we present the full density of the Gibbs energy as the superposition of the molar Gibbs potentials of the equilibrium phases, $G^\alpha(T, x_\alpha)$, in the form of:

$$G_{eq} = \frac{1}{v_m}\sum_\alpha \xi_\alpha G^\alpha(T, x_\alpha) + \sum_{\alpha<\beta} W_{\alpha\beta}\varphi_\alpha^2\varphi_\beta^2, \tag{3}$$

where the second term is similar to the well known W–potential with parameters $W_{\alpha\beta}$ which define the height of the potential barrier between the α and β states. We should note that only the Gibbs potential density of the liquid depends on the concentration, namely $G^{(1)} = G^{(1)}(T, x_1)$, whereas for the other phases the Gibbs potential densities are functions only of temperature, $G^{(2,3)} = G^{(2,3)}(T)$, and the concentrations x_2 and x_3 are fixed.

The initial non-equilibrium of the liquid state leads to the necessity of the non-equilibrium contribution to the Gibbs potential density [8]: $G_{ne} = \frac{1}{2}\Big(\beta \mathbf{J}_D^2 + \sum_\alpha \gamma_\alpha \dot{\varphi}_\alpha^2 + \sum_\alpha \sigma_\alpha \vec{\nabla}\varphi_\alpha \cdot \vec{\nabla}\varphi_\alpha\Big)$, which describes the effects associated with the relaxation of the diffusion flux $\mathbf{J}_D$, rate of the phase field change $\dot{\varphi}_\alpha$ and presence of interfaces $\frac{1}{2}\sigma_{\alpha\beta}\vec{\nabla}\varphi_\alpha\vec{\nabla}\varphi_\beta$. The set of the parameters β, γ_α, and σ_α is determined below.

4. The interface redistribution of the solute

The average concentration of the solute in all phases can be written as:

$$\bar{x} = \sum_{\alpha=1}^{3} \xi_\alpha x_\alpha. \tag{4}$$

The total quantity of the solute satisfies to the conservation law. Since the liquid is the only phase with the variable composition then differentiating Eq. (4) on time we obtain the equation describing the change of the liquid concentration caused by changing the phase composition and bulk diffusion:

$$\dot{x}_1\xi_1 = -\vec{\nabla}\cdot\mathbf{J}_D - \sum_{\alpha,\beta} x_\alpha \frac{\partial \xi_\alpha}{\partial \varphi_\beta}\dot{\varphi}_\beta. \tag{5}$$

Considering Eq. (5) as the dynamic equation for the solute in liquid we should note that the original conservation law holds at every point of space and does not involve restrictions to any of the phases. In order to avoid explicit tracking of interfaces we assume that the concentration x_1 is some function defined at each point of space at each moment of time, while the physically observable quantity is the function of $c = x_1\xi_1$. Then from Eq. (5) we can see that in the space regions without the liquid the left part of Eq. (5) vanishes, although the right side of Eq. (5) is not equal to zero. In this case Eq. (5) defines the relationship between the diffusion flux and the growth rate of the Al_2Y and Al_3Y phases at the diffuse interface of the two stoichiometric phases. But these phases have a constant composition of the solute and thus there should not be diffusion by definition. In order to remove this contradiction let us redefine Eq. (5) with the help of the Heaviside step function $\theta(\xi_1)$, which is equal to one for a positive argument and to zero for a negative or zero values of the argument

$$\dot{x}_1\xi_1 = -\Big(\vec{\nabla}\cdot\mathbf{J}_D + \theta(\xi_1)\sum_{\alpha,\beta} x_\alpha \frac{\partial\xi_\alpha}{\partial\varphi_\beta}\dot{\varphi}_\beta\Big). \tag{6}$$

However, in order to reproduce correctly Eq. (5) from Eq. (6) at each point of space we need to add to Eq. (6) the supplementary condition

$$0 = (1-\theta(\xi_1))\sum_{\alpha,\beta} x_\alpha \frac{\partial\xi_\alpha}{\partial\varphi_\beta}\dot{\varphi}_\beta. \tag{7}$$

Due to the identity $\theta(\xi_1) + (1-\theta(\xi_1)) = 1$, the sum of Eqs. (6) and (7) reproduces Eq. (5). The meaning of expression (7) is quite obvious: in the region without the liquid the growth rate of the stoichiometric phases should be constrainted. In our case, when ($\varphi_1 = 0$), condition (7) is equivalent to $(1-\theta(\xi_1))\Big(x_2\dot{\varphi}_2\partial\xi_2/\partial\varphi_2 + x_3\dot{\varphi}_3\partial\xi_3/\partial\varphi_3\Big) = 0$ or, in the absence of the liquid, we can choose $\varphi_\alpha = \varphi$ and $\varphi_\beta = 1-\varphi$. After this we immediately find that $\partial\xi_2/\partial\varphi_2 = \partial\xi_3/\partial\varphi_3 = p'(\varphi) = 6\varphi(1-\varphi)$. Then the previous condition is transformed in the following one $(1-\theta(\xi_1))p'(\varphi)(x_2-x_3)\dot{\varphi} = 0$, which implies an obvious statement $p'(\varphi)(x_2-x_3) \neq 0$ that holds within the diffuse interface between the stoichiometric phases $\dot{\varphi} = 0$.

5. The decrease of the Gibbs potential in the relaxation process

During relaxation to equilibrium monotonic decrease of the Gibbs potential takes place and taking into account relations (6) one can find that the time derivative of the Gibbs potential can be represented in the form

$$\frac{dG}{dt} = \int\Big(\mathbf{J}_D\mathbf{F}_D + \sum_\alpha \dot{\varphi}_\alpha F_\alpha + \dot{\lambda}F_\lambda\Big)\,dV \le 0. \tag{8}$$

Let us denote the bulk density of the equilibrium Gibbs potential as $\tilde{G}_\alpha = G_\alpha/v_m$ and the bulk chemical potential (more exactly the difference of the chemical potentials of the dissolved components) of the liquid phase as $\tilde{\mu} = \mu/v_m = 1/v_m\partial G^{(1)}/\partial x_1$. Then the coefficients in expression (8) are can be easily found by algebraic manipulations. We devide the products of the Gibbs potentials with the phase fields rates at the diffuse interface into two parts $\tilde{G}^\alpha\dot{\varphi}_\beta = \tilde{G}^\alpha\dot{\varphi}_\beta(\theta(\xi_1) + \theta(1-\xi_1))$ and assume that the phase field rates do not contribute to the rates of the system Gibbs potential in the regions where $\theta(\xi_1) = 0$ according to Eq. (7).

The coefficient of the change rate of the Lagrange multiplier is obviously equal to Eq. (1). Using the uncertainty of non–equilibrium thermodynamics [11] in the choice of fluxes (for the conservative order parameters) and rates (for the non–conservative order parameter), we can define the fluxes as $\mathbf{J}_D = -M_D\mathbf{F}_D$, $\dot{\lambda} = -M_\lambda F_\lambda$, $\dot{\varphi}_\alpha = -M_\alpha F_\alpha$. The choice of mobility $M_D > 0$, $M_\lambda > 0$, $M_\alpha > 0$ guarantees fulfillment of the monotone decrease of the Gibbs potential (8).

We assume that $M_\lambda \to \infty$, which assures satisfaction of Eq. (1). Expressing the mobility M_D via the diffusion coefficient in the liquid as $D_L = M_D\partial\tilde{\mu}/\partial x$, one can write the relaxation equations for the fluxes and phase fields in the form of

$$\tau_D\dot{\mathbf{J}}_D + \mathbf{J}_D = -D_L\vec{\nabla}x_1, \tag{9}$$

$$\tau_\alpha\ddot{\varphi}_\alpha + \dot{\varphi}_\alpha = M_\alpha\Big[\vec{\nabla}\big(\sigma_\alpha\vec{\nabla}\varphi_\alpha\big) - \lambda - 2\sum_{\beta\neq\alpha} W_{\alpha\beta}\varphi_\alpha\varphi_\beta^2 - \theta(\xi_1)\sum_\beta \frac{\partial\xi_\beta}{\partial\varphi_\alpha}\big(\tilde{G}^\beta - x_\beta\tilde{\mu}\big)\Big], \tag{10}$$

where $\tau_D = \beta M_D$, $\tau_\alpha = \gamma_\alpha M_\alpha$ are the relaxation times of the diffusion flux and phase fields.

6. Equilibrium conditions

Since all time derivatives and all fluxes tend to zero at equilibrium then from expression (9) it follows that the solute concentration in the liquid has a constant value $x_1 = x_0 = const$, and, respectively, $\tilde{\mu} = \tilde{\mu}_0 = const$. Then Eq. (10) can be reduced to the following relationships

$$\vec{\nabla}\left(\sigma_\alpha \vec{\nabla}\varphi_\alpha\right) - 2\sum_{\beta\neq\alpha} W_{\alpha\beta}\varphi_\alpha\varphi_\beta^2 - \theta(\xi_1)\sum_\beta \frac{\partial \xi_\beta}{\partial \varphi_\alpha}\left(\tilde{G}_0^\beta - x_\beta\tilde{\mu}_0\right) - \lambda = 0. \quad (11)$$

Next, let us consider the pairwise equilibrium of the α and β phases. Due to the lack of third phase $\varphi_\alpha = \varphi$ and $\varphi_\beta = 1 - \varphi$. Denoting $\sigma_{\alpha\beta} = \sigma_\alpha + \sigma_\beta$, and subtracting Eq. (10) for the β phase from Eq. (10) for the α phase, we obtain:

$$\vec{\nabla}\left(\sigma_{\alpha\beta}\vec{\nabla}\varphi\right) - W_{\alpha\beta}g'(\varphi) - p'(\varphi)\theta(\xi_1)\left(\tilde{G}_0^{(\beta)} - \tilde{G}_0^{(\alpha)} - (x_\beta - x_\alpha)\tilde{\mu}_0\right) = 0. \quad (12)$$

Assuming that one of the phases is liquid ($\theta(\xi_1) = 1$), we consider the case of the flat front. After integrating Eq. (12) along the normal direction for infinite limits taking into account the asymptotic boundary conditions for the phase field and the explicit form of the functions $g(\varphi) = \varphi^2(1-\varphi)^2$ and $p(\varphi) = \varphi^2(3-2\varphi)$, we obtain the thermodynamic condition of equilibrium, which corresponds to the equality to zero of the driving force of the phase transition at the interface with the liquid [12]:

$$\tilde{G}_0^{(\beta)} - \tilde{G}_0^{(\alpha)} - (x_\beta - x_\alpha)\tilde{\mu}_0 = 0. \quad (13)$$

Taking into account the equilibrium conditions from (12) one can write by analogy with paper [14], the phenomenological parameters $\sigma_{\alpha\beta}$, and $W_{\alpha\beta}$ in the form $\sigma_{\alpha\beta} = 6\chi_{\alpha\beta}\delta_{\alpha\beta}$ and $W_{\alpha\beta} = 12\chi_{\alpha\beta}/\delta_{\alpha\beta}$, where $\chi_{\alpha\beta}$ is the surface tension of the $\alpha\beta$ interface and $\delta_{\alpha\beta} = \sqrt{\sigma_{\alpha\beta}/2W_{\alpha\beta}}$ is the characteristic halfwidth of the diffusion front.

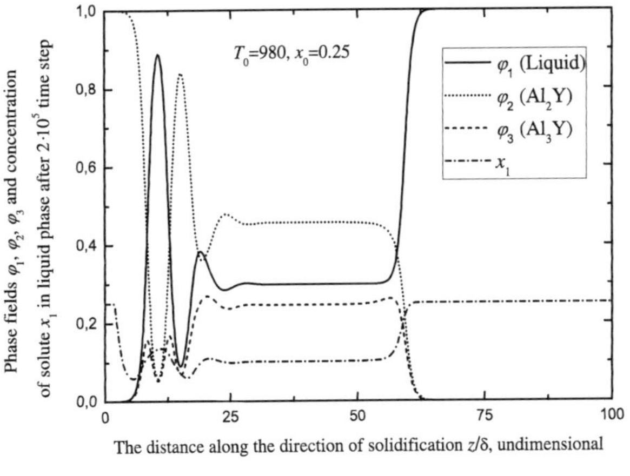

Figure 3. The dynamics of the interfaces in the initial moment of time leads to uniform values of the phase fields, which afterwards delaminate into distinct phases.

7. Numerical simulation

For simplification of numerical simulation of the equations obtained one can assume that for all phases the relaxation times, the coefficients of surface tension and the halfwidth of the diffuse interface are approximately equal. Then we can write $\tau_\alpha = \tau_0 = 10^{-11}$ s,

$\chi_{\alpha\beta} = \chi_0 = 0,87\ \mathrm{J/m^3}$, where χ_0 is equal to the surface energy of pure aluminium [13] and $\delta_{\alpha\beta} = \delta_0 = 1.2 \cdot 10^{-9}$ m. Then, taking into account the equilibrium conditions we find: $\sigma_\alpha = 3\chi_0\delta_0$, and $W_{\alpha\beta} = 12\chi_0/\delta_0$. Assuming equality of the phase fields mobilities we can present them in the form of $M_\alpha = \nu/3\chi_0\delta_0$, where $\nu = 8{\cdot}10^{-9}\ \mathrm{m^2/s}$. After the transition to dimensionless

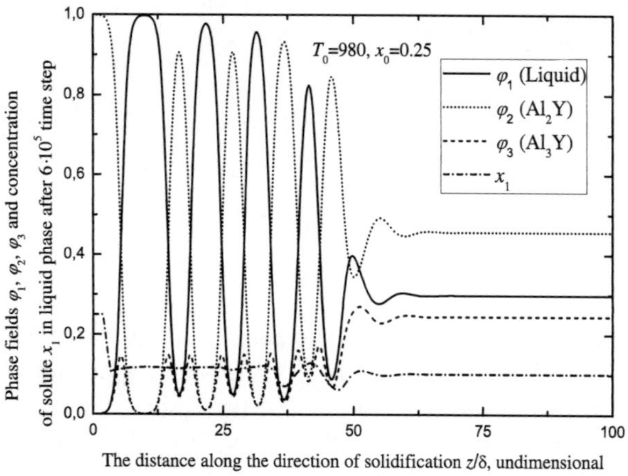

Figure 4. In subsequent moments of time the spatially nonhomogeneous structure of the phases is formed due to the driving force defined by the W–potential.

variables the set of equations obtained is solved numerically using the explicit gradient–stable algorithm [15]. Parameter λ is determined from the unit decomposition condition (1). The simulation results are presented in Fig. 3 and Fig. 4.

8. Conclusions

The simulation results show that the proposed model provides the formation mechanism of the spatially nonhomogeneous structure typical for peritectically transformations. The nucleation of the solid phase in liquid happens due to the action of the driving force (13), while the formation of interfaces is driven by the W–potential and takes place at greater time scales. Since the isothermal model in which the temperature was nearly constant was used for the study of peritectic transformations, the calculated composition of the final phase includes liquid, which is not observed in the experiments. The reason is that our calculations do not take into account the change of the temperature field in the sample up to complete solidification. This means that one should also consider other phases, such as FCC phase of aluminum, which fall out from the solution with the decreasing content of yttrium.

9. References

[1] Heulens J, Blanpain B and Moelans N 2011 *Acta Mater.* **59** 2156
[2] Park M S, Gibbons S L and Arróyave R 2012 *Acta Mater.* **60** 6278
[3] Hu S Y, Murray J, Weiland H, Liu Z K and Chen L Q 2007 *Comp. Coupl. Phase Diag. and Therm.* **31** 303
[4] Hillert M 2008 *Phase Equilibia, Phase Diagrams and Phase Transformations*, (Cambridge: Camb. Univ. Press)
[5] Luo L, Liu T, Zhang Y, Su Y, Guo J and Fu H 2016 *Acta Metall.* **52**(7) 866
[6] URL: http://cpddb.nims.go.jp/cpddb/periodic.htm from 25.01.2017
[7] Kessler D 2001 J. Crystal Growth **224** 175
[8] Cahn J W and Hilliard J E 1958 *J. Chem. Phys.* **28** 258
[9] Provatas N and Elder K 2010 *Phase–Field Methods in Materials Science and Engineering* (Weinheim: Wiley-VCH)
[10] Chen L Q 2002 *Annu. Rev. Mater. Res.* **32** 113
[11] de Groot S R and Mazur P 1962 *Non-equlibrium Thermodynamics* (Amsterdam: North-Holland)
[12] Lebedev V and Galenko P 2016 *Russ. Metallurgy* **8** 785

[13] Schmitz J, Brillo J, Egry J and Schmid-Fetzer R 2009 *Int. J. Mat. Res.* **100**(11) 1529
[14] Lebedev V G, Lebedeva A A and Galenko P K 2015 *JETP Letters* **101**(2) 136
[15] Lebedev V, Sysoeva A and Galenko P 2011 *Phys. Rev. E* **83** 026705

Acknowledgments

The work was partly supported by the Russian Foundation of Basic Researches (grant 15-02-06288) and by FASO Russia (budgetary theme number 01201157503).

IOP Conf. Series: Materials Science and Engineering **192** (2017) 012022 doi:10.1088/1757-899X/192/1/012022

Analysis of stochastic dynamics of Higgins model

L Ryashko
Ural Federal University, Lenina, 51, Ekaterinburg, 620000, Russia
E-mail: lev.ryashko@urfu.ru

Abstract. We consider the Higgins model of glycolytic oscillations. An influence of random disturbances on equilibria and cycles of this model is studied. Peculiarities of the dispersion of stochastic trajectories near the deterministic attractors of Higgins model are investigated by the direct numerical simulation and semi-analytical approach based on the stochastic sensitivity functions technique. To describe a spatial arrangement of random states, we use the confidence domains method. The results of comparative analysis of the influence of additive and parametric noise are given.

1. Introduction

At the present time, a significant interest is paid to the study of stochastic dynamics of the biochemical systems. Among a wide variety of biochemical processes, the reaction of glycolysis is one of the most important. In the present paper, we focus on the model introduced by Higgins [1]. This model describes self-sustained oscillations in the yeast glycolytic system. A limit cycle is a standard mathematical model of such oscillations. In the Higgins system, this limit cycle appears as a result of the Andronov-Hopf bifurcation. The deterministic Higgins model is well studied and plays a role of the important tutorial example [2-5].

It is well known that the random noise is an inevitable attribute of any real system. The noise can essentially change the system's dynamics [6-8]. An influence of noise on glycolysis was investigated in [9-11].

In the current paper, we study a stochastic dynamics of glycolysis on the base of Higgins model. In our theoretical analysis, we use the quasipotential method [12] and stochastic sensitivity functions technique [13-15]. This semi-analytical technique was successfully applied to the study of various noise-induced phenomena in nonlinear dynamic systems [16-19].

In Section 2, we study a bifurcation and attractors (equilibria and cycles) of the deterministic Higgins model.

In Section 3, we analyse an influence of random noise on the attractors of Higgins model. Here, we apply the stochastic sensitivity functions technique and the confidence domains method.

An impact of the parametric noise on cycles of the Higgins model is discussed in Section 4.

2. Deterministic model

In this paper, we study the glycolytic oscillations under the random disturbances on the basis of the well known Higgins model [1]

$$\begin{aligned} \dot{x} &= 1 - xy, \\ \dot{y} &= py\left(x - \tfrac{1+q}{q+y}\right). \end{aligned} \tag{1}$$

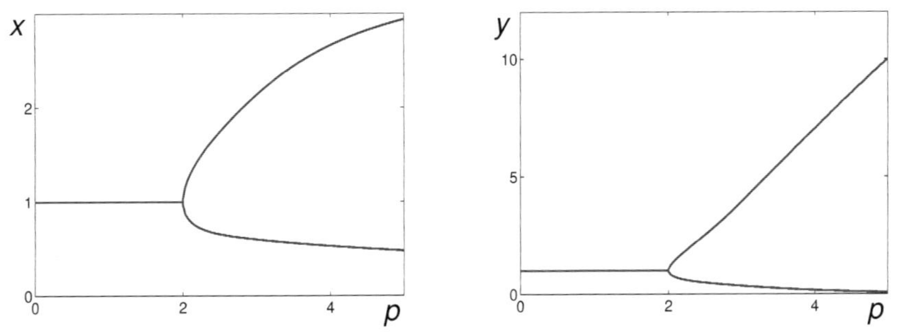

Figure 1. Bifurcation diagrams of the deterministic Higgins model (1) with $q = 1$.

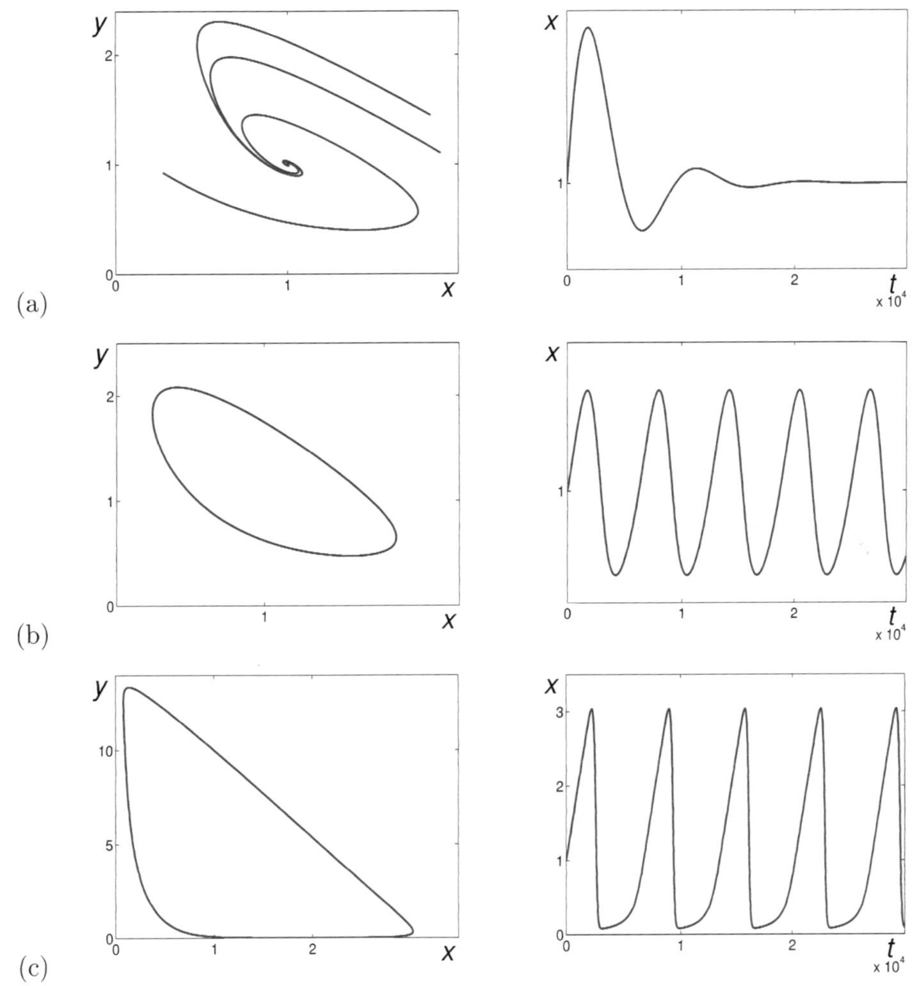

Figure 2. Phase trajectories and time series for the deterministic Higgins model (1) with $q = 1$ and $p = 1$ (a); $p = 2.2$ (b); $p = 5$ (c).

Here, p and q are the positive parameters. The system (1) is a mathematical model of the product-activated and substrate-inhibited reaction.

The parametric line $p = q + 1$ marks the Andronov - Hopf bifurcation. For $0 < p < q + 1$, system (1) has a unique stable equilibrium $\bar{x} = \bar{y} = 1$. For $p > q+1$, this equilibrium is unstable, and system (1) exhibits the stable self-sustained oscillations. In the present paper, we fix $q = 1$ and vary the parameter p.

In Figure 1, a bifurcation diagram of system (1) is presented. Here, the extreme values of x-coordinates of attractors are shown in Fig. 1a, and the extreme values of x-coordinates of attractors are shown in Fig. 1b. As parameter p increases, the amplitude of oscillations increases too, especially for y-coordinate.

For three values of parameter p, the phase portraits and time series are shown in Fig. 2. Note that close to the bifurcation point $p = 2$, the oscillations are quasi-harmonic, and with increasing p, the system exhibits relaxation oscillations.

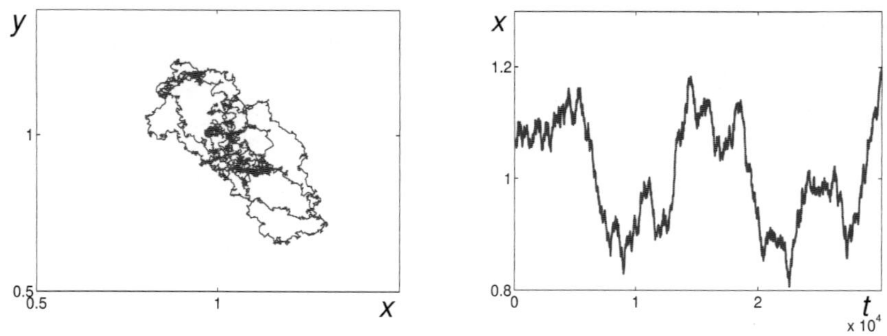

(a)

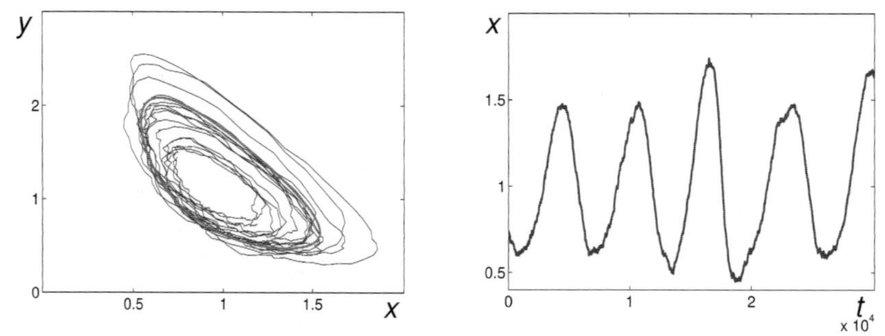

(b)

Figure 3. Phase trajectories and time series for the stochastic Higgins model (2) with $q = 1$, $\varepsilon = 0.05$, and $p = 1$ (a); $p = 2.2$ (b). By red shown is the deterministic limit cycle.

3. Stochastic model with additive noise

First, consider an influence of the additive noise modelling by the following stochastic system

$$\begin{aligned} \dot{x} &= 1 - xy + \varepsilon_1 \xi_1(t), \\ \dot{y} &= py\left(x - \frac{1+q}{q+y}\right) + \varepsilon_2 \xi_2(t), \end{aligned} \tag{2}$$

where $\xi_1(t)$ and $\xi_2(t)$ are the uncorrelated standard white Gaussian noises which describe fluctuating external forces, $\varepsilon_{1,2}$ are the additive noise intensities. In what follows, we put $\varepsilon_1 = \varepsilon_2 = \varepsilon$.

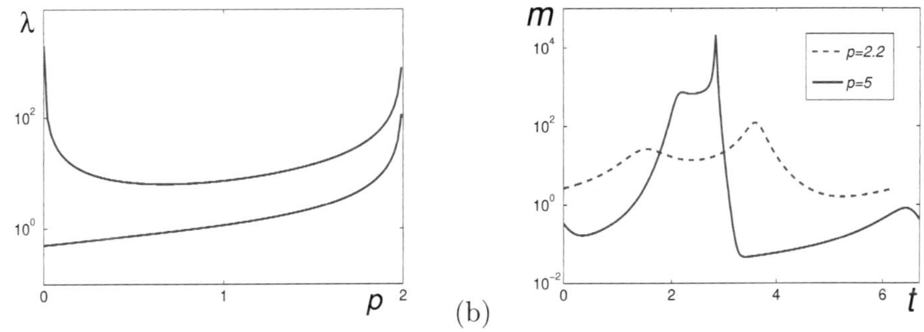

Figure 4. Stochastic sensitivity of equilibria (a) and cycles (b).

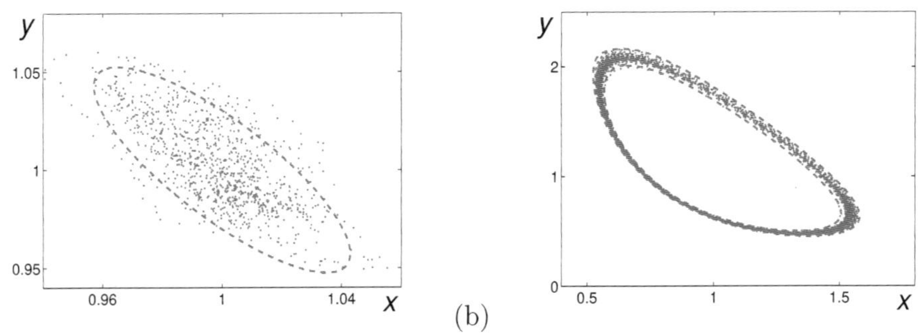

Figure 5. Random states and confidence domains for $\mathcal{P} = 0.99$: (a) ellipse for $p = 1$, $\varepsilon = 0.01$; (b) band for $p = 2.2$, $\varepsilon = 0.003$.

The trajectories of stochastically forced system (2) leave a deterministic attractor (equilibrium or cycle) and form some probabilistic distribution around it. In Fig. 3, the random trajectories and time series for model (2) with $q = 1$ and $\varepsilon = 0.05$ are plotted for the stochastically forced equilibrium with $p = 1$ (see Fig. 3a), and for the stochastically forced cycle with $p = 2.2$ (see Fig. 3b). As one can see, a dispersion of random trajectories near attractors is non-uniform.

To study this dispersion parametrically, we will apply the stochastic sensitivity functions technique.

For the equilibria, this dispersion is defined by the noise intensity and the stochastic sensitivity matrix W [13,14]. Eigenvalues λ_1, λ_2 of this matrix are the simple scalar characteristics of the stochastic sensitivity of equilibrium. In Fig. 4a, functions $\lambda_1(p) > \lambda_2(p)$ are plotted in the zone of stable equilibria for $0 < p < 2$. As can be seen, the largest $\lambda_1(p)$ unlimitedly increases near the left and right borders of this interval. In contrast to $\lambda_1(p)$, the function $\lambda_2(p)$ is monotonous and tends to infinity approaching the Andronov-Hopf bifurcation value $p = 2$. Note that the values of $\lambda_1(p)$ essentially dominate over $\lambda_2(p)$.

For the cycle, the stochastic sensitivity matrix is a T-periodic function $W(t)$. Here, T is the period of solution $(\bar{x}(t), \bar{y}(t))$ which presents a limit cycle. In the two-dimensional case, $W(t) = m(t)v(t)v^\top(t)$, where $v(t)$ is a normalized vector orthogonal to the limit cycle. The scalar function $m(t)$ is called the stochastic sensitivity function of cycle [13,14].

Using $m(t)$, we can describe some peculiarities of the dispersion along the cycle in detail. In Fig. 4b, plots of function $m(t)$ for two values of parameter p are shown. Overfalls of its values reflect the non-uniformity of stochastic bundles around cycles. Note for the larger values of p

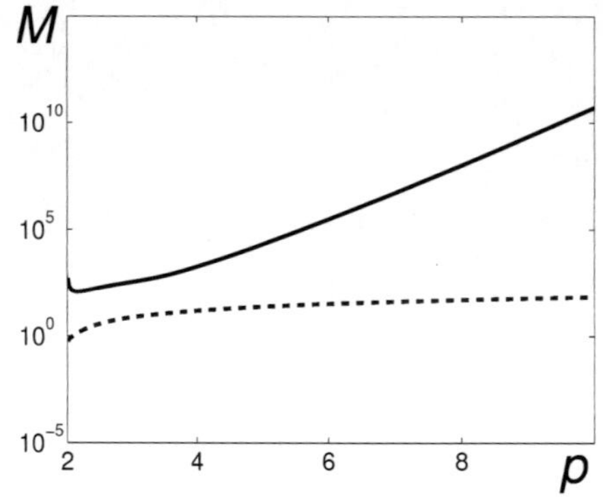

Figure 6. Stochastic sensitivity factors $M(p)$ for the system (2) (solid) and system (3) (dashed).

corresponding to the relaxation oscillations (see the solid line for $p = 5$), the overfall of values of $m(t)$ is more essential.

With the help of stochastic sensitivity function, one can the construct confidence domains around deterministic attractors. The confidence domains are simple and evident geometrical models for the description of arrangement of random states near the deterministic cycles and equilibria. In Fig. 4, along with random states of system (2), a confidence ellipse for $p = 1$, $\varepsilon = 0.01$ around the equilibrium, and confidence band for $p = 2.2$, $\varepsilon = 0.003$ around the limit cycle are shown. As one can see, these confidence domains effectively visualize a spatial arrangement of random states in the phase plane, and well agree with the results of direct numerical simulation.

As a cumulative characteristic of the response of limit cycle on random disturbances, we will use the stochastic sensitivity factor

$$M = \max_{[0,T]} m(t).$$

In Fig. 6, the function $M(p)$ for system (2) is plotted by the solid line.

4. Stochastic model with parametric noise

Consider now the influence of parametric noise on cycles of the Higgins model. Here, we focus on the random disturbances of parameter p. Consider a system

$$\begin{aligned} \dot{x} &= 1 - xy, \\ \dot{y} &= (p + \varepsilon\xi(t))y\left(x - \tfrac{1+q}{q+y}\right), \end{aligned} \tag{3}$$

obtained from system (1) by the replacement $p \to p + \varepsilon\xi(t)$. Here, $\xi(t)$ is a standard white Gaussian noise, ε is an intensity of the parametric noise.

In Fig. 4, a plot of the stochastic sensitivity factor $M(p)$ for system (3) is shown by the dashed line. Here, we can compare the response of limit cycles on the additive (solid line) and parametric (dashed line) noises. Note that the function $M(p)$ for parametric noise is monotonic. The plot of function $M(p)$ for the additive noise is different. Approaching to the bifurcation point $p = 2$, the stochastic sensitivity of the system (2) increases unlimitedly. As parameter p increases, this function decreases, passes a minimum and begins to grow sharply.

5. Conclusion

Attractors of Higgins model forced by random disturbances were studied by the stochastic sensitivity function technique and confidence domains method. An effectiveness of our approach in the analysis of peculiarities of the stochastic dynamics is demonstrated.

6. References

[1] Higgins J 1964 *Proc. Natl. Acad. Sci. U. S.* **51** 989
[2] Bhargava S C 1980 *Bull. Math. Biol.* **42** 829
[3] Gurel D and Gurel O 1983 *Oscillations in Chemical Reactions* (Berlin: Springer-Verlag)
[4] Epstein I R and Pojman J A 1998 *An Introduction to Nonlinear Chemical Dynamics: Oscillations, Waves, Patterns, and Chaos* (New York: Oxford University Press)
[5] Rubin A B 2014 *Fundamentals of Biophysics* (New Jersey: Wiley)
[6] Horsthemke W and Lefever R 1984 *Noise-Induced Transitions* (Berlin: Springer)
[7] Lindner B, Garcia-Ojalvo J, Neiman A and Schimansky-Geier L 2004 *Phys. Rep.* **392** 321
[8] Bashkirtseva I, Ryashko L and Schurz H 2009 *Chaos Solitons Fractals* **39** 72
[9] Garcia-Olivares A, Villarroel M and Marijuán P C 2000 *Biosystems* **56** 121
[10] Xu W, Kong J S and Chen P 2009 *J. Phys. Chem. C* **113** 2393
[11] Schuster P 2016 *Stochasticity in Processes Fundamentals and Applications to Chemistry and Biology* (Berlin: Springer)
[12] Bashkirtseva I and Ryashko L 2011 *Phys. Rev. E* **83** 061109
[13] Bashkirtseva I, Neiman A and Ryashko L 2013 *Phys. Rev. E* **87** 052711
[14] Bashkirtseva I and Ryashko L 2014 *Phys. A* **410** 236
[15] Ryagin MY and Ryashko LB 2004 *Int. J. Bif. Chaos* **14** 3981
[16] Bashkirtseva I and Ryashko L 2009 *Phys. Rev. E* **79** 041106
[17] Bashkirtseva I, Chen G and Ryashko L 2012 *Chaos* **22** 033104
[18] Bashkirtseva I, Ryashko L and Slepukhina E 2014 *Fluct. Noise Lett.* **13** 1450004

Acknowledgments

The work was supported by Russian Science Foundation (N 16-11-10098).

Analysis of morphology and residual porosity in selective laser melting of Fe powders using single track experiments

I V Shutov, G A Gordeev, E V Kharanzhevskiy and M D Krivilyov

Udmurt State University, Universitetskaya str. 1, building 4, 426034, Izhevsk, Russia

E-mail: shutiny@gmail.com

Abstract. Morphology and residual porosity of single tracks obtained by pulse selective laser melting (SLM) of Fe powder have been studied by metallography. Multiple cross sections of the stainless substrate with the single tracks deposited by SLM are examined and classified depending on processing parameters. A sustainable scanning strategy to reduce residual porosity is suggested for pulse laser annealing. The developed method is suitable both for improvement of processing regimes in commercial SLM machines and validation of numerical models in additive manufacturing of metal parts. The effect of the beam radius, pulse energy, its frequency and duration on a shape of the single track and its adhesion to the substrate is revealed.

1. Introduction

Additive manufacturing (AM) based on layer-by-layer fabrication of 3D parts is now one of the most rapidly developing industrial technologies. The methods of selective laser melting (SLM) and sintering (SLS) of metallic powder belongs to a most prominent AM approach in machine construction and engineering [1]. SLM/SLS like other AM methods deliver high-strength metallic parts of arbitrary shape at low level of waste and scrapped components. A vast majority of commercial 3D SLM printers use continuous laser source however in some systems pulse laser systems are implemented [2,3]. Recent studies [1,3,4] shown that pulse SLM machines have advantages in comparison with continuous ones in processing of materials with high reflectivity.

Further improvement of SLM is faced to a problem that good quality of melted or sintered layers for different powders is registered only in a narrow interval of processing parameters. Thus optimization of these parameters by the trail-and-error method is expensive and time-consuming for specific 3D prototypes. Under such conditions, analysis of simplified processing conditions observed in single track experiments allows us to receive useful qualitative dependences and quantitative estimates how the processing parameters effect on the characteristics of SLM components. Hence, the experimental study with single tracks significantly reduces costs of SLM optimization.

Computer simulation is widely used nowadays for SLM processing according to [1,5,6,7]. The developed models cannot account for all underlying physical processes in SLM due to their complexity, otherwise the computing time increases drastically. Moreover, any mathematical model requires proper validation through comparison to the results of laboratory experiments [7]. In this situation single-track experiments are very convenient. It helps the examination of powder melting and consolidation without some auxiliary effects like interference of neighboring tracks and thermal shape distortion. Additionally, morphology and depth of the molten zone, its width and other characteristics can be directly analyzed. In literature [1,5,6,7], this problem has not been solved completely therefore the aim of this paper is the study of morphology in single track SLM of Fe powder.

The Fe iron powder with low impurity content was compacted via pulse laser annealing under different modes of processing. Iron was selected since it is a well-studied and base component for steels. In what follows, morphology in terms of residual porosity and adhesion to the substrate is first analyzed by metallography of single-track processed samples. Then the effect of processing parameters of SLM on morphology is revealed.

2. Experimental technique

2.1. Processing of single-track samples

Iron powder bed on the surface of a stainless AISI 321 substrate has been processed by the pulse Yb fiber laser LRS AU-300M with a maximum power of 300 W. The ultra-dispersed Fe powder was deposited and flatten on the substrate mechanically. Due to intensive local heating, the powder is melted during processing. The liquid phase consolidates and then rapidly solidifies resulting in a single track, Fig. 1 and 2. The melted track is depicted at different magnification in Fig. 1. Each track was processed with different governing parameters including the energy E_{imp} of a laser pulse, its time duration τ_{imp}, radius of the laser beam R_b, frequency ν, and spatial distance l_{tr} between pulses. Then the instantaneous density of power is given by

$$J_{\max} = \frac{E_{imp}}{2\pi R_b^2 \tau_{imp}}, \qquad (1)$$

where J_{max} is the peak power density of laser emission. The Gauss-type spatial distribution is assumed in which the effective radius R_b holds about 80 % of delivered energy. In experiments, the following parameter intervals were studied: R_b in the range between 200 and 400 μm, E_{imp} between 0.5 and 5.1 J, τ_{imp} between 0.8 and 6 ms. The distance l_{tr} equals to R_b and the frequency varied between 30 and 100 Hz. Accuracy of E_{imp} determination was 0.1 J.

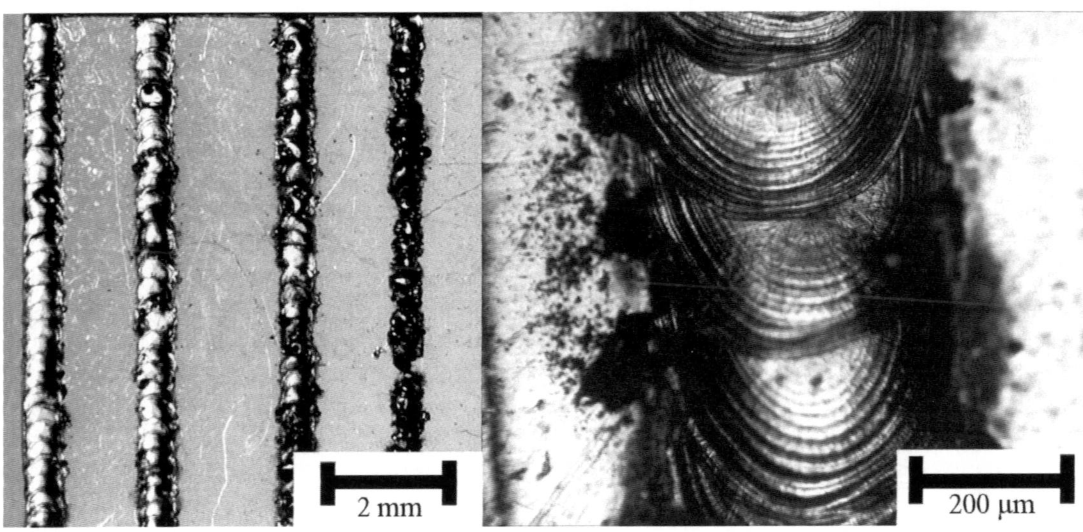

Figure 1. Top view of the molten single tracks processed with the beam radius R_b = 200 μm and pulse duration τ_{imp} = 1 ms.

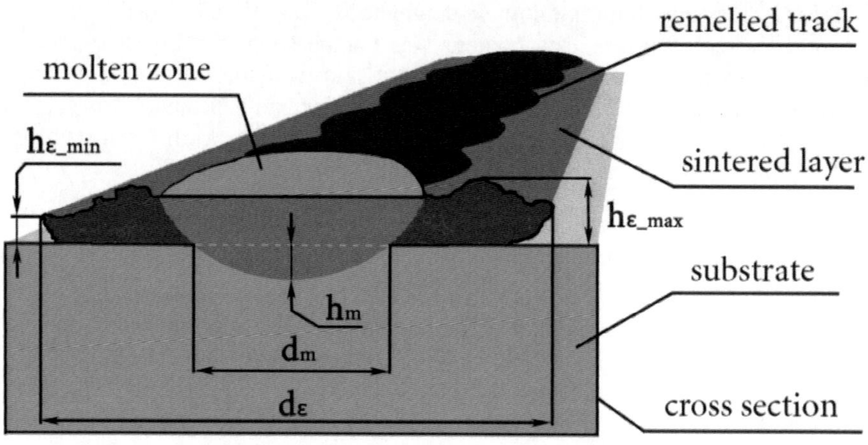

Figure 2. Sketch of morphology and characteristic distances of the molten zone after pulse annealing where h_m is the maximum depth of the remelted zone in the substrate, d_m is the width of molten powder layer, h_{ε_max} and h_{ε_min} are the maximum and minimum thickness of the sintered track correspondingly, and d_ε is the width of the sintered track.

2.2. Metallography of single tracks

The processed (Fe powder + stainless substrate) samples have been examined by metallography. The samples were fixed in epoxy resin, mechanically grinded and polished. After etching, the samples were analyzed using the optical microscope Neophot-30 Jena Zeiss. Since the samples have different porosity levels their grinding was done gradually to preserve morphology in cross sections, Fig. 2. At some processing parameters, few (between 6 and 9) cross sections were examined along the same track to reach adequate statistical variability. Other single tracks were examined only with a single cross section. Etching was done with the 5% HNO_3 solution in ethanol with the etching time between 2 and 20 seconds.

3. Results

3.1. Classification of track's morphologies

Laser melting of powder is accompanied by multiple physical phenomena that proceed at different scales and may include chemical reactions. There are the processes of ultra-rapid heating of powder and substrate, melting, solid and liquid phase sintering and consolidation, convective flow in the molten zone, capillary flow, adhesion of the powder bed to the substrate. Additionally, there are intensive evaporation of metal, surface tension, shape distortion of the substrate, rapid solidification of the molten zone and other processes. Contribution of each factor is changing depending on regimes of laser annealing. Hence their variation is a straightforward way to control the residual porosity, pore distribution in the track, shape and size of the track, adhesion to the substrate.

Based on the analysis of obtained cross sections, different morphological groups were classified depending on processing parameters and governing processes acted in separate SLM experiments, Fig. 3. Figure 3(a) corresponds to a typical shape where the pulse energy is small and no good adhesion of the remelted powder with the substrate is observed. A network of small pores is observed and the shape of the track is vague. Black color of this slice is due to the high concentration of iron and weak mixing with the substrate which contains a lot of chromium and nickel. If the pulse energy is increased, Fig. 3(b), a concave shape is registered due to powder consolidation. Melting of the substrate still does not proceed since a transition between the black track and white substrate is clearly visible. The pores are of different sizes and the black areas correspond to high concentration of small pores which are not well resolved with optical microscopy.

In Fig. 3(c), the beginning of melting of the substrate was detected. Ultimately this is caused by a higher pulse energy although the molten zone in the substrate is too small to provide effecting mixing of powder with the substrate. Large gas pores were found in some cross sections while in other sections they were absent. Finally, at $E_{imp} = 2.8$ J, Fig. 3(d) stirring of the melt becomes so intensive that good adhesion of the track to the substrate is achieved. Correspondingly, the track's color becomes lighter and residual porosity significantly decreases. At this stage small pores consolidate in larger ones since they do not have enough time to escape from the melt prior solidification. The same effect was predicted in computer simulation of powder consolidation [9, 10]. The track's shape is partially concaved that is typical for poor wetting. It was proved experimentally that such situation occurs if a high power density is combined with small pulse duration.

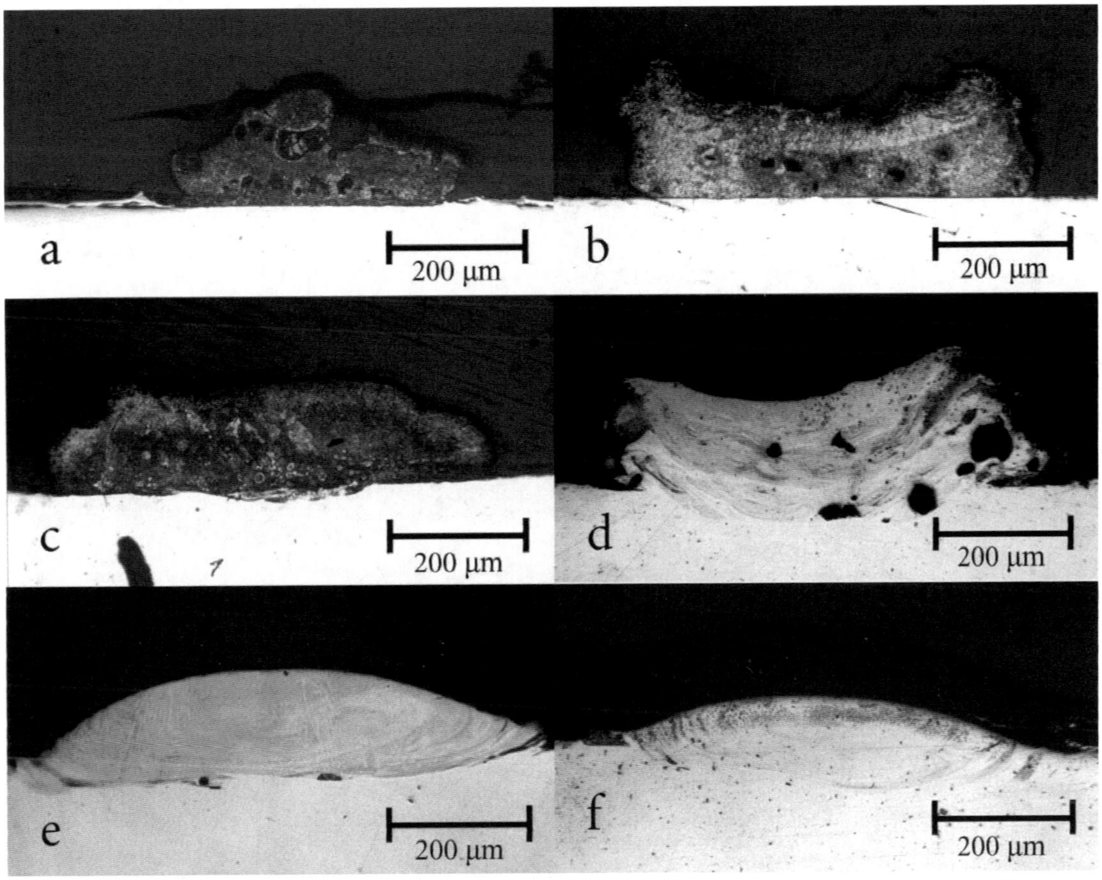

Figure 3. Morphologies of the cross sections obtained under different processing conditions in SLM of Fe powder on top of a AISI 321 substrate. The initial layer thickness before process is 150 μm. The straight-line single-track strategy is applied. The processing parameters are as follows. **a)** $E_{imp} = 0.7$ J, $\tau_{imp} = 4$ ms, $R_b = 200$ μm, ν = 50 Hz; **b)** $E_{imp} = 1.1$ J, $\tau_{imp} = 4$ ms, $R_b = 200$ μm, ν = 50 Hz; **c)** $E_{imp} = 1.6$ J, $\tau_{imp} = 4$ ms, $R_b = 200$ μm, ν = 50 Hz; **d)** $E_{imp} = 2.8$ J, $\tau_{imp} = 3.0$ ms, $R_b = 300$ μm, ν = 50 Hz; **e)** $E_{imp} = 3.6$ J, $\tau_{imp} = 1.5$ ms, $R_b = 400$ μm, ν = 50 Hz; **f)** $E_{imp} = 2.4$ J, $\tau_{imp} = 1.5$ ms, $R_b = 300$ μm, ν = 50 Hz.

Two last morphologies shown in Fig. 3(e,f) are convex. It corresponds to intensive stirring with streamlines visualized by the alternating strips along the flow streamlines, Fig. 3(d,e,f). The strips are formed by the chains of small pores which are carried by convective vortex flow in the molten zone. At such pulse energy E_{imp} and duration τ_{imp} the lowest porosity is registered with high intensity mixing of material between the powder bed and the substrate. It is logical to conclude that these morphologies comply to complete consolidation [10]. Surface tension at the melt-gas interface forms the convex shape of the track.

In Fig. 3(e), the small pores are observed at the boundary between the track and substrate since the time for gas displacement was not long enough. Comparing Figures 3(e) and (f) one may conclude that shorter pulse duration yields a higher porosity. The case (f) is obtained at a smaller effective radius R_b = 300 μm and hence at a higher power density. To summarize, the convex morphology is formed only after significant remelting of the substrate where the capillary effects and convection have sufficient time for full consolidation. Small residual porosity provided under the discussed conditions is contradicted by large evaporation of the powder material. As a result, high pulse energy may lead to melt boiling and subsequently to roughness of the processed sample. All samples examined in the study belong to one of the cases described in Fig. 3.

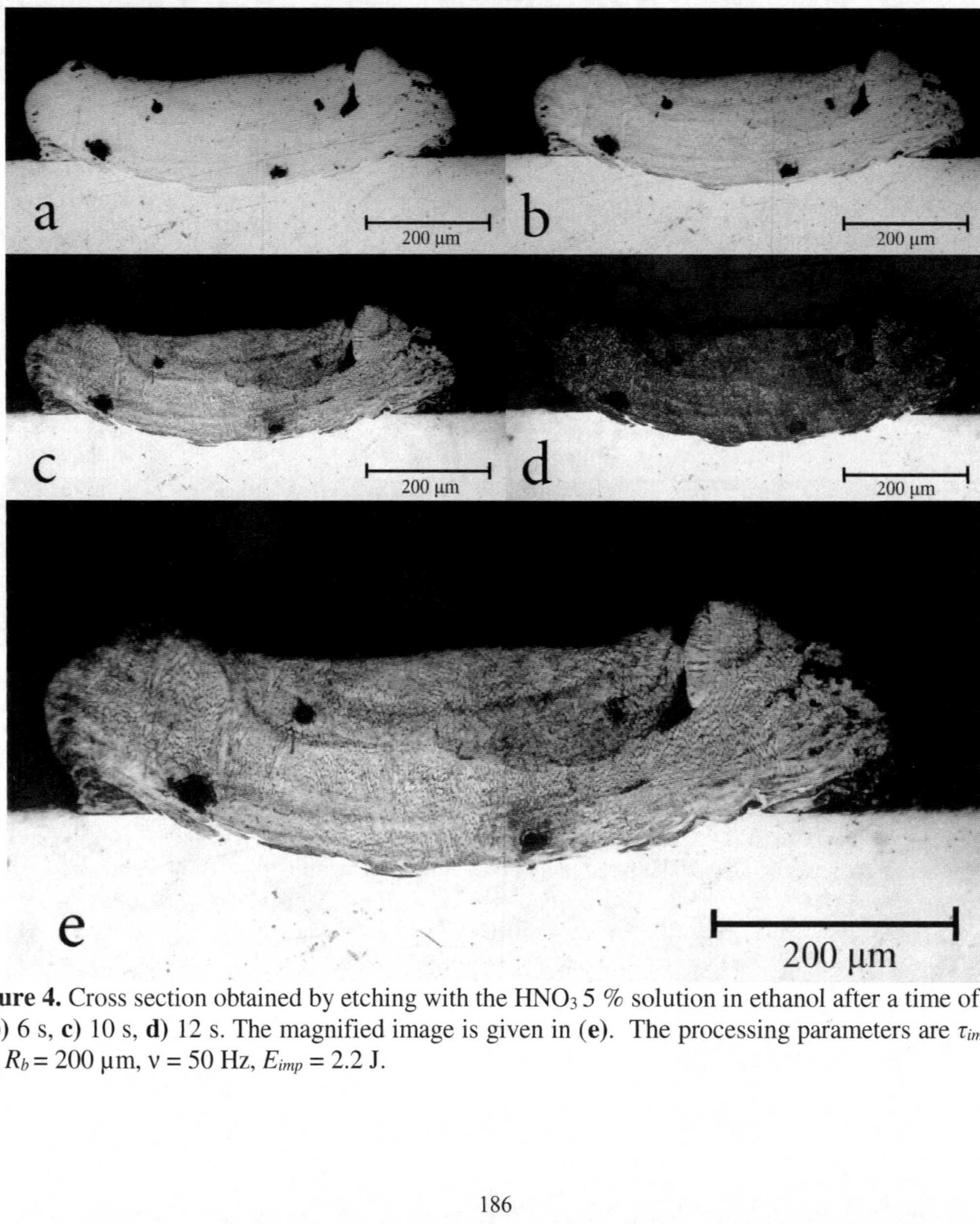

Figure 4. Cross section obtained by etching with the HNO_3 5 % solution in ethanol after a time of **a)** 4 s; **b)** 6 s, **c)** 10 s, **d)** 12 s. The magnified image is given in (**e**). The processing parameters are τ_{imp} = 4 ms, R_b = 200 μm, ν = 50 Hz, E_{imp} = 2.2 J.

IOP Conf. Series: Materials Science and Engineering **192** (2017) 012023 doi:10.1088/1757-899X/192/1/012023

The results of etching with different times are analyzed in Fig. 4. In Fig. 4(e), a pattern typical for metallographic analysis is observed. At this stage, its interpretation is ambiguous. The first explanation is that this pattern in fact represents the traces of small pores which deepen during etching and lead to large roughness of the surface. That areas do not reflect light and are seen as black in optical microscopy. On the other hand, this pattern can be a result of chemical inhomogeneity due to convective mixing of the molten Fe powder and the substrate. The second argument is supported by the fact that the patterns are observed only in the regimes (c), (d) and (e) in Fig. 3 where melting of the substrate occurs. At high convective stirring where the track is enriched by material from the substrate the pattern was not found, Fig. 3(f). At small and large etching times the pattern was not registered, Fig. 4(a) and Fig. 4(d), correspondingly.

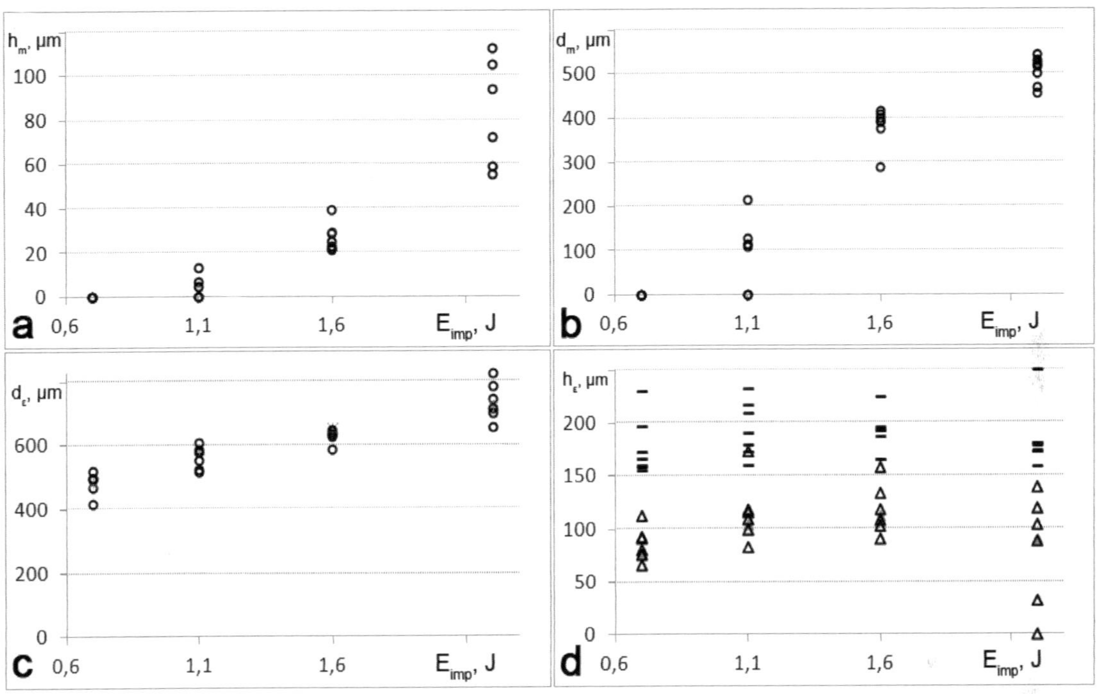

Figure 5. Dependence of the characteristic distances of the track as a function of the pulse energy E_{imp} in experiments (symbols «°», «-», «Δ» in the plot) in different cross sections of the same track. The processing parameters were R_b= 200 μm, τ_{imp} = 4 ms, ν = 50 Hz, $l_{tr}=R_b$. The notation of distances h_m , d_m, h_ε and d_ε is given in Figure 2. Here «-» and «Δ» are the maximum and minimum values of h_ε, correspondingly.

3.2. Effect of the pulse energy

The results of examination of the samples are shown in Fig. 5. Increase of E_{imp} results in nonlinear growth of the track depth h_m and quasi linear growth of the track width d_m, Fig. 5(a,b). At the smallest pulse energy $E_{imp} \leq 2.1$ J, weak adhesion to the substrate is combined with a wide molten track. The average track height first slightly grows and then starts to decrease if E_{imp} increases, Fig. 5(d). This observation is explained by combination of two opposite processes. On one side, the higher volume of molten powder gets a larger convex shape due to capillary forces. On other side, the local porosity ε decreases with higher E_{imp} hence the relative density grows. Additionally, at high E_{imp} > 2.1 J, intensive evaporation of material occurs. Powder burning was registered both experimentally that was predicted in our numerical simulations [8].

Thus, a conclusion was drawn that the increase of pulse energy E_{imp} yields better adhesion of the track with the substrate and hence good efficiency of SLM processing. However excessive E_{imp} leads to powder burning. At R_b= 200 μm, τ_{imp} = 4 ms, ν = 50 Hz, $l_{tr}=R_b$ the optimal regime occurs at processing with E_{imp} between 1.6 and 2.1 J. In other processing regimes at R_b= 300 μm, ν = 50 Hz, $l_{tr}=R_b$, the beginning of sintering and good adhesion was registered at $E_{imp} \leq 1$ J and 1.3 J $\leq E_{imp} \leq 2$ J. The recommended pulse duration τ_{imp} is between 0.8 and 4 ms. At R_b = 400 μm, 30 Hz ≤ ν ≤50 Hz, $l_{tr}=R_b$, powder sintering starts at $E_{imp} \leq 2$ J and good adhesion is provided at 3 J $\leq E_{imp} \leq 3.6$ J with τ_{imp} between 0.8 and 4 ms. The general tendency is as follows. Increase of τ_{imp} requires higher E_{imp} for initiation of adhesion since heat is effectively removed from the molten zone to the substrate.

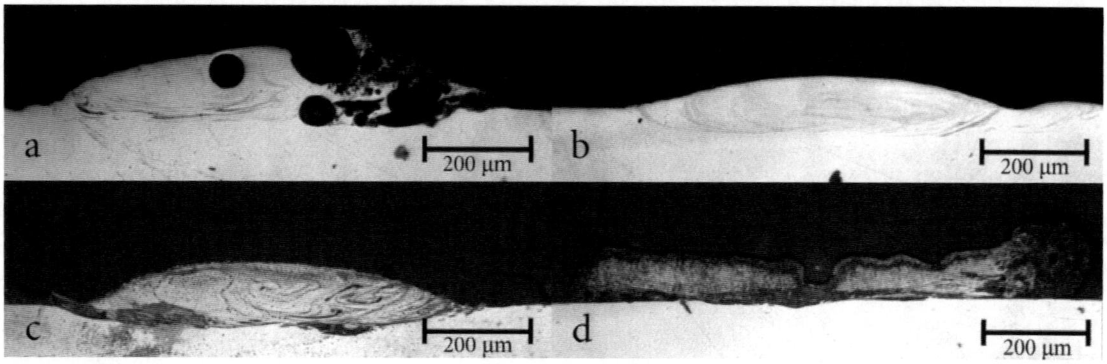

Figure 6. Macrostructure of samples processed by SLM with the pulse energy E_{imp} between 3.4 and 3.5 J, R_b = 400 μm, ν = 50 Hz. The pulse durations τ_{imp} are **a)** 0.8, **b)** 1, **c)** 1.5, and **d)** 4 ms.

3.3. Effect of the pulse duration

The pulse duration τ_{imp} is also an important governing parameter like the pulse energy E_{imp}. Equation (1) shows that the pulse power density can be increased through both E_{imp} and τ_{imp}. The phenomenological model suggested in literature [9,10,11] assumes Arrhenius-type kinetics of powder consolidation. Hence, strictly speaking, powder sintering starts before melting. Around the melting temperature, intensity of consolidation exponentially increases as a function of temperature. High temperature in the annealing zone after pulse heating exists during a time period which covers the pulse itself and a short time immediately after the pulse. Thus it is necessary to account for both opposite effects. Sufficiently large time τ_{imp} is desired to complete consolidation. At the same time, small time τ_{imp} leads to increase of the power density and processing temperature in the molten zone.

Figure 6 demonstrates the described effect as interplay of two factors. At small τ_{imp} the power density is high however powder has not completed its consolidation. Large pores are seen in Fig. 6 (a) which corresponds to the results of numerical simulation [10]. With the increase of τ_{imp} in Fig. 6 (b,c), the power density reduces which results in a higher temperature and almost complete consolidation with low residual porosity. Further increase of τ_{imp} provides good sintering of the powder but there is no sufficient adhesion to the substrate required in SLM, Fig. 6(d). In [3,12] it was shown that pulse laser annealing allows us to avoid the balling effect. In our study, the balling effect sometimes occurred if the pulse duration was large, Fig. 7(a). Once the pulse duration was reduced, the absence of the balling effect was observed, Fig. 7(b).

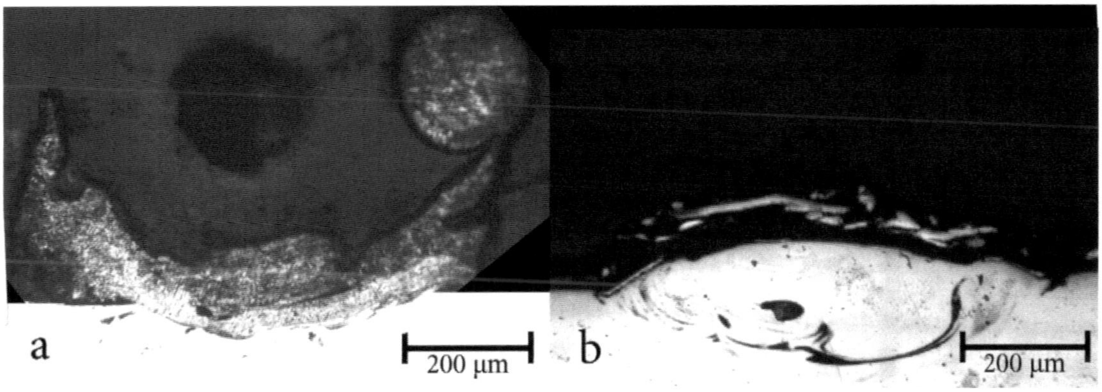

Figure 7. Cross sections of the single tracks obtained after processing with $R_b = 200$ μm, ν = 50 Hz. **a)** $\tau_{imp} = 4$ ms, $E_{imp} = 2.2$ J, the balling effect is registered. **b)** $\tau_{imp} = 2$ ms, $E_{imp} = 1.8$ J, no balling effect.

To summarize this discussion on selection of optimal processing regimes, if the parameter E_{imp} is fixed then it is suggested to use a lower pulse duration τ_{imp}. It helps to increase the laser energy power density. However the value τ_{imp} has a natural lower limit below which powder burning occurs or residual porosity becomes too high. Sometimes it is possible to fulfill both conditions for the high energy density and the duration when the molten zone exists. Although such regime has a risk of balling.

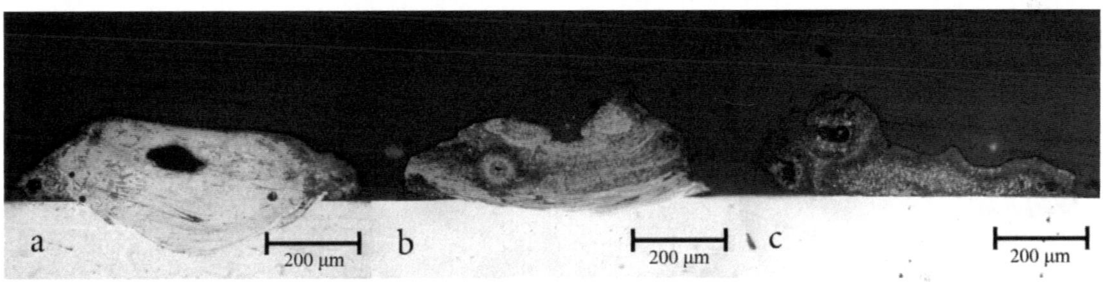

Figure 8. Cross sections of the single tracks obtained after processing with $\tau_{imp} = 4$ ms, ν = 50 Hz: **a)** $R_b = 200$ μm, $E_{imp} = 2.2$ J; **b)** $R_b = 300$ μm, $E_{imp} = 2.5$ J; **c)** $R_b = 400$ μm, $E_{imp} = 2.8$ J.

3.4. Effect of the beam radius

The beam radius R_b is another governing parameter of selective laser melting. It influences an area where heat is pumped into the powder. Therefore the width d_ε of the sintered track directly depends on R_b, Fig. 8. Decrease of R_b at the fixed E_{imp} facilitate increase of the peak and average power density inside the laser beam and hence improve efficiency of SLM. The discussed tendency was confirmed in experiments, Fig. 8. Deep melting of the substrate is given in Fig. 8(a) at $R_b = 200$ μm. Increase to $R_b = 300$ μm, Fig. 8(b) results in reduction of the distance h_m. At $R_b = 400$ μm, Fig. 8(c) no adhesion is registered and the track has high residual porosity and irregular shape.

In some cases, processing with small R_b and high pulse duration τ_{imp} leads to a controversial morphology. On the periphery of the track porous pattern was etched while closer to the middle of the molten zone good adhesion was found like in Figs. 3 (c,d), 7 (a), and 8 (a). Our explanation is that the effect is due to a significant difference in processing conditions between the edge and center of the laser beam. For efficient and uniform processing of metal powder with a wide laser beam, the applied power should be increased nonlinearly according to Eq. (1). As was discussed before, such scenario may results to overheating of the surface, shape deformation and long relaxation of the thermal field between impulses.

Thus, the beam radius has to be optimized according to processing conditions and specification of the SLM machine. For higher speed of processing, the beam radius should be increased. In SLM of thin-wall and high precision parts it is recommended to decrease R_b to fulfill the imposed technological requirements. At the same time, the pulse duration τ_{imp} is limited from above in order to prevent sintering at the edge of the track where local porosity is high and adhesion with the substrate is poor.

4. Conclusions

Analysis of the impact of governing SLM parameters on morphology of single tracks is performed. The optimal strategy in selection of the beam radius, pulse energy, and pulse duration is developed through the combined experimental and numerical study of Fe samples processed by a pulse laser.

The performed examination of samples shows that the resulting morphology is closely linked to the processing regimes. All images of cross sections of single tracks have been classified depending on the parameters of laser annealing. The high pulse energy combined with a larger beam radius and a short pulse duration leads to uniform convective mixing in the molten zone with a low residual porosity with good adhesion to the substrate.

In processing with the small pulse energy and large pulse duration, powder consolidation is weak and morphology of the track is irregular. The gradual increase of pulse energy and beam radius were combined with the reduced pulse duration facilitates capillary flow and the convex morphology. Increase of the pulse energy alone extends adhesion to the substrate and is considered as useful. However this parameter has a limit from above beyond which burning of material occurs. For a given pulse energy it is suggested to reduce the pulse duration as far as it possible before burning starts. A key indicator that the pulse duration is too long is the balling effect.

The beam radius has been selected depending on a particular problem and specification of a SLM machine. A large beam radius is better in terms of residual porosity. At the same time, manufacturing of thin wall parts requires a small beam radius and a short pulse duration.

5. References

[1] Gibson I, Rosen D and Stucker B 2015 *Additive Manufacturing Technologies* (New York: Springer Science+Business Media) 656
[2] Olakanmi E O, Cochrane R F and Dalgarno K W 2015 *Progress in Materials Science* **74** 401–77
[3] Demir A G, Colombol P and Previtali B 2017 *Int.* J. *Adv. Manuf. Technol.* (*in print*)
[4] Regenfuss P, Streek A, Hartwig L, Klötzer S, Th. Brabant, Horn M, Ebert R and Exner H 2007 *Rapid Prototyp.* J. **13** (4) 204–12
[5] King W E, Anderson A T, Ferencz R M et al. 2015 *Applied Physics Reviews* **2** (4) 041304
[6] Yap C Y, Chua C K, Dong Z L et al. 2015 *Applied Physics Reviews* **2**(4) 041101
[7] Gordeev G A, Ankudinov V E, Krivilyov M D and Kharanzhevskiy E V 2012 *IOP Conf. Series:Materials Science and Engineering* **27** 012079
[8] Gordeev G A, Krivilyov M D, Ankudinov V E, Kharanzhevskiy E V and Bogdanov A A 2016 *Proc. of the Second International Conference* «Additive technologies: today and tomorrow» (Moscow: VIAM) 15
[9] Krivilyov M D, Ankudinov V E, Gordeev G A, Reshetnikov S M, Kharanzhevskiy E V and Galenko P K 2016 *Proc. of the Second International Conference* «Additive technologies: today and tomorrow» (Moscow: VIAM) 15
[10] Krivilyov M D, Mesarovic S Dj and Sekulic D P 2017 J*ournal of Materials Science* 4155-63
[11] Dong L, Makradi A, Ahzi S et al. 2009 J*ournal of materials processing technology* (209) 700–6
[12] Kruth J P, Froyen L, J Van Vaerenbergh, Mercelis P, Romboutsb M and Lauwers B 2004 J*ournal of Materials Processing Technology* **149** 616–22

Acknowledgments
The authors express their acknowledgement to A. Bogdanov for processing of the samples, A. Sklyamina and A. Mosin for metallographic analysis, N. Orlova and V. Ankudinov for useful discussions. This work was supported financially by the Russian Foundation for Basic Research project No. 16-38-00839 mol_a and the Ministry of Education and Science of Russia (project No. 11.6513.2017/BP). The research was financially supported by the UMNIK program of Innovation Support Foundation.

IOP Conf. Series: Materials Science and Engineering **192** (2017) 012024 doi:10.1088/1757-899X/192/1/012024

Stochastic sensitivity analysis: theory and numerical algorithms

I Bashkirtseva

Ural Federal University, Lenina, 51, Ekaterinburg, 620000, Russia

E-mail: irina.bashkirtseva@urfu.ru

Abstract. A problem of the analysis of nonlinear dynamic systems forced by the random disturbances is considered. For the description of probabilistic distributions around the deterministic attractors, a theory of the quasipotential and its approximations based on the stochastic sensitivity functions (SSF) technique is suggested. Constructive methods for the computation of SSF for the equilibria and limit cycles are presented and discussed. We show how SSF can be used for the visual description of probabilistic distributions in the form of confidence domains.

1. Introduction

Currently, the stochastic dynamic systems are widely used for the modeling of processes in various fields of science [1]. In nonlinear systems, inevitably presented random disturbances can cause some unexpected and complex probabilistic phenomena [2, 3]. A direct numerical simulation is still a main tool of the investigation of these phenomena. However, in parametric studies, this method is extremely time-consuming. So, the analytical methods of the probabilistic analysis attract attention of many researchers.

In this paper, for the analysis of nonlinear stochastic systems, we set forth a new constructive approach based on the stochastic sensitivity functions and confidence domains.

Consider a general stochastic system of nonlinear differential equations

$$dx = f(x)dt + \varepsilon\sigma(x)dw, \tag{1}$$

where x is an n-dimensional vector, $f(x)$ is an n-vector function, $\sigma(x)$ is an $n \times m$-matrix function, $w(t)$ is an m-dimensional Wiener process, and ε is a scalar parameter of the noise intensity.

It is supposed that the corresponding deterministic system (1) with $\varepsilon = 0$ has an exponentially stable attractor A. It means that for the small neighbourhood D of the attractor A, there exist constants $K > 0$, $l > 0$ such that for any solution $x(t)$ of the deterministic system with $x(0) = x_0 \in D$ the following inequality holds

$$||\Delta(x(t))|| \leq Ke^{-lt}||\Delta(x_0)||.$$

Here, $\Delta(x) = x - \gamma(x)$ is a vector of a deviation of the point x from the attractor A, $\gamma(x)$ is a point of the attractor A that is nearest to x.

IOP Conf. Series: Materials Science and Engineering **192** (2017) 012024 doi:10.1088/1757-899X/192/1/012024

As a result of stochastic forcing, the random trajectories of system (1) form some flow with the probability density function $\rho(t,x,\varepsilon)$. This function is governed by the Kolmogorov-Fokker-Planck equation [4]

$$\frac{\partial\rho}{\partial t} = L\rho, \quad L\rho = \frac{\varepsilon^2}{2}\sum_{i,j=1}^{n}\frac{\partial^2}{\partial x_i \partial x_j}(a_{ij}\rho) - \sum_{i=1}^{n}\frac{\partial}{\partial x_i}(f_i\rho), \quad a_{ij} = [\sigma\sigma^\top]_{ij}.$$

If a character of the transient is unessential and the main interest is focused on the stable stationary regime, then it is possible to restrict a stochastic analysis by the study of the stationary density function $\rho(x,\varepsilon)$. This function is a solution of the stationary Kolmogorov-Fokker-Planck equation

$$L\rho = 0.$$

The analysis of this equation is a very difficult problem technically even for the 2-dimensional case. So, the asymptotics based on the quasipotential

$$v(x) = -\lim_{\varepsilon\to 0}\varepsilon^2 \ln \rho(x,\varepsilon)$$

are commonly used [5].

The quasipotential is connected with some variational problem of the minimization of the action potential and governed by the Hamilton-Jacobi equation

$$\left(f(x), \frac{\partial v}{\partial x}\right) + \frac{1}{2}\left(\frac{\partial v}{\partial x}, \sigma(x)\sigma^\top(x)\frac{\partial v}{\partial x}\right) = 0$$

with conditions $v|_A = 0$, $v|_{D\setminus A} > 0$.

The solution of this equation is still difficult problem, therefore in a small neighbourhood of the attractor A, a first quadratic approximation of the function $v(x)$ is used [6]

$$v(x) \approx \frac{1}{2}(\Delta(x), \Phi(\gamma(x))\Delta(x)), \; \Phi(x) = \frac{\partial^2 v}{\partial x^2}(x).$$

In what follows, we will present constructive methods of such approximations for equilibria and cycles.

2. Stochastic sensitivity of equilibria

Let the attractor of the deterministic system be an equilibrium $\bar{x}$. In this case, the quadratic approximation $v(x) \approx \frac{1}{2}(x-\bar{x}, \Phi(\bar{x})x - \bar{x})$ of the quasipotential gives the following Gaussian approximation of $\rho(x,\varepsilon)$ in a neighbourhood of the equilibrium $\bar{x}$

$$\rho(x,\varepsilon) \approx N\exp\left(-\frac{(x-\bar{x}, W^{-1}(x-\bar{x}))}{2\varepsilon^2}\right), \; W = \Phi^{-1}(\bar{x})$$

with the mean $\bar{x}$, the covariance matrix $C = \varepsilon^2 W$, and the normalization constant, N. Here, the matrix W is a solution of the equation

$$FW + WF^\top + S = 0, \tag{2}$$

where $F = \frac{\partial f}{\partial x}(\bar{x})$ is the Jacobi matrix of the deterministic system at $\bar{x}$ and $S = \sigma(\bar{x})\sigma^\top(\bar{x})$. For the exponentially stable equilibrium $\bar{x}$, the eigenvalues of the Jacobi matrix F have the negative real parts, and the matrix equation 2 has a unique solution being the stochastic sensitivity matrix of the equilibrium [7].

For the low-dimensional cases, it is convenient to find the solution of system (2) by direct numerical methods. In the case of systems with large dimensions, the iterative methods are more suitable. As an example, consider the iterative method associated with the procedure of the stabilization of solutions for the corresponding system of differential equations.

The matrix W (required stochastic sensitivity function of the equilibrium $\bar{x}$) as a solution of equation (2) can be obtained by the following limit procedure.

Let matrix function $V(t)$ be an arbitrary solution of the differential equation

$$\dot{V} = FV + VF^{\top} + S$$

on the interval $[0, +\infty)$. If the equilibrium is exponentially stable then the matrix $V(t)$ tends to the matrix W as $t \to +\infty$

$$\lim_{t \to +\infty} V(t) = W.$$

Note that the convergence rate depends on degree of the equilibrium stability.

To visualize a spatial probabilistic distribution of random states around the equilibrium $\bar{x}$, one can construct a confidence ellipsoid:

$$\Big(x - \bar{x}, W^{-1}(x - \bar{x})\Big) = \varepsilon^2 K(P),$$

where P is a fiducial probability. The function $K(P)$ is an inverse function to $P(K)$:

$$P(K) = \frac{\Phi_n(K)}{\Phi_n(\infty)}, \qquad \Phi_n(K) = \int_0^{\sqrt{K}} e^{-\frac{t^2}{2}} t^{n-1} dt.$$

In the one-dimensional case ($n = 1$),

$$P(K) = \sqrt{\frac{2}{\pi}} \int_0^{\sqrt{K}} e^{-\frac{t^2}{2}} dt = \mathrm{erf}\left(\sqrt{\frac{K}{2}}\right), \quad \mathrm{erf}(x) = \frac{2}{\sqrt{\pi}} \int_0^{x} e^{-t^2} dt,$$

and the corresponding confidence interval $(\bar{x} - r, \bar{x} + r)$ is defined by $r = \varepsilon\sqrt{2\mu}\,\mathrm{erf}^{-1}(P)$, where the stochastic sensitivity μ can be found from the explicit formula:

$$\mu = -\frac{\sigma^2(\bar{x})}{2f'(\bar{x})}.$$

For the 3σ-rule, it holds that $r = 3\varepsilon\sqrt{\mu}$.

In the two-dimensional case ($n = 2$),

$$P(K) = 1 - e^{-\frac{K}{2}}, \qquad K(P) = -2\ln(1 - P).$$

The equation of the confidence ellipse can be written as

$$\frac{\beta_1^2}{\mu_1} + \frac{\beta_2^2}{\mu_2} = \varepsilon^2 K(P), \qquad \beta_1 = (x - \bar{x}, v_1), \;\; \beta_2 = (x - \bar{x}, v_2). \tag{3}$$

Here, μ_1, μ_2 are the eigenvalues, and v_1, v_2 are the normalized eigenvectors of the stochastic sensitivity matrix W. These vectors define the directions of the confidence ellipse axis, and μ_1, μ_2 define the values of corresponding semi-axis.

In the three-dimensional case ($n = 3$),

$$P(K) = \sqrt{\frac{2}{\pi}} \left[\int_0^{\sqrt{K}} e^{-\frac{t^2}{2}} \, dt - \sqrt{K} e^{-\frac{K}{2}} \right] = \text{erf}\left(\sqrt{\frac{K}{2}} \right) - \sqrt{\frac{2K}{\pi}} \text{e}^{-\frac{K}{2}}.$$

The confidence ellipsoid can be written in the following form

$$\frac{\beta_1^2}{\mu_1} + \frac{\beta_2^2}{\mu_2} + \frac{\beta_3^2}{\mu_3} = \varepsilon^2 K(P), \tag{4}$$

where $\beta_1 = (x - \bar{x}, v_1)$, $\beta_2 = (x - \bar{x}, v_2)$, $\beta_3 = (x - \bar{x}, v_3)$. Here, μ_1, μ_2, μ_3 are the eigenvalues, and v_1, v_2, v_3 are the normalized eigenvectors of the stochastic sensitivity matrix W.

For $n = 4$, we have $P(K) = 1 - e^{-\frac{K}{2}} (1 + 0.5K)$.

3. Stochastic sensitivity of limit cycles

Now, let the attractor of the deterministic system be an exponentially stable limit cycle defined by a T-periodic solution $\bar{x}(t)$, $\bar{x}(t+T) = \bar{x}(t)$. Let Π_t be a hyperplane orthogonal to the cycle at the point $\bar{x}(t)$. For the Poincare section Π_t in the neighbourhood of the point $\bar{x}(t)$, the quadratic approximation

$$v(x)|_{\Pi_t} \approx \frac{1}{2} (x - \bar{x}(t), \Phi(\bar{x}(t))(x - \bar{x}(t)))$$

of the quasipotential gives the approximation $\rho_t(x, \varepsilon)$ of the probability density $\rho(x, \varepsilon)|_{\Pi_t}$ in the Gaussian form

$$\rho_t(x, \varepsilon) = N \exp\left(-\frac{(x - \bar{x}(t), W^+(t)(x - \bar{x}(t)))}{2\varepsilon^2} \right), \quad W(t) = \Phi^+(\bar{x}(t))$$

with the mean $\bar{x}(t)$ and covariance matrix $C(t) = \varepsilon^2 W(t)$. Here, the sign "+" means the pseudoinversion.

For the exponentially stable limit cycle, the matrix function $W(t)$ is a unique solution of the boundary problem [8]

$$\dot{W} = F(t)W + WF^\top(t) + P(t)S(t)P(t), \tag{5}$$

$$W(0) = W(T), \tag{6}$$

$$W(t)r(t) \equiv 0. \tag{7}$$

Here,

$$F(t) = \frac{\partial f}{\partial x}(\bar{x}(t)), \quad S(t) = \sigma(\bar{x}(t))\sigma^\top(\bar{x}(t)), \quad r(t) = f(\bar{x}(t)), \quad P(t) = P_{r(t))},$$

and $P_r = I - rr^\top / r^\top r$ is a projection matrix onto the hyperplane that is orthogonal to the vector r.

3.1. How to calculate the stochastic sensitivity of two-dimensional cycle

For the 2D-case, the stochastic sensitivity matrix $W(t)$ can be written as $W(t) = \mu(t)P(t)$. The scalar function $\mu(t) > 0$ is a T-periodic stochastic sensitivity function satisfying the following boundary problem [7]

$$\dot{\mu} = a(t)\mu + b(t), \quad \mu(0) = \mu(T) \tag{8}$$

with the T-periodic coefficients

$$a(t) = p^\top(t)(F^\top(t) + F(t))p(t) \text{ , } b(t) = p^\top(t)S(t)p(t), \quad P(t) = p(t)p^\top(t),$$

where $p(t)$ is a normalized vector orthogonal to $f(\bar{x}(t))$. An explicit formula for the solution $\mu(t)$ of problem (8) is given by

$$\mu(t) = g(t)(c + h(t)),$$

where

$$g(t) = \exp\left(\int_0^t a(s)ds\right), \quad h(t) = \int_0^t \frac{b(s)}{g(s)}ds, \qquad c = \frac{g(T)h(T)}{1 - g(T)}.$$

The value $M = \max \mu(t),\ t \in [0, T]$ plays an important role in the analysis of stochastic dynamics near the limit cycle. We shall consider M as a *stochastic sensitivity factor* of the cycle.

Note that the function $\mu(t)$ allows us to construct a confidence band around the deterministic cycle. For the line Π_t that is orthogonal to the cycle at the point $\bar{x}(t)$, the boundaries $x_{1,2}(t)$ of the confidence band can be written in an explicit parametric form:

$$x_{1,2}(t) = \bar{x}(t) \pm \varepsilon\sqrt{2\mu(t)}\, \mathrm{erf}^{-1}(P)p(t). \tag{9}$$

The confidence bands are sufficiently simple and evident models for the spatial description of random states around the deterministic cycle.

3.2. How to calculate the stochastic sensitivity of three-dimensional cycle

In the three-dimensional case ($n = 3$), a singular decomposition of the symmetric non-negative defined 3×3-matrix $V(t)$ looks like

$$V(t) = \lambda_1(t)v_1(t)v_1^\top(t) + \lambda_2(t)v_2(t)v_2^\top(t) + \lambda_3(t)v_3(t)v_3^\top(t),$$

where $\lambda_1(t) \geq \lambda_2(t) \geq \lambda_3(t)$ are the eigenvalues, and $v_1(t), v_2(t), v_3(t)$ are the normalized eigenvectors of matrix $V(t)$.

From condition (7) it follows that for any t, the matrix $V(t)$ is degenerated (a distribution of intersection points is concentrated in a hyperplane Π_t). It means that $\lambda_3(t) \equiv 0$, and the appropriate eigenvector $v_3(t) = r(t)/\|r(t)\|$ is tangent to the cycle. It allows us to write the decomposition of the matrix $V(t)$ as follows

$$V(t) = \lambda_1(t)v_1(t)v_1^\top(t) + \lambda_2(t)v_2(t)v_2^\top(t). \tag{10}$$

Here $V(t)$ is determined by the scalar functions $\lambda_1(t), \lambda_2(t)$ and vectors $v_1(t),\ v_2(t)$. In the case of non-degenerate noises, the functions $\lambda_1(t), \lambda_2(t)$ are strictly positive and determine for any t a dispersion of random trajectories around cycle along vectors $v_1(t), v_2(t)$. The values $\lambda_1(t),\ \lambda_2(t)$ determine the size and $v_1(t), v_2(t)$ determine the directions of the dispersion ellipse axes. The equation of this ellipse in plane Π_t looks like

$$(x - \bar{x}(t))^\top W^+(t)(x - \bar{x}(t)) \;=\; -2\varepsilon^2 \ln(1 - P),$$

where P is a fiducial probability. A set of these confidence ellipses along the deterministic limit 3D-cycle forms a confidence torus (see details in [9]).

Denote by $u_1(t),\ u_2(t)$ some orthonormal basis of the plane Π_t. One can easily find this basis if the T-periodic solution $\bar{x}(t)$ is known (see Remark). The eigenvectors $v_1(t),\ v_2(t)$, can be represented by rotation of basis $u_1(t),\ u_2(t)$ with some angle $\varphi(t)$

$$\begin{aligned} v_1(t) &= u_1(t)\cos\varphi(t) + u_2(t)\sin\varphi(t), \\ v_2(t) &= -u_1(t)\sin\varphi(t) + u_2(t)\cos\varphi(t). \end{aligned} \tag{11}$$

Thus, the decomposition (10), (11) allows us to express an unknown solution of the system (5)-(7) by means of three scalar functions $\lambda_1(t)$, $\lambda_2(t)$, $\varphi(t)$.

Denote

$$P_1(t) = v_1(t) \cdot v_1^\top(t), \quad P_2(t) = v_2(t) \cdot v_2^\top(t).$$

Remark that $P_i(t)$ $(i = 1, 2)$ are the projective matrices

$$P_i v_i = v_i, \quad P_i v_j = 0 \ (i \neq j). \tag{12}$$

Rewrite the decomposition (10) as follows

$$V(t) = \lambda_1(t) \cdot P_1(t) + \lambda_2(t) \cdot P_2(t).$$

Theorem 1. [8] *The matrix $V(t)$ is a solution of system (5),(6) if and only if the scalar functions $\lambda_1(t)$, $\lambda_2(t)$, $\varphi(t)$ of decomposition (10), (11) satisfy the following system*

$$\begin{aligned} \dot{\lambda}_1 &= \lambda_1 v_1^\top [F + F^\top] v_1 + v_1^\top S v_1 \\ \dot{\lambda}_2 &= \lambda_2 v_2^\top [F + F^\top] v_2 + v_2^\top S v_2 \\ (\lambda_1 - \lambda_2)\dot{\varphi} &= \lambda_2 v_1^\top F v_2 + \lambda_1 v_1^\top F^\top v_2 + v_1^\top S v_2 - (\lambda_1 - \lambda_2)\dot{u}_1^\top u_2. \end{aligned} \tag{13}$$

As we see, a construction of the solution $V(t)$ of system (5),(6) on the basis of a singular decomposition is reduced to the solution of system (13) for three scalar functions. The matrix $W(t)$ (required stochastic sensitivity function of a cycle) of the solution of system (5)-(7) can be obtained by the following limit procedure.

Theorem 2. [8] *Let the T-periodic matrix $W(t)$ be a solution of system (5)-(7). Let $\lambda_1(t)$, $\lambda_2(t)$, $\varphi(t)$ be an arbitrary solution of system (13) on the interval $[0, +\infty)$. Put $V(t) = \lambda_1(t) \cdot P_1(t) + \lambda_2(t) \cdot P_2(t)$, where $P_i(t) = v_i(t)v_i^\top(t)$ with the vector functions $v_i(t)$ obtained from (11). Then the matrix $V(t)$ tends to the matrix $W(t)$ as $t \to +\infty$*

$$\lim_{t \to +\infty} (V(t) - W(t)) = 0.$$

Remark. For the computation of system (13) right parts, it is necessary to know the vector functions $u_1(t), u_2(t)$ and $\dot{u}_1(t)$. To find these, the following method is suggested.

Rewrite the initial 3D-system $\dot{x} = f(x)$ and its T-periodic solution $x = \bar{x}(t)$ by coordinates

$$\dot{x}_1 = f_1(x_1, x_2, x_3), \ \dot{x}_2 = f_2(x_1, x_2, x_3), \ \dot{x}_3 = f_3(x_1, x_2, x_3),$$

$$\bar{x}(t) = (\bar{x}_1(t), \bar{x}_2(t), \bar{x}_3(t))^\top.$$

A tangent vector $r(t) = (f_1(t), f_2(t), f_3(t))^\top$ has the coordinates $f_i(t) = f_i(\bar{x}(t))$, $i = 1, 2, 3$. One can choose the vectors $u_1(t), u_2(t)$ of the orthonormal basis for the plane Π_t in the following form

$$u_1 = g_1 \cdot \begin{pmatrix} -f_2 \\ f_1 \\ 0 \end{pmatrix}, \quad u_2 = g_2 \cdot \begin{pmatrix} -f_1 f_3 \\ -f_2 f_3 \\ f_1^2 + f_2^2 \end{pmatrix},$$

where

$$g_1 = \left(f_1^2 + f_2^2\right)^{-\frac{1}{2}}, \quad g_2 = \left(f_3^2\left(f_1^2 + f_2^2\right) + \left(f_1^2 + f_2^2\right)^2\right)^{-\frac{1}{2}}.$$

As a result of such choice the following formula holds

$$\dot{u}_1 = \dot{g}_1 \cdot \begin{pmatrix} -f_2 \\ f_1 \\ 0 \end{pmatrix} + g_1 \cdot \begin{pmatrix} -\dfrac{\partial f_2}{\partial x_1} & -\dfrac{\partial f_2}{\partial x_2} & -\dfrac{\partial f_2}{\partial x_3} \\ \dfrac{\partial f_1}{\partial x_1} & \dfrac{\partial f_1}{\partial x_2} & \dfrac{\partial f_1}{\partial x_3} \\ 0 & 0 & 0 \end{pmatrix} \cdot \begin{pmatrix} f_1 \\ f_2 \\ f_3 \end{pmatrix},$$

where

$$\dot{g}_1 = -\left(f_1^2 + f_2^2\right)^{-\frac{3}{2}} \cdot \left(f_1 \cdot \left(\frac{\partial f_1}{\partial x_1} f_1 + \frac{\partial f_1}{\partial x_2} f_2 + \frac{\partial f_1}{\partial x_3} f_3\right) + f_2 \cdot \left(\frac{\partial f_2}{\partial x_1} f_1 + \frac{\partial f_2}{\partial x_2} f_2 + \frac{\partial f_2}{\partial x_3} f_3\right)\right).$$

4. A development of the theory. Applications

In continuous-time dynamic systems, along with equilibria and limit cycles, the invariant tori are an important type of attractors. An extension of SSF technique for the case of tori is given in [10].

For the discrete-time systems, SSF technique was elaborated for such regular attractors as equilibria, cycles [11] and closed invariant curves [12]. For the description of probabilistic distributions near chaotic attractors, SSF approach is presented in [13].

Some constructive abilities of the SSF technique and confidence domains method were demonstrated in applications to the wide range of noise-induced phenomena: noise-induced transitions between attractors [7, 14]; stochastic deformations of oscillatory regimes in period-doubling bifurcations [15, 16]; backward stochastic bifurcations [17]; noise-induced chaos-order transitions [18].

Our approach clarified probabilistic mechanisms of these stochastic phenomena in the population systems [7, 19], neuron activity [20, 21, 22], and climate dynamics [14, 23].

The SSF technique was successfully used to solve the engineering problems of stabilization of stochastically forced operating modes and controlling of chaos [24, 25].

5. References

[1] Moss F and McClintock P V E 1989 *Noise in Nonlinear Dynamical Systems* (Cambridge: Cambridge University Press)
[2] Horsthemke W and Lefever R 1984 *Noise-Induced Transitions* (Berlin: Springer)
[3] Anishchenko V S, Astakhov V V, Neiman A B, Vadivasova T E and Schimansky-Geier L 2007 *Nonlinear Dynamics of Chaotic and Stochastic Systems. Tutorial and Modern Development* (Berlin/Heidelberg: Springer-Verlag)
[4] Gardiner C 2009 *Stochastic Methods. A Handbook for the Natural and Social Sciences* (Berlin: Springer)
[5] Freidlin M I and Wentzell A D *Random Perturbations of Dynamical Systems* (New York: Springer)
[6] Milshtein G N and Ryashko L B 1995 *PMM J. Appl. Math. Mech.* **59** 47
[7] Bashkirtseva I and Ryashko L 2011 *Chaos* **21** 047514
[8] Bashkirtseva I and Ryashko L 2004 *Math. Comput. Simulat.* **66** 55
[9] Ryashko L, Bashkirtseva I, Gubkin A and Stikhin P 2009 *Math. Comput. Simulat.* **80** 256
[10] Bashkirtseva I and Ryashko L 2016 *Electron. J. Diff. Equat.* **2016** 1
[11] Bashkirtseva I, Ryashko L and Tsvetkov I 2010 *Dyn. Cont., Discr. Impul. Syst.: Ser. A: Math. Analysis* **17** 501
[12] Bashkirtseva I and Ryashko L 2014 *Physica A* **410** 236
[13] Bashkirtseva I and Ryashko L 2015 *Int. J. Bifurcat. Chaos* **25** 1550138
[14] Alexandrov D V, Bashkirtseva I A and Ryashko L B 2014 *Tellus A* **66** 23454
[15] Ryagin MY and Ryashko LB 2004 *Int. J. Bifurcat. Chaos* **14** 3981
[16] Bashkirtseva I, Chen G and Ryashko L 2010 *Int. J. Bifurcat. Chaos* **20** 1439
[17] Bashkirtseva I, Ryashko L and Stikhin P 2013 *Int. J. Bifurcat. Chaos* **23** 1350092
[18] Bashkirtseva I and Ryashko L 2017 *Physica A* **467** 573

[19] Ryashko L and Bashkirtseva I 2011 *Front. Life Sci.* **5** 141
[20] Ryashko L and Bashkirtseva I 2011 *Phys. Rev. E* **83** 061109
[21] Bashkirtseva I, Neiman A and Ryashko L 2015 *Phys. Rev. E* **91** 052920
[22] Bashkirtseva I, Ryashko L and Slepukhina E 2015 *Nonlin. Dynam.* **82** 919
[23] Alexandrov D V, Bashkirtseva I A and Ryashko L B 2015 *Eur. Phys. J. B* **88** 304
[24] Bashkirtseva I, Chen G and Ryashko L 2012 *Commun. Nonlin. Sci. Numer. Simulat.* **17** 3381
[25] Bashkirtseva I, Chen G and Ryashko L 2017 *J. Franklin Inst.* **354** 1658

Acknowledgments
The work was supported by Russian Science Foundation (N 16-11-10098).

IOP Conf. Series: Materials Science and Engineering **192** (2017) 012025 doi:10.1088/1757-899X/192/1/012025

Boundary integral approach for elliptical dendritic paraboloid as a form of growing crystals

E A Titova[1], D V Alexandrov[1] and P K Galenko[2]

[1] Department of Theoretical and Mathematical Physics, Laboratory of Multi-Scale Mathematical Modeling, Ural Federal University, Ekaterinburg 620000, Russian Federation
[2] Friedrich-Schiller-Universität-Jena, Physikalisch-Astronomische Fakultät, 07743 Jena, Germany

E-mail: ekatitova@mail.ru

Abstract. A method based on the boundary integral approach to the phase interface propagation is used to obtain the Horvay and Cahn solutions. An elliptical paraboloid as one of the crystal (dendritic) shapes growing under the steady state conditions is derived from the boundary integral.

1. Introduction

The dynamics of phase interfaces plays a very important role in different phase and structural transformations and leads to various types of pattern formation in materials [1–7]. The Stefan-like problems occurring in the presence of moving boundary problems [8–28] may be mentioned as an important class of such evolutionary tasks.

The boundary integral method represents one of the useful approaches to describe the time-dependent moving boundary problems. This technique was developed for the first time by Nash [29] and Nash and Glicksman [30] in their pioneering papers. This method leans upon a single integral equation for the interface function, which is derived on the basis of the Green's function technique. Langer and Turski [31,32] developed the boundary integral method to describe a chemical diffusion problem in two-phase solid/liquid, solid/solid and fluid/fluid systems. Recently, their theory was extended for the description of binary non-isothermal mixtures by Alexandrov and Galenko in [33]. Note that the integral equation for the interface function represents the basis for stability and selection problems [34–40].

It is well-known that in many cases the growing dendrites are non-symmetrical in their shapes and cannot be described by means of the Ivantsov solutions [39,41]. In this case, one can use the Horvay and Cahn solutions describing the non-symmetrical shape of a dendritic tip [42–44]. The present paper connects the Horvay and Cahn solutions with the boundary integral theory [33]. Namely, we discuss below how to derive the Horvay and Cahn solutions describing the elliptical paraboloid from the boundary integral method [33].

IOP Conf. Series: Materials Science and Engineering **192** (2017) 012025 doi:10.1088/1757-899X/192/1/012025

2. Equation of the interface motion and its analysis for elliptical paraboloid

Horvay and Cahn [42] obtained a form of an isothermal dendrite that has an elliptical cross section. The surface of the dendrite in dimensionless coordinates is given by

$$\frac{x^2}{\omega - b} + \frac{y^2}{\omega + b} = \omega - 2z, \tag{1}$$

where $2D_T/V = \rho/P_T$ is the length scale and $P_T = \rho v/2D_T$ is the growth Péclet number (V and ρ represent the dendrite tip velocity and diameter of its any cross section (or the average value of dendritic tip curvature radii), and D_T is the thermal diffusivity). The interface corresponds to $\omega = P_T$. Here b determines the aspect ratio of the elliptical cross-section of dendrite. For $b = 0$ the cross-section is circular; we will assume $|b| < P_T$. Eq. (1) is actually a symmetrized version of the corresponding expression given by Horvay and Cahn, with dimensionless radii of curvature $P_T - b$ in the $x-z$ plane and $P_T + b$ in the $y-z$ plane. Thus, the average dimensionless radius of curvature is P_T [44]. In dimensional form this equations transforms to

$$\frac{x_d^2}{\omega_d - b_d} + \frac{y_d^2}{\omega_d + b_d} = \omega_d - 2z_d, \tag{2}$$

where $\omega_d = \omega\rho/P_T$ and $b_d = b\rho/P_T$ and subscript d designates the dimensional variables and parameters. An arbitrary deformed solid-liquid interface then propagates in the z-direction and is defined by the function $z_{\text{interface}} = \zeta(x, t)$, where t stands for the time. The dimensionless undercooling as a function of P_T in the three-dimensional case can be written out as [33]

$$\Delta - \frac{d_c}{\rho}K - \beta V\left(1 + \frac{\partial\zeta(x,t)}{\partial t}\right) = I_\zeta^T, \tag{3}$$

where (in dimensionless variables)

$$\begin{aligned} I_\zeta^T &= P_T^{3/2}\int_0^\infty \frac{d\tau}{(2\pi\tau)^{3/2}} \int_{-\infty}^{\infty}\int_{-\infty}^{\infty} d^2x_1 \left[1 + \frac{\partial\zeta(x_1, t-\tau)}{\partial t}\right] \\ &\times \exp\left\{-\frac{P_T}{2\tau}\left[|x - x_1|^2 + (\zeta(x,t) - \zeta(x_1, t-\tau) + \tau)^2\right]\right\}. \end{aligned} \tag{4}$$

Here $\Delta = (T_M - T_\infty)c_p/Q$ is the dimensionless undercooling, T_M is the freezing temperature of the planar front, T_∞ is the liquid phase temperature far from the moving interface, Q is the latent heat released per unit volume of the solid phase, c_p is the specific heat β is the kinetic coefficient, variables x, x_1 and ζ are measured in units of ρ whereas variables t and τ are measured in units of ρ/V. In dimensional coordinates, when the elliptical paraboloid does not change its shape and moves with the constant velocity V (i.e. $\partial\zeta(x,t)/\partial t = 0$), Eq. (4) takes the following form

$$\begin{aligned} I_{\zeta d}^T &= \left(\frac{P_T}{2\pi\rho}\right)^{3/2}\frac{1}{\sqrt{V}} \\ &\times \int_0^\infty \frac{d\tau}{\tau^{3/2}} \int_{-\infty}^{\infty}\int_{-\infty}^{\infty} d^2x_1 \exp\left\{-\frac{P_T}{2\tau}\frac{1}{\rho V}\left[|x - x_1|^2 + (\zeta(x,t) - \zeta(x_1, t-\tau) + \tau V)^2\right]\right\}. \end{aligned} \tag{5}$$

Here $I_{\zeta d}^T$ denotes the integration over dimensional coordinates. Taking into account expressions (2) and (5) one can get

$$I_{\zeta d}^T = \left(\frac{P_T}{2\pi\rho}\right)^{3/2}\frac{1}{\sqrt{V}}$$

$$\times \int_0^\infty \frac{d\tau}{\tau^{3/2}} \int_{-\infty}^{\infty}\int_{-\infty}^{\infty} \exp\left\{-\frac{P_T}{2\tau}\frac{1}{\rho V}\left[(x-x_1)^2+(y-y_1)^2\right.\right.$$

$$\left.\left.+\left(\frac{x_1^2-x^2}{2(\omega_d-b_d)}+\frac{y_1^2-y^2}{2(\omega_d+b_d)}+\tau V\right)^2\right]\right\} dx_1 dy_1. \quad (6)$$

Replacing two integration variables τ and y_1 by ω_1 and z_1 as

$$\tau = \frac{(x-x_1)^2}{2\omega_1}, \quad y - y_1 = (x-x_1)z_1, \quad (7)$$

we get

$$I_{\zeta d}^T = \left(\frac{P_T}{\pi\rho}\right)^{3/2} \frac{1}{2\sqrt{V}} \int_0^\infty \frac{d\omega_1}{\sqrt{\omega_1}} \int_{-\infty}^{\infty} \exp(-\frac{P_T\omega_1}{\rho V}(1+z_1^2))dz_1$$

$$\times \int_{-\infty}^{\infty} \exp\left\{-\frac{P_T\omega_1}{\rho V}\left(-\frac{x_1+x}{2(\omega_d-b_d)} - \frac{z_1(2y-z_1(x-x_1))}{2(\omega_d+b_d)} + \frac{V(x-x_1)}{2\omega_1}\right)^2\right\} dx_1. \quad (8)$$

Integration of the right-hand side of Eq. (8) over x_1 leads to

$$I_{\zeta d}^T = \frac{P_T(\omega_d+b_d)}{\pi\rho} \int_0^\infty \frac{\exp(-\frac{P_T\omega_1}{\rho V})d\omega_1}{\omega_1} \int_{-\infty}^{\infty} \frac{\exp(-\frac{P_T\omega_1 z_1^2}{\rho V})dz_1}{z_1^2 + (\omega_d+b_d)(\frac{V}{\omega_1}+\frac{1}{\omega_d-b_d})}. \quad (9)$$

Now it is possible to integrate the right-hand side of the Eq. (9) over z_1

$$I_{\zeta d}^T = \frac{P_T\sqrt{\omega_d+b_d}}{\rho} \int_0^\infty \frac{\exp\left(\frac{P_T}{V\rho}(\omega_d+b_d)\left(V+\frac{\omega_1}{(\omega_d-b_d)}\right) - \frac{P_T\omega_1}{V\rho}\right)}{\sqrt{\omega_1}\sqrt{V+\frac{\omega_1}{(\omega_d-b_d}}}$$

$$\times \operatorname{erfc}\left(\sqrt{\frac{P_T}{V\rho}(\omega_d+b_d)\left(V+\frac{\omega_1}{\omega_d-b_d}\right)}\right) d\omega_1. \quad (10)$$

Here it has been taken into consideration that [45]

$$\int_0^\infty \frac{\exp(-\mu^2u^2)du}{u^2+\beta^2} = \frac{\pi}{2\beta}\operatorname{erfc}(\beta\mu). \quad (11)$$

Replacing the variable ω_1 by $u = \frac{1}{\sqrt{V}}\sqrt{V+\frac{\omega_1}{\omega_d-b_d}}$, we obtain

$$I_{\zeta d}^T = \frac{2P_T}{\rho}\sqrt{\omega_d^2-b_d^2}\exp(\frac{P_T}{\rho}(\omega_d-b_d))J(\omega_d), \quad (12)$$

where

$$J(\omega_d) = \int_1^\infty \frac{\exp\left(\frac{P_T}{\rho}2b_d u^2\right)\operatorname{erfc}\left(u\sqrt{\frac{P_T}{\rho}(\omega_d+b_d)}\right)du}{\sqrt{u^2-1}}. \quad (13)$$

Further, applying the method of differentiation and integration of $J(\omega_d)$ and keeping in mind that

$$\int_1^\infty \frac{\exp(-\mu u)\mathrm{d}u}{\sqrt{u-1}} = \sqrt{\frac{\pi}{\mu}}\exp(-\mu), \tag{14}$$

we get

$$I^T_{\zeta d} = \frac{P_T}{\rho}\sqrt{\omega_d^2 - b_d^2}\exp\left(\frac{P_T\omega_d}{\rho}\right)\int_{\omega_d}^\infty \frac{\exp\left(-\frac{P_T}{\rho}t\right)\mathrm{d}t}{\sqrt{t^2-b_d^2}}. \tag{15}$$

At the interface, $\omega_d = \rho$, one obtains

$$I^T_{\zeta d} = \sqrt{P_T^2 - b_d^2\frac{P_T^2}{\rho^2}}\exp(P_T)G(P_T), \tag{16}$$

where

$$G(P_T) = \int_{P_T}^\infty \frac{\exp(-y)\mathrm{d}y}{\sqrt{y^2 - b_d^2\frac{P_T^2}{\rho^2}}}, \tag{17}$$

and

$$\frac{c_p(T_M - T_\infty)}{Q} = I^T_{\zeta d}. \tag{18}$$

This relation shows that the undercooling $\Delta T = T_M - T_\infty$ connects two unknown dendrite parameters: ρ and V of the paraboloid. Our solution (16) transforms to the Horvay-Cahn two-fold solution [44] after returning to the dimensionless value $b = b_d P_T/\rho$

$$I^T_\zeta = I^T_{\zeta d} = \sqrt{P_T^2 - b^2}\exp(P_T)\int_{P_T}^\infty \frac{\exp(-y)dy}{\sqrt{y^2-b^2}}. \tag{19}$$

For $b = 0$, the cross-section is circular and one recovers the Ivantsov's solution [39, 41].
The thermal field can be expressed as [41, 44]

$$T_d(\omega) = (T_M - T_\infty)\frac{G(\omega) - G(P_T)}{G(P_T)}. \tag{20}$$

The function $\omega = \omega(x, y, z)$ is given by Eq. (1). This equation has three roots, but only one of them is positive at all possible values of parameters. It can be found from the Cardano's formula

$$\omega(x,y,z) = \frac{2z}{3} + \sqrt[3]{\frac{-q+\sqrt{q^2+\frac{4p^3}{27}}}{2}} + \sqrt[3]{\frac{-q-\sqrt{q^2+\frac{4p^3}{27}}}{2}}, \tag{21}$$

where

$$q(x,y,z) = \frac{4}{3}zb^2 - bx^2 + by^2 - \frac{16}{27}z^3 - \frac{2}{3}z(x^2+y^2), \tag{22}$$

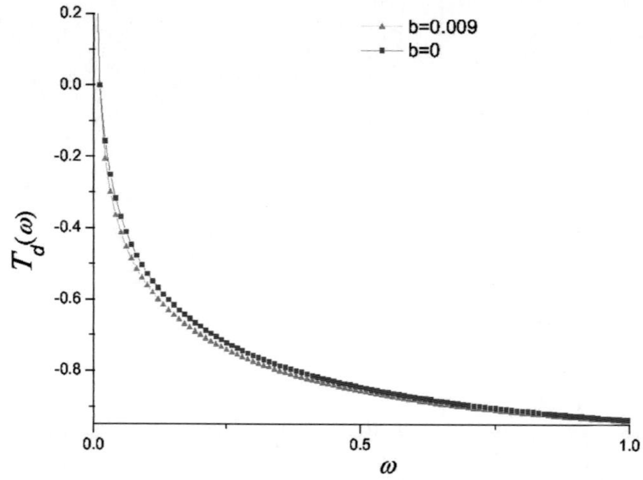

Figure 1. The temperature as a function of dimensionless variable ω at different values of parameter b.

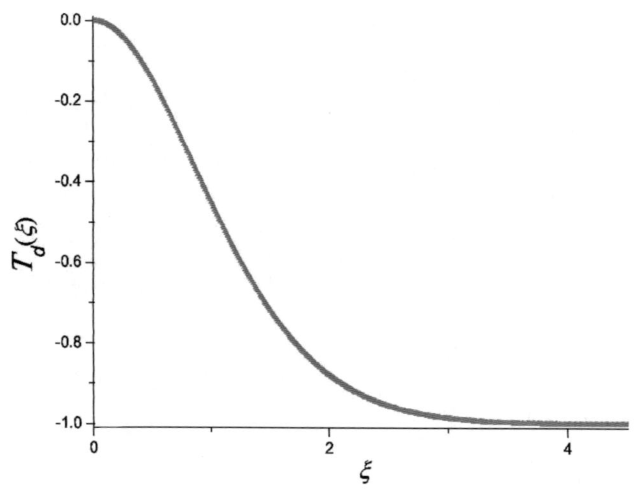

Figure 2. The temperature as a function of parabolic variable ξ at the fixed values of $b = 0.009$, $\eta = 1$ and $\varphi = \pi$.

and

$$p(x, y, z) = -\left(\frac{4}{3}z + b^2 + x^2 + y^2\right). \tag{23}$$

IOP Conf. Series: Materials Science and Engineering **192** (2017) 012025 doi:10.1088/1757-899X/192/1/012025

The parabolic coordinates are often used to describe the dendritic growth. The relations between the parabolic coordinates (ξ, η, φ) and the Cartesian coordinates (x, y, z) are

$$x = \xi\eta\cos\varphi, \quad y = \xi\eta\sin\varphi, \quad z = \frac{1}{2}(\xi^2 - \eta^2). \tag{24}$$

For a steady-state isothermal dendrite growing in the z-direction, the dendrite surface has the form $\xi = f(\eta, \varphi)$. The elliptical paraboloid in the parabolic coordinate system is given by

$$\xi^2\eta^2\left[\frac{\cos^2\varphi}{\omega - b} + \frac{\sin^2\varphi}{\omega + b}\right] = \omega - \xi^2 + \eta^2. \tag{25}$$

This relation defines ω as a function of the parabolic coordinates

$$\omega(\xi, \eta, \varphi) = \frac{\xi^2 - \eta^2}{3} + \sqrt[3]{\frac{-q_1 + \sqrt{q_1^2 + \frac{4p_1^3}{27}}}{2}} + \sqrt[3]{\frac{-q_1 - \sqrt{q_1^2 + \frac{4p_1^3}{27}}}{2}}, \tag{26}$$

where

$$q_1(\xi, \eta, \varphi) = \frac{2(\eta^2 - \xi^2)^3}{27} + \frac{(\eta^2 - \xi^2)(b^2 + \xi^2\eta^2)}{3} + b^2(\xi^2 - \eta^2) - \xi^2\eta^2 b(\cos^2\varphi - \sin^2\varphi), \tag{27}$$

and

$$p_1(x, y, z) = -\left(\xi^2\eta^2 + b^2 + \frac{(\eta^2 - \xi^2)^2}{3}\right). \tag{28}$$

3. Conclusions

Figures 1 and 2 demonstrate the solution given by Eq. (20) for the ice-water system ($T_M = 0$°C, $T_\infty = -1$°C, $P_T = 0.01$).

Let emphasize in conclusion the main output of our study. By leaning upon the previously developed theory of the thermo-solutal boundary-integral method [33], we have sewed together the Horvay and Cahn analytical solutions derived for the elliptical paraboloid [42] and the boundary integrals given by expressions Eq. (3) and Eq. (4) [33]. The temperature distribution in the liquid phase for the elliptical paraboloid is analyzed and illustrated as well.

4. References

[1] Chalmers B 1961 *Physical Metallurgy* (New York: John Wiley)
[2] Flemings M 1974 *Solidification Processing* (New York: McGraw Hill)
[3] Kurz W and Fisher D J 1989 *Fundamentals of Solidification* (Aedermannsdorf: Trans. Tech. Publ.)
[4] Mullins W W and Sekerka R F 1964 *J. Appl. Phys.* **35** 444–451
[5] Alexandrov D V and Ivanov A O 2000 *J. Crystal Growth* **210** 797–810
[6] Alexandrova I V, Alexandrov D V, Aseev D L and Bulitcheva S V 2009 *Acta Physica Polonica A* **115** 791–794
[7] Lee D and Alexandrov D V 2010 *Int. J. Pure Appl. Math.* **58** 381–416
[8] Hills R N, Loper D E and Roberts P H 1983 *Q. J. Appl. Math.* **36** 505–539
[9] Fowler A C 1985 *IMA J. Appl. Math.* **35** 159–174
[10] Worster M G 1986 *J. Fluid Mech.* **167** 481–501
[11] Mansurov V 1990 *Mathl. Comput. Modelling* **14** 819–821
[12] Alexandrov D V 2001 *J. Crystal Growth* **222** 816–821
[13] Alexandrov D V 2004 *Int. J. Heat Mass Trans.* **47** 1383–1389
[14] Aseev D L and Alexandrov D V 2006 *Doklady Physics* **51** 291–295
[15] Alexandrov D V and Aseev D L 2005 *J. Fluid Mech.* **527** 57–66
[16] Aseev D L and Alexandrov D V 2006 *Acta Mater.* **54** 2401–2406
[17] Alexandrov D V and Malygin A P 2006 *Int. J. Heat Mass Trans.* **49** 763–769

[18] Alexandrov D V and Malygin A P 2006 *Dokl. Earth Sciences* **411** 1407–1411
[19] Alexandrov D V, Aseev D L, Nizovtseva I G, Huang H-N and Lee D 2007 *Int. J. Heat Mass Trans.* **50** 3616–3623
[20] Alexandrov D V, Nizovtseva I G, Malygin A P, Huang H-N and Lee D 2008 *J. Phys.: Condens. Matter* **20** 114105
[21] Alexandrov D V and Nizovtseva I G 2008 *Int. J. Heat Mass Trans.* **51** 5204–5208
[22] Alexandrov D V, Ivanov A A and Malygin A P 2009 *Acta Physica Polonica A* **115** 795–799
[23] Alexandrov D V and Malygin A P 2012 *Int. J. Heat Mass Trans.* **55** 3196–3204
[24] Alexandrov D V and Malygin A P 2013 *J. Phys. A: Math. Theor.* **46** 455101
[25] Alexandrov D V and Nizovtseva I G 2014 *Proc. R. Soc. A* **470** 20130647
[26] Alexandrov D V and Malygin A P 2014 *Modelling Simul. Mater. Sci. Eng.* **22** 015003
[27] Alexandrov D V 2014 *J. Phys. A: Math. Theor.* **47** 125102
[28] Alexandrov D V 2014 *Phil. Mag. Lett.* **94** 786–793
[29] Nash G E 1974 *NRL Report* **7679** May 1974
[30] Nash G E and Glicksman M E 1974 *Acta Metall.* **22** 1291–1299
[31] Langer J S and Turski L A 1977 *Acta Metall.* **25** 1113–1119
[32] Langer J S 1977 *Acta Metall.* **25** 1121–1137
[33] Alexandrov D V and Galenko P K 2017 *Physica A* **469** 420–428
[34] Barber M N, Barbieri A and Langer J S 1987 *Phys. Rev. A* **36** 3340–3349
[35] Brener E A and Mel'nikov V A 1991 *Adv. Phys.* **40** 53–97
[36] Alexandrov D V, Galenko P K and Herlach D M 2010 *J. Crystal Growth* **312** 2122–2127
[37] Alexandrov D V and Galenko P K 2013 *Phys. Rev. E* **87** 062403
[38] Alexandrov D V and Galenko P K 2013 *J. Phys. A: Math. Theor.* **46** 195101
[39] Alexandrov D V and Galenko P K 2014 *Physics-Uspekhi* **57** 771-786
[40] Alexandrov D V and Galenko P K 2015 *Phys. Chem. Chem. Phys.* **17** 19149-19161
[41] Ivantsov G P 1947 *Dokl. Akad. Nauk USSR* **58** 567–570
[42] Horvay G and Cahn J W 1961 *Acta Metal* **9** 695–705
[43] Ananth R and Gill W 1989 *J. Fluid Mech.* **208** 575–593
[44] McFadden G B, Coriell S R and Sekerka R F 2000 *J. Cryst. Growth* **208** 726–745
[45] Gradshteyn I S and Ryzhik I M 2014 *Table of Integrals, Series, and Products*, (San Diego: Academic Press)

Acknowledgments

This work was supported by the Russian Science Foundation (grant no. 16-11-10095).

IOP Conf. Series: Materials Science and Engineering **192** (2017) 012026 doi:10.1088/1757-899X/192/1/012026

Numerical simulation of heat transfer and melting of Fe-based powders in SLM processing

V Ankudinov, G A Gordeev and M D Krivilyov

Udmurt State University, Universitetskaya str. 1, building 4, 426034, Izhevsk, Russia

E-mail: `vladimir@ankudinov.org`

Abstract. Modelling of selective laser melting (SLM) of Fe and stainless steel powders has been performed to analyze unsteady heat transfer in a porous medium under conditions of rapid phase transformations. The pulsed laser heat source develops high temperature gradients that lead to large solidification velocities, which can be used in acquisition of complex microstructures in SLM manufacturing. First, the effective thermal conductivity as a function of the local porosity which is a dynamically varying parameter in SLM was evaluated numerically. It showed a high role of heat transfer through the gas phase in powders with low thermal conductivity, i.e. in stainless steel. At the porosity 65% and above, the mechanism of heat transfer drastically changed and a linear dependence of thermal conductivity on porosity frequently used in literature becomes incorrect. Second, the obtained dependence was accounted for modelling of thermal fields and powder consolidation in the powder bed. The results of simulation agree well with the obtained experimental data and show the importance of correct dynamical evaluation of the thermophysical properties in SLM processes.

1. Introduction

Selective laser melting of metallic powders is a complex process. Therefore its computer simulation is important in modern material sciences and technology. Selective laser sintering (SLS) and melting (SLM) are the successfully developed techniques which allow us to modify phase and structural characteristics in a thin surface layer [1, 2] located on the substrate or formerly processed powder layers. Extension of this method towards processing of ultrafine powders through pulsed laser annealing was done for roughness reduction of SLM produced parts [2, 3]. The key feature of pulsed laser powder annealing is enermous local heating of the surface and large thermal gradients and velocities during the treatment. Due to high velocities of the solidification fronts, melting and crystallization of ultrafine (submicron and nanoscale) particles proceed under highly nonequilibrium conditions. Several literature sources estimate the cooling rates of the melting pool during the laser processing as very high [4, 5]. Both sintering and remelting process of ultrafine powders could be used as the suitable techniques for synthesis of metastable metallic materials [6] with desired phase composition. A characteristic time of the cycle "heating-melting-solidification" or "heating-sintering" depends on the annealing mode and follows from the processing parameters. In pulsed lasers which may lead to high melting/solidification velocity, the duration of a single impulse is about $\tau \sim 10^{-6} \div 10^{-3}$ s. Hence, in such short times it is difficult to control the dynamics of the remelting process and proper selection of the optimal processing parameters is vital. The objective of this study is the development of a comprehensive model and then verification analysis of laser sintering and melting process of ultrafine powders. In the paper, the results of experimental and computational examination are used to reveal the effect of dynamic thermal properties of the powder media under high energy pulsed laser treatment on the dynamics of melting and solidification during SLM processing.

Transport of heat through a porous medium is of great interest in relation with its regular and stochastic properties since in SLM the effective thermal properties control the dynamics of melting and solidification [5, 7]. The structure of a porous media is complex and irregular and it is defined by a wide distribution of morphology and size of pores [8]. Due to the fact that powder in SLM is spheroidized, the simplest approximation of particles by spheres can be used. However, this approximation somehow simplifies the results of analysis for real industrial systems [9]. Thus, the detailed prediction of the effective thermal conductivity of heterogeneous media requires consideration of the shape and size distribution, conductivity of each particle, atmosphere and heat exchange between the metal and gas phases [10]. Calculation of the effective heat conductivity for a case of stochastic filling of the powder can be made by combination of the classic packed bed modelling [8, 10] and the stochastic random filling model [9]. This approach is presented in this article.

The computing of SLM processing considering the whole powder layer including such effects as compacting and remelting of powder under the high-energy laser irradiation is represented by a wide range of models [2, 5, 7]. They all can be reduced to three classes in description of laser processing: (i) models of the arbitrary powder bed that directly account for partial melting and mass transfer coupled to heat conduction (implemented by the finite volume method); (ii) models with the stochastic distribution of particles which incorporate the phase state of the system using a state functions to determine the molten domain (by the discrete element or cellular automata methods); (iii) models which consider the powder in the approximation of a continuous medium assuming shrinkage of the powder as a continuous function (by the finite element method, models with a moving boundary).

In our work, the approximation of a continuous medium is used for modelling of heating of the powder bed, sintering and consolidation [11, 12, 13] during SLM of iron and stainless steel (CL20ES) powder by the pulsed infrared optic fiber YAG:Nd^{3+} laser. To control the effective thermal conductivity k_{eff} a state function $k_{eff}(\varepsilon)$ is used where ε is the local porosity. The function $k_{eff}(\varepsilon)$ is calculated separately for Fe and stainless steel powders under different atmosphere conditions (Vac, Air, Argon) suitable for SLM manufacturing.

2. Computer simulation of local heat transfer in the powder bed

In our work, we directly model heat transfer in the metallic powder beds using stochastic filling of the representative volume with a number (up to 10^3) particles. The numerical model gives us a possibility to calculate the dynamics of heat propagation in a metal powder with the given porosity ε treated in inert atmosphere or in vacuum. The model is allowing the variation of the fractional composition of particles, their size distribution and relative density of the bed.

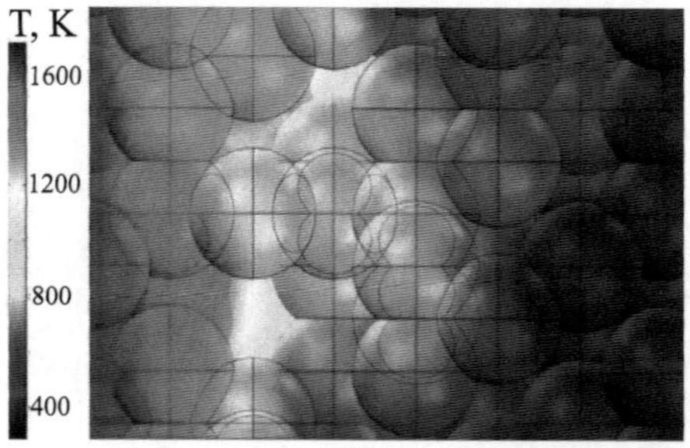

Figure 1. Instantaneous thermal field T in the ultrafine Fe powder bed calculated for the particles with a radius R_p=1.5 μm in argon and the porosity ε = 0.65. The time is $t = 3 \cdot 10^{-5}$ s and the evaluated effective heat conductivity for this case is $k_{eff} = 3$ W/(m · K).

IOP Conf. Series: Materials Science and Engineering **192** (2017) 012026 doi:10.1088/1757-899X/192/1/012026

2.1. Model

The powder bed was modelled by random placement of spherical particles according to the size and contact positions verified by scanned electron microscopy. The average radius of the particles in ultrafine iron powder was found about $R_p = 1.5$ μm and for stainless steel $R = 45$ μm assuming Gauss-type size distribution [1, 5, 14]. The specified porosity is achieved by removal of random particles from the bed till the target average density (or porosity ε) is attained. The spheres intersect each other by $10 - 20\%$ of the radius R_p to model heat contacts through each pair of particles.

For construction of a mixed two-phase system "gas-metal", the transient heat conduction equation is used. The two-phase problem with heat sources/sinks was solved using the finite element method. There is no need to take into consideration the effect of heat irradiation in the system of particles because of its small contribution to the overall energy balance. The convection mechanism of heat exchange was also neglected for the specified capillary systems [9] due to a pulse mode of processing ($\tau = 10^{-3}$ s for the used laser system). Thus, the time scale of heat annealing of the powder was significantly shorter than a characteristic time of effective gas flows through the powder. Then the boundary value problem for heat conduction in the powder bed is

$$C_p^{\text{gas}} \rho^{\text{gas}} \frac{\partial T}{\partial t} = k^{\text{gas}} \nabla^2 T, \quad C_p^{\text{metal}} \rho^{\text{metal}} \frac{\partial T}{\partial t} = k^{\text{metal}} \nabla^2 T, \tag{1}$$

$$T|_{t=0} = T_{amb}, \quad T|_{z=0} = T_m, \quad \mathbf{n} \cdot k \nabla T = 0 \text{ at other boundaries,} \tag{2}$$

where C_p is the specific heat capacity, ρ is density, k is the thermal conductivity for each phase. For the interphases between gas and metal a continuity of heat flows and temperatures was defined. The temperature at the boundary $z = 0$ was selected as the melting temperature $T_m^{\text{Fe}} = 1808$ K and $T_m^{\text{stainless steel}} = 1713$ K. In Figure 1 the instantaneous thermal field calculated in the powder bed is presented.

2.2. Evaluation of the effective thermal conductivity

The inverse problem of heat conduction [15, 16] was solved in the self-similar approximation which was used to get the effective thermal diffusivity coefficient α_{eff} as follows

$$\frac{\partial T}{\partial t} = \alpha_{eff} \nabla^2 T, \quad \text{which leads to a solution} \quad T(z,t) = T_m - (T_m - T_{amb}) \operatorname{erfc}\left(\frac{z}{\sqrt{4\alpha_{eff}\, t}} \right). \tag{3}$$

In performed simulations, we take a number of numerically calculated profiles at fixed times in the mixed gas-metal domain. Then a minimization Levenberg-Marquardt procedure to the functional $T^{\text{num}} - T^{\text{analyt}}$ was applied. Figure 2 depicts both the calculated temperature probes at different locations of the domain and the fitted semi-similar solution. In this manner the effective diffusivity coefficient for the specific times can be approximated. Then α_{eff} is averaged in time and the effective thermal conductivity for the two-phase powder is obtained by

$$k_{eff}(\varepsilon) = \alpha_{eff} \left(C_p^{\text{gas}} \cdot \rho^{\text{gas}}\, \varepsilon + C_p^{\text{metal}} \cdot \rho^{\text{metal}} (1 - \varepsilon) \right) \tag{4}$$

for each set of heat profiles for given ε. Since this task is devoted to evaluation of the effective thermal conductivity as a function of geometric characteristics of powder, the heat capacity C_p of both gas and metal is here assumed to be temperature independent. Dependence of k_{eff} on temperature is introduced in the model below where the macroscopic model of sintering/melting is considered.

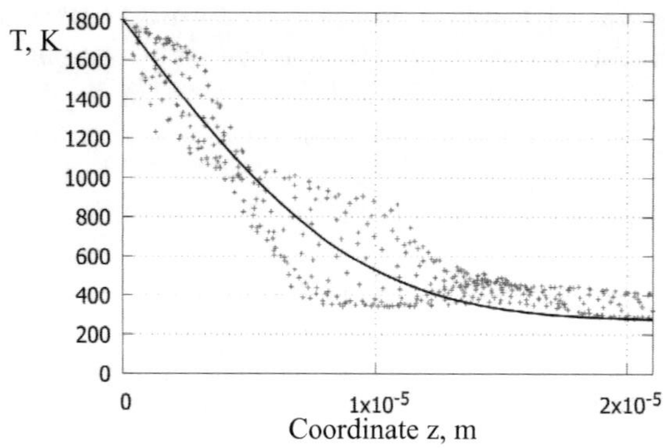

Figure 2. Temperature field (denoted by the color dots) calculated numerically for Fe powder in air with the porosity $\varepsilon = 0.65\,(65\%)$ at time $t = 10^{-5}$ s. Propagation of a heat wave experiences thermal resistance on gas inclusions which leads to deterioration in effective heat transfer. The black line is the approximated curve plotted for the self-similar solution (Eq. 3). The obtained effective thermal diffusivity is $\alpha_{eff} = 2.265 \cdot 10^{-6}$ m^2/s compared to the $\alpha_{\mathrm{Fe}} = 2.85 \cdot 10^{-6}$ m^2/s.

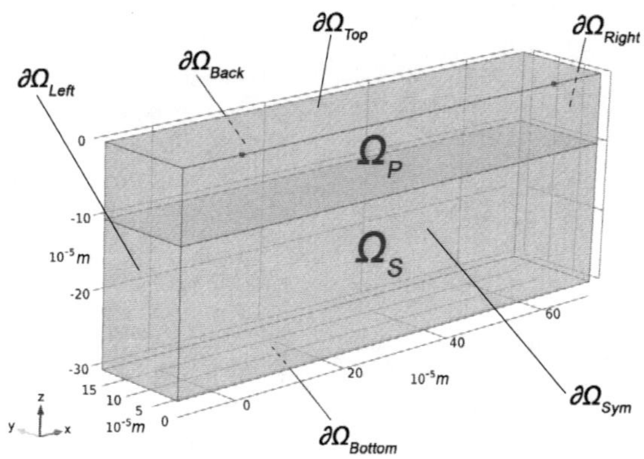

Figure 3. Computational domain for simulation of the powder bed melting. The top subdomain corresponds to the powder bed with a height of $h_0 = 100 \div 150$ µm. The bottom subdomain represents the stainless substrate. In SLM the previously processed layer acts as the substrate. The finite element mesh contains up to 10^7 degrees of freedom.

3. Simulation of sintering and melting of the powder bed

Mathematical description of underlying in SLM thermophysical phenomena [12, 13] is provided by the heat conduction equation adapted for enthalpy of phase transitions [15, 16, 17] and a surface heat source F due to laser annealing. Two conjugated variables temperature T and specific enthalpy H of the metal fraction are used, Eqs. (5) and (6). In addition, Eq. (7) [18] for the kinetics of local porosity ε is assumed. Then the given system of equations is as follows:

$$\frac{\partial T}{\partial t} = \frac{\partial T}{\partial H}\frac{\partial H}{\partial t}, \tag{5}$$

$$\frac{\partial H}{\partial t}(1-\varepsilon) = \nabla \cdot (k_{eff}(\varepsilon)\nabla T) + F, \tag{6}$$

$$\left.\frac{\partial \varepsilon}{\partial t}\right|_{\Omega_P} = -\varepsilon A \exp\left(\frac{E_a}{RT}\right), \tag{7}$$

$$\Delta Z|_{\Omega_P} = \frac{h_0(1-\varepsilon)}{1 - \int_z \varepsilon dz/h_0} - h_0, \tag{8}$$

where Ω_P is the powder subdomain (Fig. 3), A is the exponential factor describing the dynamics of consolidation of the powder, E_a is the energy of activation of viscous flow of liquid metal. The displacement ΔZ along the z axis defines an actual position of the top boundary according to local consolidation of the powder. Its definition includes the initial thickness h_0 of the powder layer and the integral on local porosity ε along z-direction in the powder. The effective conductivity $k_{eff}(\varepsilon)$ given in Eq. (6) represents a correlation between the thermal conductivity and local structure of the powder bed. Equation (7) is the Arrhenius-type [18] kinetic equation which is often used for description of viscoplastic flow in multiphase systems with phase transitions. In our model it is used for description of the dynamics of the powder porosity [12]. Equation (8) defines the unsteady position of the top boundary $\partial\Omega_{Top}$ in time, Fig. 3. At $\partial\Omega_{Top}$, the boundary heat source F is implied to reproduce laser heating of the surface conjugated to the convective and radiate cooling. The computer model has been implemented using the commercial software COMSOL Multiphysics (licence No. 1056903) in the weak form. The processing parameters [12, 13, 19] were selected for the laser facility BULAT LRS which was used for SLM processing.

4. Results and discussions

The dependence of the effective thermal conductivity on gas atmosphere is presented in Fig. 4. The calculated points are fitted by a polynomial function which depends on the local porosity ε. Then this function is included in the macroscopic model in which heating, melting and solidification of the powder bed are simulated in the approximation of a continuous medium. It allows us to account for the effect of dynamically changed porosity in powder consolidation.

The effect of gas atmosphere for powders with low conductivity like stainless steel was analyzed. In powders with relatively small porosity or in powders with relatively high thermal conductivity the presence of air or protective atmosphere in pores is insignificant. The influence of atmosphere can not be neglected if the initial porosity ε achieves a value 0.6. It happens due to the percolation effect in powders with high porosity $\varepsilon > 0.7$ where thermal contacts between particles are not sufficient. As a result, quality of the fabricated SLM parts can be controlled by reduction of the initial porosity by mechanical milling of the powder.

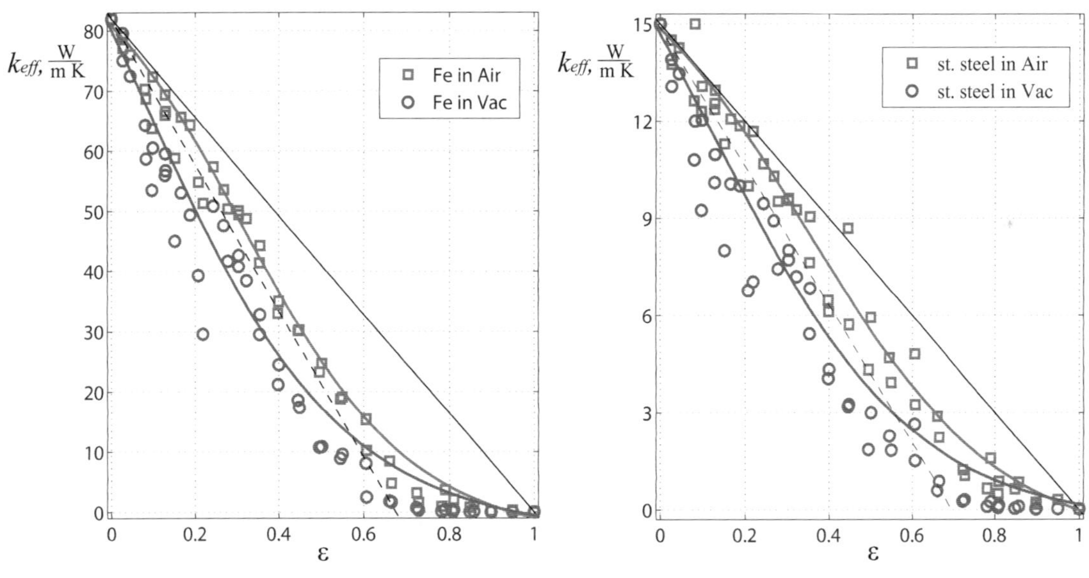

Figure 4. Dependence of the effective thermal conductivity k_{eff} on the local porosity ε calculated for a) Fe powder in vacuum (green curve and circles) and air (red curve and circles); b) stainless steel CL20ES powder in vacuum (green curve and circles) and air (red curve and circles). The dashed line represents the theoretical percolation limit for the medium, the black thin line represents the geometrical linear interpretation of the porosity.

A comparison of the calculated profiles in the melted powder with optical images taken in similar full-scale experiments is given in Fig. 5. Introduction of the effective thermal characteristics such as the effective thermal conductivity $k_{eff}(\varepsilon)$ has made it possible to achieve good qualitative and quantitative agreement with the experiment [19].

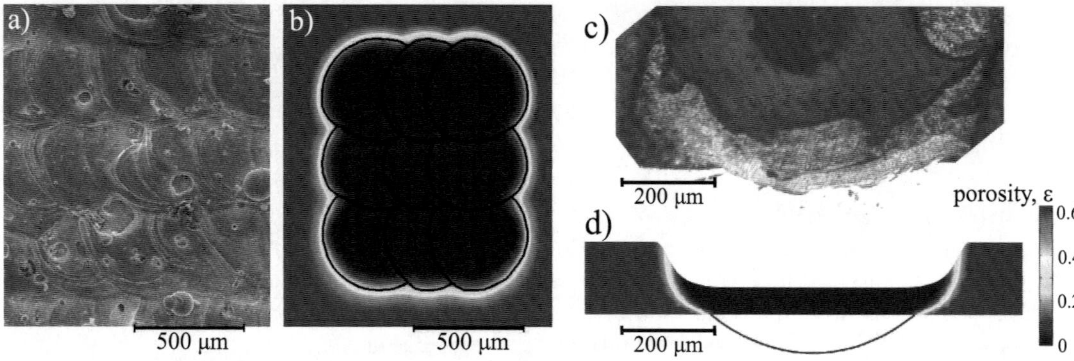

Figure 5. Comparison of the optical images (subfigures *a* and *c*) of the sample with the results of numerical simulation (subfigures *b* and *d*). Good quantitative agreement is achieved both in the track patterns and cross sections for Fe powder. The processing parameters for SLM of the stainless steel substrate were as follows. The frequency of the laser pulses is $\nu = 50$ Hz and the initial porosity is $\varepsilon = 0.6$. a,b) Top view of the melted powder layer, pulse energy is $E_{imp} = 3.4$ J, pulse duration $\tau = 1.5 \cdot 10^{-3}$ s. c,d) Cross sections, $E_{imp} = 2.2$ J, $\tau = 4 \cdot 10^{-3}$ s.

5. Conclusions

The effective thermal conductivity k_{eff} as a function of porosity ε for Fe and stainless steel powders were calculated in a series of computer simulations. The obtained relations can be used for modeling of powder consolidation in selective laser sintering/melting. It is also useful for estimation of the effect of porosity on partial melting.

The purpose of this work was the development of a mathematical model and study of melting and consolidation of the powder bed. In a series of computations, the underlying in SLM phenomena were analyzed for the industrially relevant Fe-based powders. Good agreement with the experiment data ensures that these data are useful for improvement of quality of the melted tracks. Further optimization of the whole process to receive the highest speed of fabrication at decent quality of the compacts and adhesion to the substrate was suggested.

6. References

[1] Haranzhevskiy E, Danilov D, Krivilyov M and Galenko P 2004 *Materials Sci. Eng.* A **375–77** pp 502–506
[2] Kumar S 2003 *JOM* **55** (10) pp 43–47
[3] Viswanathana V, Lahab T, Balanib K, Agarwalb A and Seala S 2006 *Mater. Sci. Eng.* R **54** (5-6) pp 121-285
[4] Kharanzhevskiy E and Krivilyov M 2011 *The Physics of Metals and Metallography* **111** (1) pp 53–61
[5] King W E, Anderson A T and Ferencz R M et al. 2015 *Applied Physics Reviews* **2** (4) p 041304
[6] Herlach D, Galenko P and Holland-Moritz D 2007 *Metastable Solids from Undercooled Melts Materials Series* (Amsterdam: Pergamon)
[7] Yap C Y, Chua C K and Dong Z L et al. 2015 *Applied Physics Reviews* **2** (4) p 041101
[8] Hsu C-T, Cheng P and Wong K W 1994 *Journal of Porous Media* **37** (1) pp 19–38
[9] Kaviany M 1995 *Principles of Heat Transfer in Porous Media* (Springer New York) p 726
[10] Zehnder P and Schlünder E U 1970 *Chemie. Ingr.-Tech* **42** pp 933–941
[11] Gordeev G A, Ankudinov V E, Krivilyov M D, Kharanzhevskiy E V 2012 *Mat. Sci. and Eng.* **27** p 012079
[12] Gordeev G A, Krivilyov M D, Ankudinov V E, Kharanzhevskiy E V and Bogdanov A A 2016 *Proc. of the Second Int. Conf. "Additive technologies: today and tomorrow"* VIAM, Moscow p 15

[13] Krivilyov M D, Ankudinov V E, Gordeev G A, Reshetnikov S M, Kharanzhevskiy E V and Galenko P K 2016 *Proc. of the Second Int. Conf. "Additive technologies: today and tomorrow"* VIAM, Moscow p 15
[14] Gusarov A 2010 *Quantum Electronics* **40** (5) pp 451–9
[15] Carslaw H S and Jaeger J C 1984*Conduction of heat in solids* (Oxford, Clarendon Press) p 510
[16] Bird R, Stewart W and Lightfoot E 2002 *Transport phenomena* (Wiley) p 905
[17] Borisov V 1962 *Dokl. Acad. USSR* **142** (3) pp 581–83
[18] Dong L, Makradi A and Ahzi S 2009 *Journal of materials processing technology* **209** pp 700–6
[19] Krivilyov M D, Gordeev G A, Ankudinov V E, Shutov I V, Ipatov A G, Matveeva U U and Kharanzhevskiy E V 2017 *Proc. of the Third Int. Conf. "Additive technologies: today and tomorrow"* VIAM, Moscow p 14

Acknowledgments
The authors wish to acknowledge I. V. Shutov for the metallographic analysis of sintered samples and A. A. Bogdanov for laser processing of the samples. This work was financially supported by the Russian Foundation for Basic Research (project No. 16-38-00839 mol_a) and Ministry of Education and Science of Russia (project No. 11.6513.2017/BP). The research is financially supported by the UMNIK program of the Innovation Support Foundation of Russian Federation.

IOP Conf. Series: Materials Science and Engineering **192** (2017) 012027 doi:10.1088/1757-899X/192/1/012027

Computation material science of structural-phase transformation in casting aluminium alloys

V M Golod[1] and L Yu Dobosh[1,]

[1]Peter the Grate St.Petersburg Polytechnic University, Institute of Metallurgy, Mechanical Engineering and Transport, Department of Metallurgical and Casting Technologies, 195251, Saint-Petersburg, Russia

E-mail: lpi2015@mail.ru

Abstract. Successive stages of computer simulation the formation of the casting microstructure under non-equilibrium conditions of crystallization of multicomponent aluminum alloys are presented. On the basis of computer thermodynamics and heat transfer during solidification of macroscale shaped castings are specified the boundary conditions of local heat exchange at mesoscale modeling of non-equilibrium formation the solid phase and of the component redistribution between phases during coalescence of secondary dendrite branches. Computer analysis of structural - phase transitions based on the principle of additive physico-chemical effect of the alloy components in the process of diffusional - capillary morphological evolution of the dendrite structure and the o of local dendrite heterogeneity which stochastic nature and extent are revealed under metallographic study and modeling by the Monte Carlo method. The integrated computational materials science tools at researches of alloys are focused and implemented on analysis the multiple-factor system of casting processes and prediction of casting microstructure.

1. Introduction

Reliable technology foresight of casting technology for manufacturing high-quality products, as well as for the successful synthesis of casting alloys with the desired structure and properties is provided by the development of adequate mathematical models produced in the field of ***computational thermodynamics*** (CTD) and ***computational heat transfer*** (CHT) for the analysis of the conditions of solidification of castings, also on the basis of ***computational materials science*** models (CMS) in order to predict the processes of formation of the microstructure of multicomponent alloys with a dendritic morphology that can be integrated into the modern theory of casting processes.

2. Problems and prospects of CTD models implementation

Using a number of specialized software systems (FactSage - since 1976, ThermoCalc - since 1981, Polytherm - since 2000, etc.) allows to implement in technological developments the powerful mathematical models of equilibrium thermodynamic analysis of phase transformations of various physico-chemical nature. These models should reflect the crystalline structure of alloys and processes occurring during phase transitions, taking into account the redistribution of components between phases, as well as the composition and amount of solid phase under equilibrium and non-equilibrium crystallization of stable and metastable phases. The mathematical apparatus of CTD for metal systems is based on a universal model of subregular solutions used for quantitative description of the physico-chemical processes under the phase transformations based on the synthesis of metal physics, materials science and computer technology [1-3].

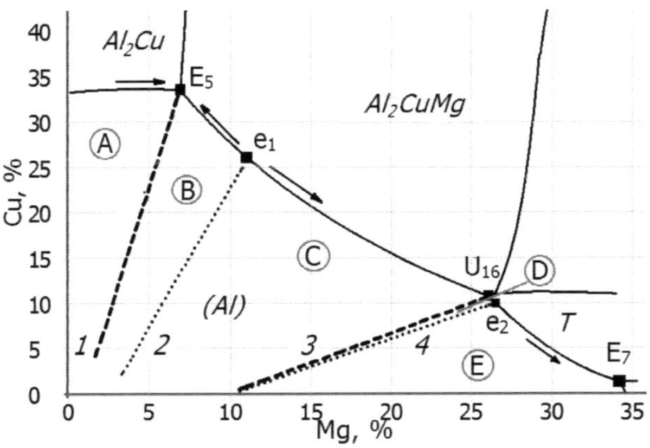

Figure 1. Areas (A-E) formation of various structural components and the corresponding dividing lines (*1-4*) under crystallization of Al-Cu-Mg alloys.

Figure 1 shows the calculated fragment of liquidus surface for ternary system Al-Cu-Mg [4], indicating areas and boundaries of areas of the various structural components, which is the basis for modeling of its crystallization. Modeling of phase transformation provides the ability to calculate the temperature functions for coefficients $k(t)$ and $p(t)$, which is dictated by the considerable scale of the it's changes taking place during the crystallization (figure 2.) - from 30 to 70-90%. The appropriate correction of algorithm of used CTD software allows to calculate the course of phase transformation not only under equilibrium conditions, but also at the complete suppression of diffusion in the solid phase (Scheil model) [3-5], including the fraction of the releasing phases, temperature functions of specific heat of phases $c_m(t)$ and the latent heat of crystallization $L_m(t)$.

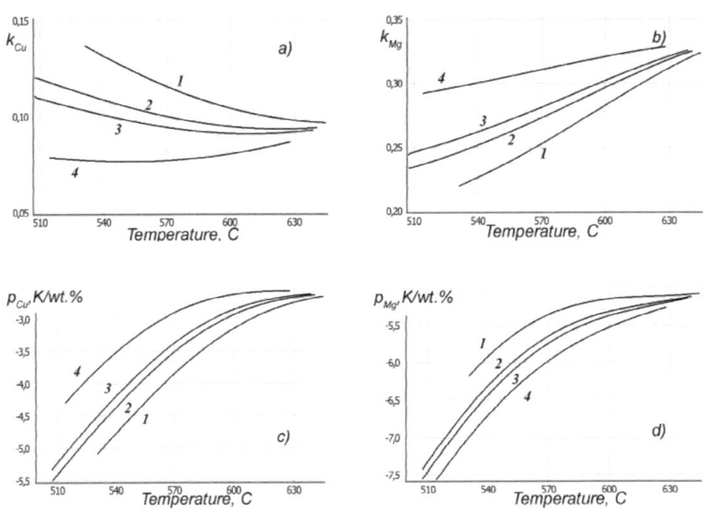

Figure 2. Effect of the composition of alloys Al - 4,5% Cu - Mg on changing the distribution coefficients k_{Cu} (*a*)- k_{Mg} (*b*) and slope of liquidus p_{Cu} (c) - p_{Mg} (d) in the temperature range of equilibrium crystallization of the primary solid phase α-Al at Mg,% = 0.8 (*1*); 1.6 (*2*); 2.0 (*3*); 4.0 (*4*).

Further development of tools and experimental techniques CTD will eliminate the existing at the moment a significant (up to 5-10 wt.%) number of mismatching lines and critical points, delimiting the phases formation region (including - for the most important systems on the basis of iron and aluminum) [5], as well as realization of the thermodynamic support users of a certain group of alloys (silumins, tin bronzes, stainless steels, etc.), for which a local adaptation of a computational thermodynamic model can be made cheaper and more accurate.

3. The multi-level synthesis of CHT tools

The analysis of the complex processes that occur during the formation of the casting of any configuration is carried out by generation for the macrosystem the net of small mesoelements, which are characterized by a uniform temperature distribution (the grid number of Bi << 1) and the concentration of residual melt. Solution the problem of casting solidification is based on the analysis of its heat exchange with the environment – the mold or ingot (figure 3), which is described by the Fourier equation:

$$c_m \frac{\partial t}{\partial \tau} = \lambda_m \nabla^2 t + \frac{\partial Q_m}{\partial \tau}, \tag{1}$$

where τ – time; c_m, Q_m, λ_m – volumetric heat capacity, heat of crystallization and thermal conductivity of the alloy. For further solution it is necessary to define the initial and boundary conditions of heat transfer and thermal properties of materials [3].

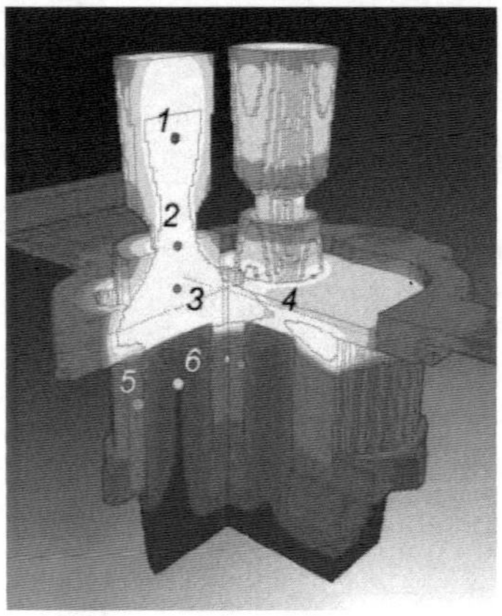

Figure 3. Distribution of temperature during the casting solidification (alloy AK12, weight - 45 *kg*) in sand mold with registration the local heat transfer conditions in the control points of casting (*1-5*) and mould (*6*).

After splitting the macro-object by spatial grid with a increments $\Delta x \times \Delta y \times \Delta z$ for three coordinate axes the calculation of the effective heat flux for the *i*-th selected mesoelement of the three-dimensional geometric macromodel is based on the local heat balance equation:

$$q_i = \left(\frac{V_i}{F_i} \right) \frac{Q_m}{(\tau_{LS})_i} = \frac{R_i Q_m}{(\tau_{LS})_i}, \tag{2}$$

where the value of q_i is to be defined by averaging the results of numerical calculation of heat transfer in the modeling of solidification by time τ_{LS} for mesoelement, which is selected for the purpose of local analysis of the crystallization process in a given domain of R_i size.

Dataware for heat calculation of casting solidification, based on thermodynamic modeling of phase transformation of alloy involves the additional determination of the temperature functions of the density $\rho_m(t)$ and the thermal conductivity $\lambda_m(t)$, which is defined by generalizing of known experimental data using neural network models [6], as well as their diffusion coefficients in liquid D_L and solid D_S phase. Thermal properties of the moulding materials $c_f(t)$ and $\lambda_f(t)$, which technological characteristics are caused by grain and mineralogical composition, humidity of mixture and its degree of compaction is determined by computer-automated procedure for solving the inverse problem based on data of thermal analysis of mould heating during the solidification test of experimental sample [6].

4. Mesoscale model of non-equilibrium releasing of the solid phase

Progress of the non-equilibrium crystallization of multi-component (i = 2, ... K) alloy describes the system (3a-3c) which interconnected the changes in the concentration of the components in melt, fraction of the released primary solids phase $m(t)$ and temperature $t(\tau)$ [7-8]:

$$\begin{cases} \dfrac{dC_i^L}{dm} = \dfrac{C_i^L(1-k_i)}{1-m(1-\sigma_i k_i)}; \text{ (a)} \\ \dfrac{dm}{dt} = \dfrac{1}{\sum_2^K \dfrac{p_i C_i^L(1-k_i)}{1-m(1-\sigma_i k_i)}}; \text{ (b)} \\ q d\tau = -Rc_m dt + RL_m dm, \text{ (c)} \end{cases} \quad (3)$$

where $\sigma_i = 2\alpha_i/(1+2\alpha_i)$; $\alpha_i = 8D_i^S \tau_{LS}/\lambda_2^2$ [9] – the degree of suppression of diffusion in the solid phase, estimated on the basis of adjacent calculation of secondary dendrite arm spacing $\lambda_2(\tau)$ of the primary solids phase.

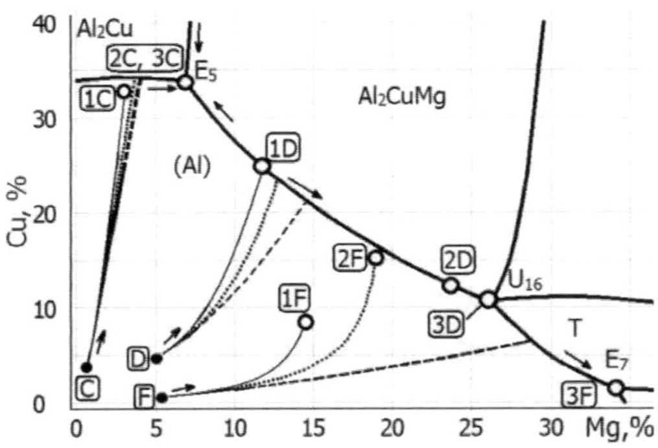

Figure 4. Solidification paths of C, D and F alloys on the liquidus surface of ternary system Al-Cu-Mg at various intensity of diffusion in the solid phase:
$D_i^S = \infty$; $\sigma_i = 1$ (*1*);
$\infty > D_i^S > 0$; $1 > \sigma_i > 0$ (*2*);
$D_i^S = 0$; $\sigma_i = 0$ (*3*).

The paths of crystallization of various alloys (figure 4) with similar initial composition (0.5-5% Cu; 1-5% Mg) demonstrate the radical difference of the trajectories and the final reached structure (the fractions and compositions of the resulting eutectic - E_5, U_{16} or E_7), depending on the diffusion conditions in solid phase and the morphology of the dendritic structure that has a significant influence on the course of the casting processes and the mechanical properties of the alloy. The results of numerical calculation using system of non-equilibrium models (3a - 3c) for estimating the rate of primary solid phase formation (figure 5) and changing the composition of the liquid phase (figure 6) are compared with the data of thermodynamic calculation (CTD) under equilibrium conditions ($D_S = \infty$) and complete suppression of diffusion in the solid phase ($D_S = 0$ - Scheil model).

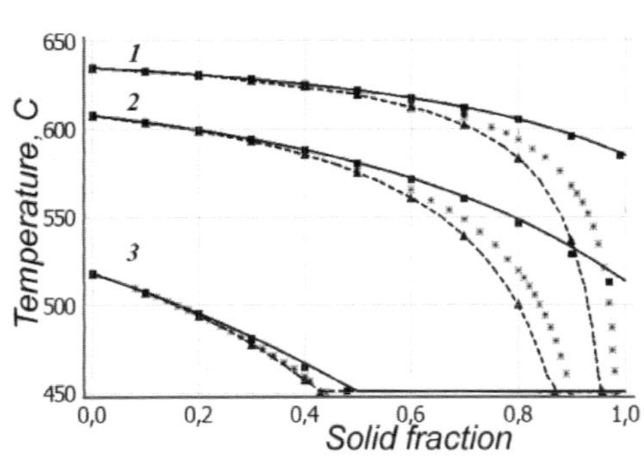

Figure 5. Changes temperature of Al-Mg alloy with 5 (*1*), 10 (*2*) and 25 (*3*)% Mg under different crystallization conditions: lines - CTD, $D_S = \infty$; dotted lines - CTD, $D_S = 0$; points - system of models (3a-3c) for σ = 1.0 (■); σ = 0,5 (*); σ = 0.0 (▲).

Evident coincidence between the numerical calculation results according to the system of models (3), presented in figure 5-6 by points, with CTD data under the same assumptions (σ = 1.0 and σ = 0.0) indicates the adequacy of both system components – (3a) and (3b), and the possibility of their use for predicting the phase transformation.

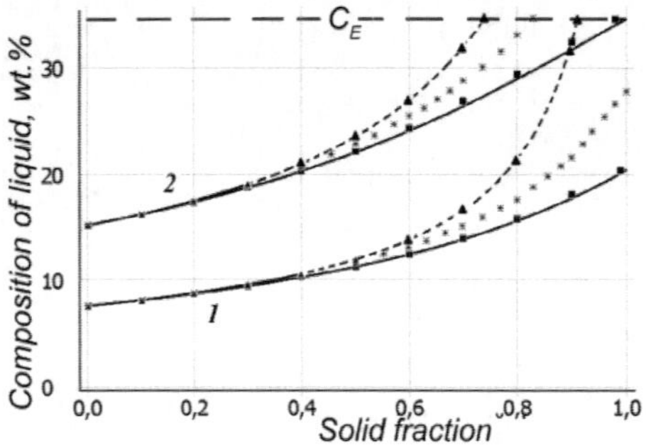

Figure 6. Changes of liquid phase composition of Al-Mg alloy with 8 (*1*) and 15 (2)% Mg under different crystallization conditions: lines - CTD, $D_S = \infty$; dotted lines - CTD, $D_S = 0$; points - system of models (3a-3c) for σ = 1.0 (■); σ = 0,5 (*); σ = 0.0 (▲).

An important feature of the above systems of models (3a-3c) is the using in the modeling of principles: *independent* action of intra-phase and inter-phase diffusion of components (3a - model) and *additive* influence of components on rate of formation of the solid phase (3b - model) with account the dependence from temperature of thermodynamic (*p*, *k*) and kinetic (D_i^S) parameters.

5. Mesoscale model of dendritic structure evolution

Mesoscale equiaxed crystallization of multicomponent alloys develop step-by-step by formation the solid phase during the creation and growth of the ensemble of dendritic crystallites in *interdendritic* melt up to their closing (first stage) followed by capillary-diffusion coalescence in the volume of *intradendritic* liquid phase (second stage) under the influence of the Gibbs-Thomson effect. The initial process of formation of dendrite in the period of maximum supercooling of the melt and its subsequent recalescence leads to origination the ensemble of differently oriented and non-uniformly distributed secondary branches that differ significantly in length, diameter and curvature of the surface in array with a variety of dendrite arm spacing λ_0 [10]. The dendritic morphology, formed during the evolution of the primary non-homogeneous massif during a competitive implementation of various coalescence mechanisms [11], leads to a continuous increase the average size of the secondary dendrite space $\lambda_2(\tau)$ and their scope $\sigma_\lambda(\tau)$ up to the moment of exhaustion of intradendritic liquid phase at $\tau = \tau_{LS}$.

The course of evolution of the dendritic structure, evaluated by changes in the most important parameter – averaged $\lambda_2(\tau)$, describes an equation based on a generalized model [12]:

$$\frac{d(\lambda_2^3)}{d\tau} = \frac{1}{\varphi_j \sum_K \frac{p_i(1-k_i)C_i^L}{\Gamma_i D_i^L}}, \tag{4}$$

where Γ_i – Gibbs-Thomson coefficient for the *i*-th component of the alloy, φ_j – adopted rate of coalescence mechanism. Analysis of a large array of experimental data (more than 110, *R*=0.98) of the averaged (finite) values of secondary dendritic spacing λ_K for binary and ternary aluminum alloys (figure 7) in comparison with the calculation on the equation (4) allow to carry out its statistical calibration and set the effective value of the coefficient φ for alloys of different composition, established the predominant mechanism of coalescence – radial dissolution/thickening of the branches ($\varphi \approx \varphi_{II}$ [12]).

The set of equations (3) and (4) forms a system model ALSYS [16], which allows to consider quantitatively the relationships of physical and chemical nature in the course of consistent non-equilibrium crystallization of multicomponent aluminum alloys and reveal the influence of thermodynamic, thermal, diffusion and capillary factors on the morphology and volume fraction of the dendrite structure (with further modeling of the second stage of alloys crystallization – formation of binary and ternary eutectics).

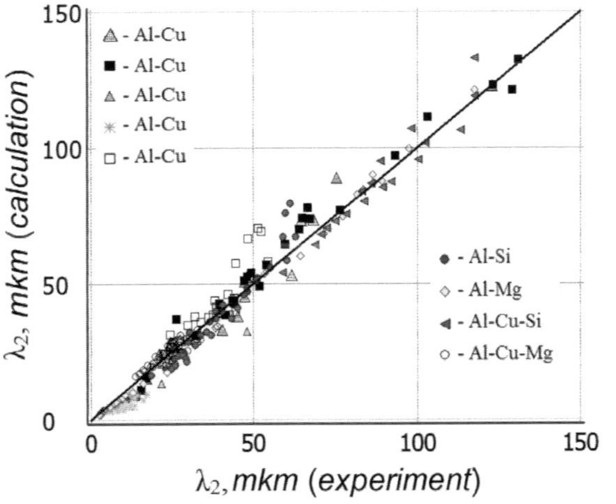

Figure 7. Comparison of calculated and experimental values λ_2 [13-15 etc.] of binary and ternary aluminum alloys using after calibration testing the different coalescence models I-IV [12] respectively various φ for alloy groups: $\varphi_I \approx \varphi_{III}$ (Al-Cu); φ_{II} (Al-Cu-Si, Al-Si; Al-Mg);
φ_{IV} (Al-Cu-Mg).

Complex diagram of system modeling the characteristics of crystallization process of ternary alloy Al-5% Cu-5% Mg, shown in figure 8 for different conditions of the heat removal, shows the relationship of the time of successive stages of the crystallization process (τ_α, τ^b_E, τ^t_E), release kinetics and the fraction of solid phase (figure 8,a-b), the composition of the residual liquid phase (figure 8, c-d), as well as the intensity and completeness of diffusion processes (figure 8, e-f) and – as a result – the evolution of the parameters (λ_2, λ^b_E, λ^t_E) of the microstructure (figure 8, g) from the rate of cooling at different stages of solidification.

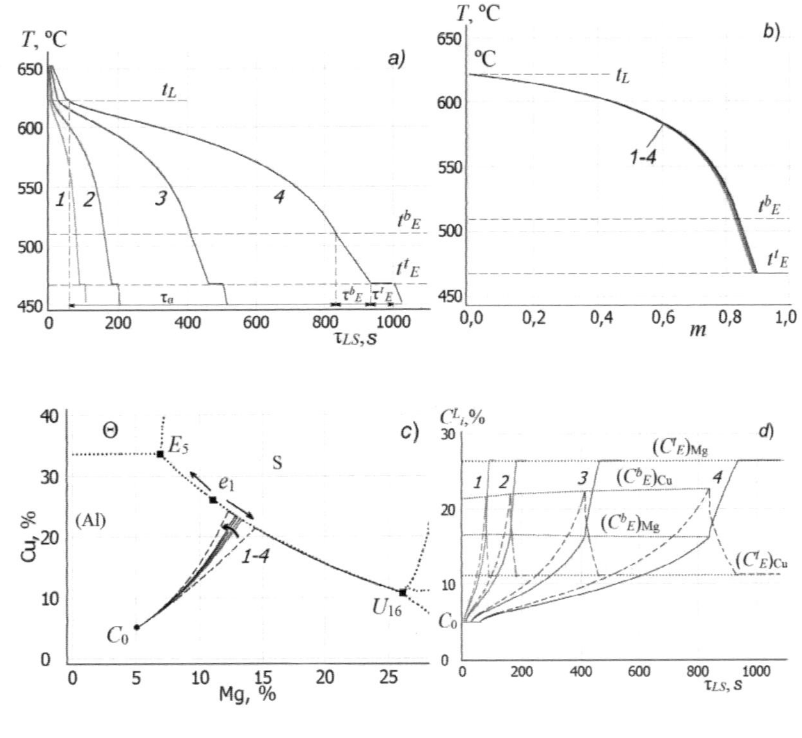

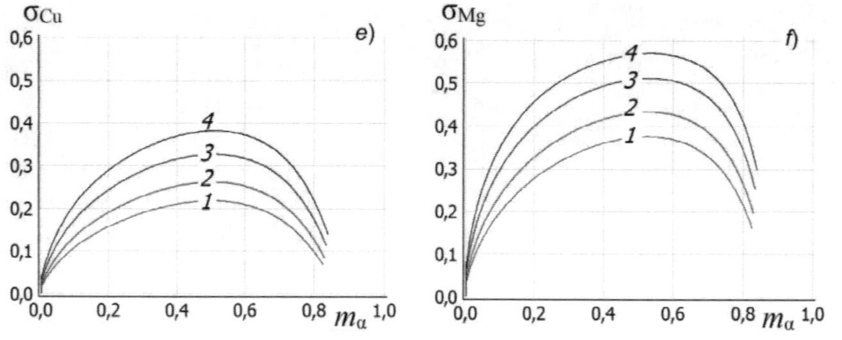

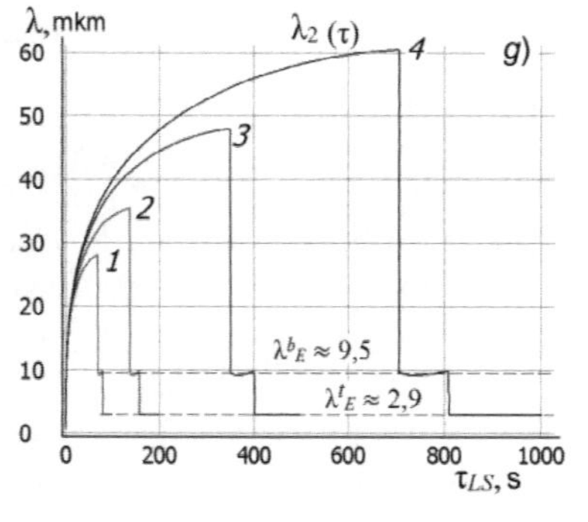

Figure 8. Effect of local crystallization time of the alloy Al-5% Cu-5% Mg (τ_{LS}, c: 1-100; 2-200; 3-500; 4-1000) on character of thermal curves (*a*), part of solid phase (*b*), composition of liquid phase (*c-d*), intensity of solid phase diffusion (*e-f*) and parameters of microstructure (*g*).

The adequacy of the system model to predict the dendritic structure confirmed by the results of metallographic control (figure 9), which shows a linearized dependence $\lambda_2 = K\tau_{LS}^n$ of local solidification time τ_{LS} (in a wide range from 5.2 to 10^3s) for averaged values of dendrite arm spacing $(\lambda_2)_{av}$ of investigated cast samples.

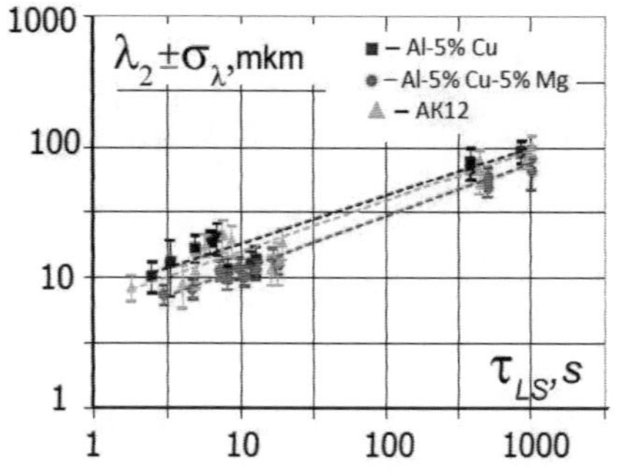

Figure 9. Dependence of the secondary dendrite spacing (averaged statistical data $(\lambda_2)_{av} \pm\sigma_\lambda$ of microstructure studied on the cross section of the cast samples) for alloys Al-5% Cu, Al-5% Cu-5% Mg and AK12 from a local solidification time τ_{LS}.

The data obtained are characterized by similar values of exponents (n=0.37-0.40) at relatively high values of the correlation coefficient (R≈0.94-0.98) and large scale of λ_2 values reaching 18-30% of $(\lambda_2)_{av}$.

6. Local inhomogeneity of the dendrite structure

The mentioned character of the inhomogeneity of the dendritic structure are represented by histograms distribution of end dendritic arm spacings (figure 10), which show significant local range on the scale of a separated dendrite and its nearest neighbors. This inhomogeneity of secondary arms caused by intense coalescence, develops between the growing side branches with the cessation of growth and their dissolution on the interface area of increased curvature.

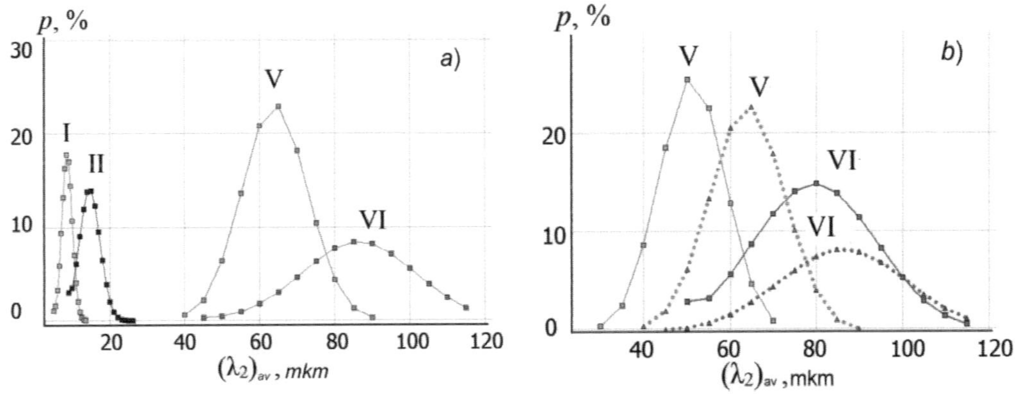

Figure 10. The frequency distribution of dendritic arm spacing in castings of different thicknesses (I-II: 10-15 mm; V-VI: 40-70 mm) in metal (I-II) or sand (V-VI) molds a) structure of axial zones; b) structure of surface (solid line) or centre (dotted line) zones.

Modeling the evolution of the dendritic structure was carried out using the Monte Carlo method [11] by calculation of competitive serial realization of various steps of coalescence in a statistical array with inhomogeneous distribution of morphological parameters of the crystallizing system [10]. Its result shows – in the accordance with experimental data (figure 10) – that the increasing of solidification and coalescence time causes growth of a mean value $(\lambda_2)_{av}$ and the limiting values $(\lambda_2)_{max}$ and $(\lambda_2)_{min}$ with sequential increase their scope, which is the driving force for repeated and continuous growth of these parameters.

7. Conclusion

System computer analysis of the inhomogeneous initial ensemble of secondary branches and its coalescence during crystallization shows that the stochastic processes of nucleation and evolution of dendritic branches in the interdendritic and intradendritic melt play a decisive role in shaping the final morphology and dispersity of the dendritic structure of alloys and non-equilibrium rate of releasing the solid phase under redistribution of components between the phases. The results of mesoscale analysis done in the framework of macroscale multi-dimensional configuration of the solidified casting (figure 3) allow to quantify the nature of the casting processes, that developed in the period of liquid to solid transformation for prediction the intensity and localization of casting defects [10].

8. References

[1] Hillert M and Staffanson L-I The regular solution model for stoichiomertric phases and ionic melts 1970 *Acta Chem. Scand.* **24** 3618-3626

[2] *Materials Modeling Series: SGTE bookcase. Thermodynamics* 1996 (London: Inst. of Materials)

[3] Golod V M, Savel'ev K D and Basin A S 2008 *Modeling and Computer Analysis of Crystallization of Multicomponent Iron-based Alloys* (Saint Petersburg: Polytechnic Univ Press)

[4] Golod V M, Savel'ev K D, Dobosh L Yu and Orlova I G 2014 Computational thermodynamics – the tool of foundry technology (review and forecast) *Metallurgy of Mechanical Engineering* **6** 27-33

[5] Golod V M and Savel'ev K D 2010 *Computational Thermodynamics in Materials Science* (Saint Petersburg Polytechnic Univ Press)

[6] Golod V M 2008 Computer analysis of foundry technology. Problems of its information support and adaptation to the conditions of production *Bull. Udmurt University: Physics. Chemistry* **1** 67-87

[7] Dobosh L Yu and Golod V M 2015 Prediction of microstructure parameters of ternary aluminum alloys on the basis of modeling non-equilibrium crystallization *Scientific and Technical Bull.* **3** 92-101

[8] Golod V M, Dobosh L Yu and Savel'ev K D 2014 Assessing the adequacy of the model of additive components influence on the crystallization of aluminum alloys *Proc. 10-th Int. Conf. Foundry today and tomorrow* 366-376

[9] Ohnaka J 1986 Mathematical analysis of solute redistribution during solidification with diffusion in solid phase *Trans. ISIJ* **26** 1045-1051

[10] Golod V M and Emel'yanov K I 2014 System analysis of morphological evolution of the dendritic steel structure *Ferrous Materials* **4** 49-54

[11] Golod V M, Emel'yanov K I and Orlova I G 2013 The dendritic heterogeneity of cast steel: an overview of problems and computer analysis (Part 3) *Ferrous Materials* **11** 18-25

[12] Han Q, Hu H and Zhong X 1997 Models for the isothermal coarsening of secondary dendrite arms in multicomponent alloys *Metall. Mater. Trans.* B **28** 1185-1187

[13] Bouchard D and Kirkaldy JS 1997 Prediction of dendrite arm spacings in unsteady and steady-state heat flow of unidirectionally solidified binary alloys *Metall. Mater. Trans. B* **28** 651-663

[14] Rodrigues J R P, Melo M L N M and Santos R G 2010 Effect of magnesium content on thermal and structural parameters of Al-Mg alloys directionally solidified *J. Mater. Sci.* **45** 2285-2295

[15] Berkdemir A and Gündüz M 2009 Effect of growth rate and Mg content on dendrite tip characteristics of Al–Cu–Mg ternary alloys *Appl. Phys. A* **96** 873-886

[16] Dobosh L Yu and Golod V M 2013 The influence of composition multicomponent aluminum alloys on the secondary arm spacing of dendrites *Foundryman of Russia* **3** 35-39

IOP Conf. Series: Materials Science and Engineering **192** (2017) 012028 doi:10.1088/1757-899X/192/1/012028

Solidification kinetics of a Cu-Zr alloy: ground-based and microgravity experiments

P K Galenko [1,2], R Hanke [1], P Paul [1], S Koch [1], M Rettenmayr [1], J Gegner [3], D M Herlach [3,4], W Dreier [5] and E V Kharanzhevski [5]

[1] Friedrich-Schiller-Universität Jena, Otto-Schott-Institut für Materialforschung, D-07743 Jena, Germany
[2] Ural Federal University, Laboratory of Multi-Scale Mathematical Modeling, 620002 Ekaterinburg, Russia
[3] Deutsches Zentrum für Luft- und Raumfahrt, Institut für Materialphysik im Weltraum, 51170 Köln, Germany
[4] Ruhr-Universität, Physikalisch-Astronomische Fakultät, 44810 Bochum, Germany
[5] Udmurt State University, Faculty of Physics and Energetics, 426037 Izhevsk, Russia

E-mail: peter.galenko@uni-jena.de

Abstract. Experimental and theoretical results obtained in the MULTIPHAS-project (ESA-European Space Agency and DLR-German Aerospace Center) are critically discussed regarding solidification kinetics of congruently melting and glass forming $Cu_{50}Zr_{50}$ alloy samples. The samples are investigated during solidification using a containerless technique in the Electromagnetic Levitation Facility [1]. Applying elaborated methodologies for ground-based and microgravity experimental investigations [2], the kinetics of primary dendritic solidification is quantitatively evaluated. Electromagnetic Levitator in microgravity (parabolic flights and on board of the International Space Station) and Electrostatic Levitator on Ground are employed. The solidification kinetics is determined using a high-speed camera and applying two evaluation methods: "Frame by Frame" (FFM) and "First Frame – Last Frame" (FLM). In the theoretical interpretation of the solidification experiments, special attention is given to the behavior of the cluster structure in $Cu_{50}Zr_{50}$ samples with the increase of undercooling. Experimental results on solidification kinetics are interpreted using a theoretical model of diffusion controlled dendrite growth.

1. Introduction

A valuable progress in the investigation of glass forming metals and alloys has been made since their discovery by Turnbull [3]. The development of containerless processing techniques, especially using different methods of levitation, provided a new ability to reach deep undercoolings and to investigate solidification into the crystalline or glassy state [1]. A favourable advantage of the containerless methods is the prevention of heterogeneous nucleation, which in turn allows for a better investigation of the effects of high undercoolings and rapid growth conditions on the solidification of metals and alloys.

The present work is devoted to a systematic view on recent results obtained regarding solidification of the congruently melting and glass forming $Cu_{50}Zr_{50}$ alloy. Wang et al. [4] showed that CuZr exhibits a maximum growth velocity at an undercooling of ~200 K and a steep decline at higher undercoolings, with the according deviation from the Arrhenius law in the dynamic viscosity where the structure of the liquid changes, as pointed out by Kelton et al. [5]. Furthermore, Kobold [6] finds an additional growth path with lower velocities than the ones found by Wang et al., with the question arising where this comes from. Motivation for the present research is that a theoretical model provided by H. Wang et al. [7] for solidification kinetics does not sufficiently explain all experimental data points obtained by Q. Wang et al. [4] and thus the need for a more accurate model arises. In order to study the peculiarities of $Cu_{50}Zr_{50}$, ground-based and microgravity experiments have recently been carried out in parabolic flights.

2. Experimental methods

Electrostatic and electromagnetic levitation techniques, where the heterogeneous nucleation may effectively be prevented (at least from the outer surface of the sample), are employed to investigate the kinetics of dendritic solidification under the conditions of normal gravitational forces (1g) and microgravity (μg).

Electromagnetic levitation [8] is used in terrestrial laboratories at DLR Köln and Jena University (see Figure 1). The temperature is measured contactless by a two-colour pyrometer with an accuracy of ±5 K. An alternating electromagnetic field is causing both a levitation force and heating of the sample due to eddy currents. No extra heating source is required. Levitation and heating cannot be controlled separately. Therefore, a helium gas stream is needed to transfer the heat from the sample to the environment.

The limitation of coupled heating and levitating is overcome if electromagnetic levitation is applied in microgravity. Using the concept of two independently working coil systems, a quadrupole field for positioning, and a dipole field for heating, positioning and heating can be controlled separately at an increase of the efficiency from 1 to 30 % [9]. An Electromagnetic Levitator (EML) is in use on board the International Space Station.

To measure the dendrite growth velocity, a high-speed camera is used that records the advancement of the intersection line of the solidification front with the sample surface with 500 frames/sec [10].

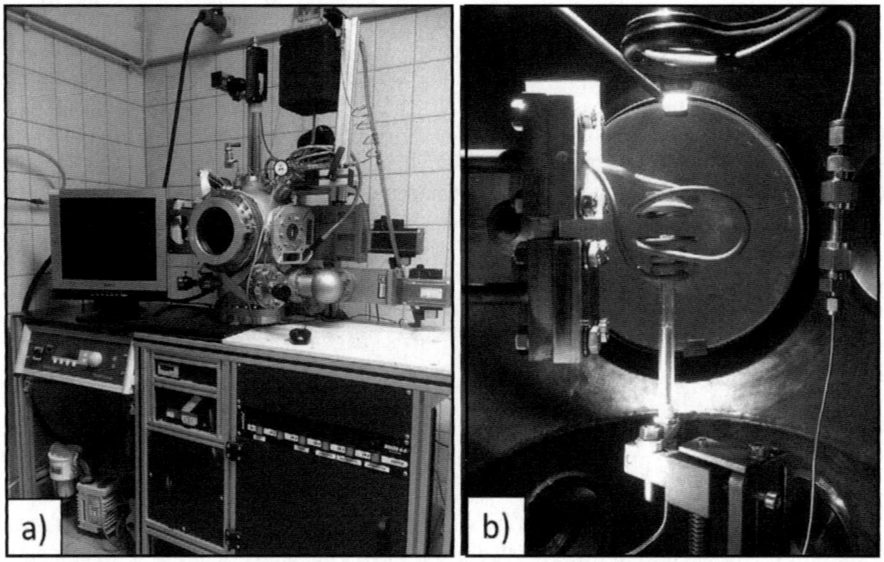

Figure 1. Experimental setup of electromagnetic levitation facility at Otto Schott Institute of Materials Research at Friedrich Schiller University Jena. (a) A general view of EML device. (b) Sample holder with a cooling system and an alternating current coil.

3. Methods of evaluation of solidification kinetics

The experimental solidification velocity for an initial undercooling is evaluated by recalescence fronts. Under fronts of recalescence we understand an envelope of growing dendrite tips, nucleation points or corners of crystalline patterns having specific crystallographic orientation. Figure 2 shows snapshots of a propagating recalescence front, defined by an envelope of the dendrite growth area.

Two different methods for evaluation of crystal growth kinetics are used. These are: evaluation with the Frame-by-Frame Method (FFM) and the First Frame-Last Frame Method (FLM). The FLM measures the growth velocity as a function of droplet radius R and the time interval Δt_{tot} from the beginning of dendrite growth until its end:

$$v_{FLM} = \frac{2R}{\Delta t_{tot}}. \quad (1)$$

Proper measurement is achieved by setting up the camera and a mirror to record images of a large part of the surface area of the sample. Details can be further examined by mapping these images onto a 3d spherical model [6]. Therefore, the FFM is used for the estimation of growth velocity as a function of the evolving growth area Δr (estimated as a spherical area) during the interval between two frames Δt_{fr}:

$$v_{FFM} = \frac{\Delta r}{\Delta t_{fr}}. \quad (2)$$

One can obviously suppose that the calculated growth velocities using FLM and FFM should be (almost) the same if only one nucleation center is observed (as shown in Fig. 2). Four videos with recalescence occuring at various undercoolings were analyzed. As shown in Table 1, indeed, measurements with just one recalescence (nucleation) center give approximately the same growth velocity measured by these two methods. In contrast, three recalescence centers yield a large difference in the measured velocities.

Table 1. Data of microgravity investigations: measured dendrite growth velocities (evaluated by FLM and FFM) in undercooled $Cu_{50}Zr_{50}$ alloy samples processed in the Electromagnetic Levitator during parabolic flights.

No. experiment	Undercooling (K)	Recalescence centers (nuclei points)	Velocity by FLM (mm/s)	Velocity by FFM (mm/s)
00	95	1	8,61	9,77
01	170	1	13,61	11,50
04	145	1	21,85	17,68
05	195	3	31,71	11,69

4. Results

Data sets for dendrite growth velocities in CuZr samples have been obtained under microgravity conditions in parabolic flights (Airbus A310 Zero-G). During these flights, the airplane is accelerating periodically until it reaches an angle of 47 degrees, then stopping acceleration and starting the descent and microgravity phase for about 20 seconds. This flight pattern results in parabolae - hence the name - with microgravity phases persisting for each parabola between the reversal points of the flight curve. As a result, during every 20-22 seconds, measurements of dendrite growth velocity are obtained.

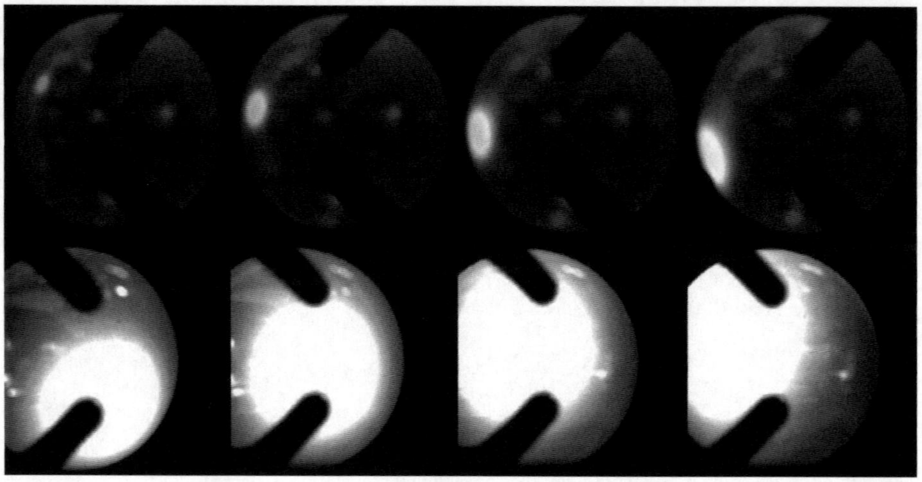

Figure 2. Different stages of solidification front evolving from one nucleation center (bright spot on top left) at an undercooling of 145 K. Top left: a few milliseconds after the start of growth. Top right: after 7 ms. Bottom left: after 76 ms. Bottom right: after 84 ms.

The growth velocities obtained via FLM and FFM are compared with previous studies in Fig. 3. Wang et al. [4] and Kobold [6] find a maximum in velocity for $Cu_{50}Zr_{50}$ at undercoolings in the region of 200 K as well as a steep decline of velocities at undercoolings above 250 K. Furthermore, Kobold [6] obtained a second growth regime (see blue squares for the low velocity sequence of data in Fig. 3) which is also confirmed by our present measurements (see three points obtained by both FLM and FFM in the region of undercooling 170-195K). Therefore, to explain this second behavior, the model of the diffusion-limited growth should be extended for prediction of metastable phase formation taking into account the process of phase selection.

For the undercooling of 95 K, FLM and FFM agree with other data points. This also applies to FFM at 145 K, although the FLM value appears to be somewhat too high. At 170 K, both values reside near the lower growth regime. At the undercooling of 195 K there is a strong deviation between FLM and FFM. The FLM value is nearly three times as high as the FFM value. Evaluation of the corresponding video shows three recalescence centers that lead to assuming too fast solidification if FLM is utilized. However, using the FFM, the recently measured velocity is close to the lower data points (see blue squares in Fig. 3).

Experimental data on growth kinetics represent a special interest for our theoretical advancement in solidification kinetics. Using a dendrite growth model based on the diffusion-limited approach to glass forming alloys, Wang et al. [7] provided a comparison of their calculation with experimental data. Figure 3 shows this comparison (see the solid line as predicted by Wang et al.). It can be seen that the modeling results describe the experimental data quite well up to the maximum velocity around the undercooling of 210-220 K. The velocity maximum in Fig. 3 is consistent with the point where gradual freezing of atoms and the decrease of their mobility starts such that the transition to sluggish growth kinetics occurs.

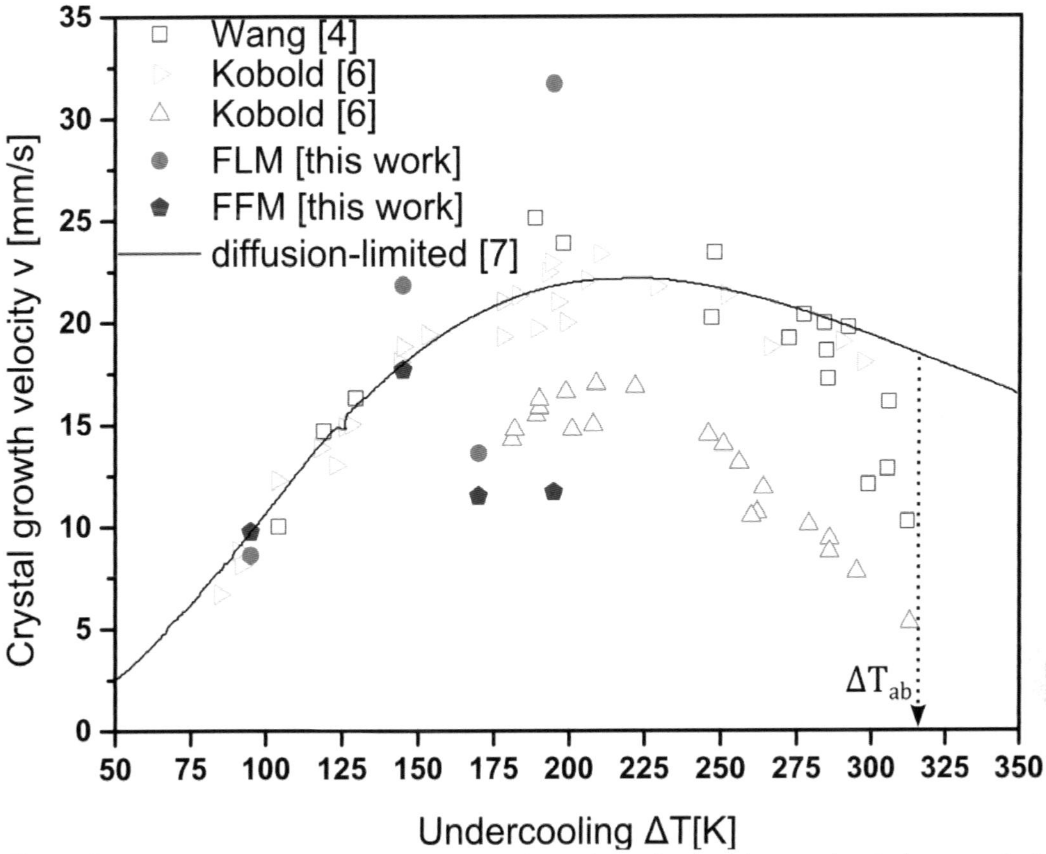

Figure 3. Solidification kinetics in $Cu_{50}Zr_{50}$. Experimental data for dendrite growth velocity were obtained by different authors (see insert). Theoretical model is given by Wang et al. [7] using a diffusion-limited approach. Results of the present work have been obtained using high-speed camera in an Electromagnetic Levitator under microgravity conditions during parabolic flights. They are shown for two different evaluation methods of recalescence fronts velocities: Frame by Frame Method (FFM) and First Frame-Last Frame Method (FLM). The critical undercooling ΔT_{ab} (or range of undercooling around ΔT_{ab}) shows the undercooling where an abrupt decrease of the solidification velocity in experimental data occurs – other than in the theoretical curve showing monotonic and gradual decrease of the velocity.

As is shown in Fig. 3, at undercoolings above the one with the velocity peak, the theoretical curve increasingly overestimates the experimental data of Wang et al. [4]. At the undercooling $\Delta T_{ab} \approx 320K$, experimentally measured velocities abruptly decrease. Note that such an abrupt transition has also been confirmed in other glass-forming Zr-Cu-Ni alloys with primary solidification of the CuZr-phase [11].

5. Discussion

As we have shown in the previous section, the solidification kinetics of the glass forming CuZr alloy exhibits several distinguished features that are not described theoretically so far (see Fig. 3). These features are related to the clearly observable velocity peak, the second sequence of experimental data for lower velocity values (in comparison with the main and complete data for the upper velocity sequence), and the abrupt decrease of the velocity at very high undercoolings. In the present section, we only discuss them in the light of experimental and modeling data for the dynamic viscosity. Theoretical predictions on the basis of these data will be presented and discussed in forthcoming publications.

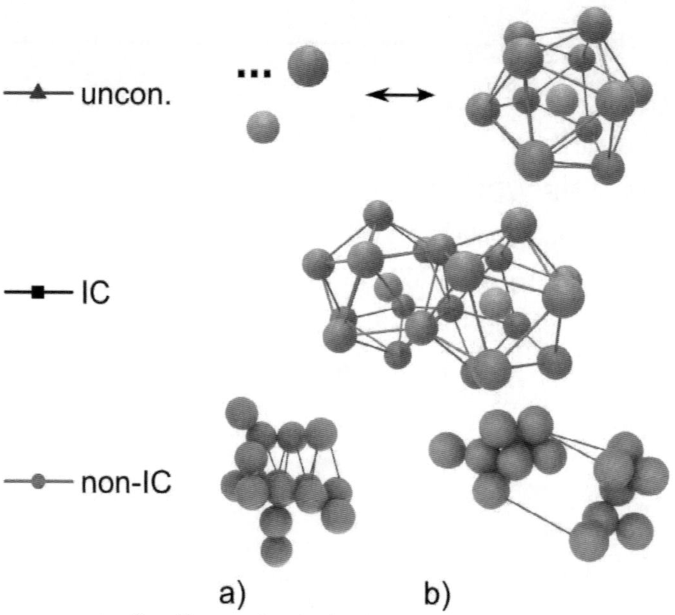

Figure 4. Cluster structures in $Cu_{50}Zr_{50}$ melt obtained *via* molecular dynamics simulations and Voronoi tesselation analysis [12]. Top: "uncon." structures are represented by unconnected single Zr or Cu atoms (left) and by single clusters (right). Middle: IC structures are represented by interpenetrating clusters – five atoms are shared between two shells. Bottom: non-IC structures are represented by vertex-, edge- or face-sharing clusters with a) pinning and/or b) bridging connection. Note that icosahedral clusters are unique structural elements of every non-IC-structure which thus is represented by a net of interpenetrating clusters.

5.1. Viscosity behaviour

Figure 4 depicts atomistic structures obtained in molecular dynamics simulations [12]. They are represented by isolated atoms or clusters that one can consider as a completely disordered structure of the overheated to slightly undercooled $Cu_{50}Zr_{50}$ melt (top of Fig. 4). In the deeply undercooled $Cu_{50}Zr_{50}$ melts, in Ref. [12] a double-connected or multiple-connected cluster structure was found (middle and bottom row of Fig. 4). These connected structures already show features of the ordering process in the undercooled melt, with a gradual transition from a "strong" liquid phase to "fragile" liquid phase (so-called fragility transition). Dominant structures in the liquid states are of controversy. If a coordination number of 12 is set, then icosahedral clusters as Kelton et al. via molecular simulations [12] suggest can be presumed present. This however was not confirmed experimentally by Holland-Moritz et al. [13] who find the nearest-neighbor coordination number Z_{NN} to be 13,8 and therefore the corresponding structure cannot be of icosahedras primarily but an aggregate of different other polyhedral structures. For the present consideration, it is not important which type of cluster structure exists in the undercooled melt. The present analysis takes into account clusterization and net of cluster construction as a process leading to glass formation. Critical points for beginning, developing and freezing of net of clusters are important for our understanding of the viscosity behaviour and kinetics of solidificiation.

Figure 5 (top) shows the structural ordering in CuZr melts as a content in percent amongst clusters. These results were obtained using molecular dynamic simulations in different states, particularly the overheated state $T>T_l$ state, at the melting temperature $T=T_l$, in the supercooled state $T<T_l$ and in a glassy state, $T\leq T_g$ by Kelton et al. [12]. These authors found three distinguishable structures, which are a) unconnected structures constantly building and collapsing from Cu and Zr atoms, b) interpenetrating clusters, where five atoms coincide on two shells (IC) and c) vertex-, edge-, or face-sharing polyhedrals with a pinning and/or bridging connections between Cu atoms (non-IC).

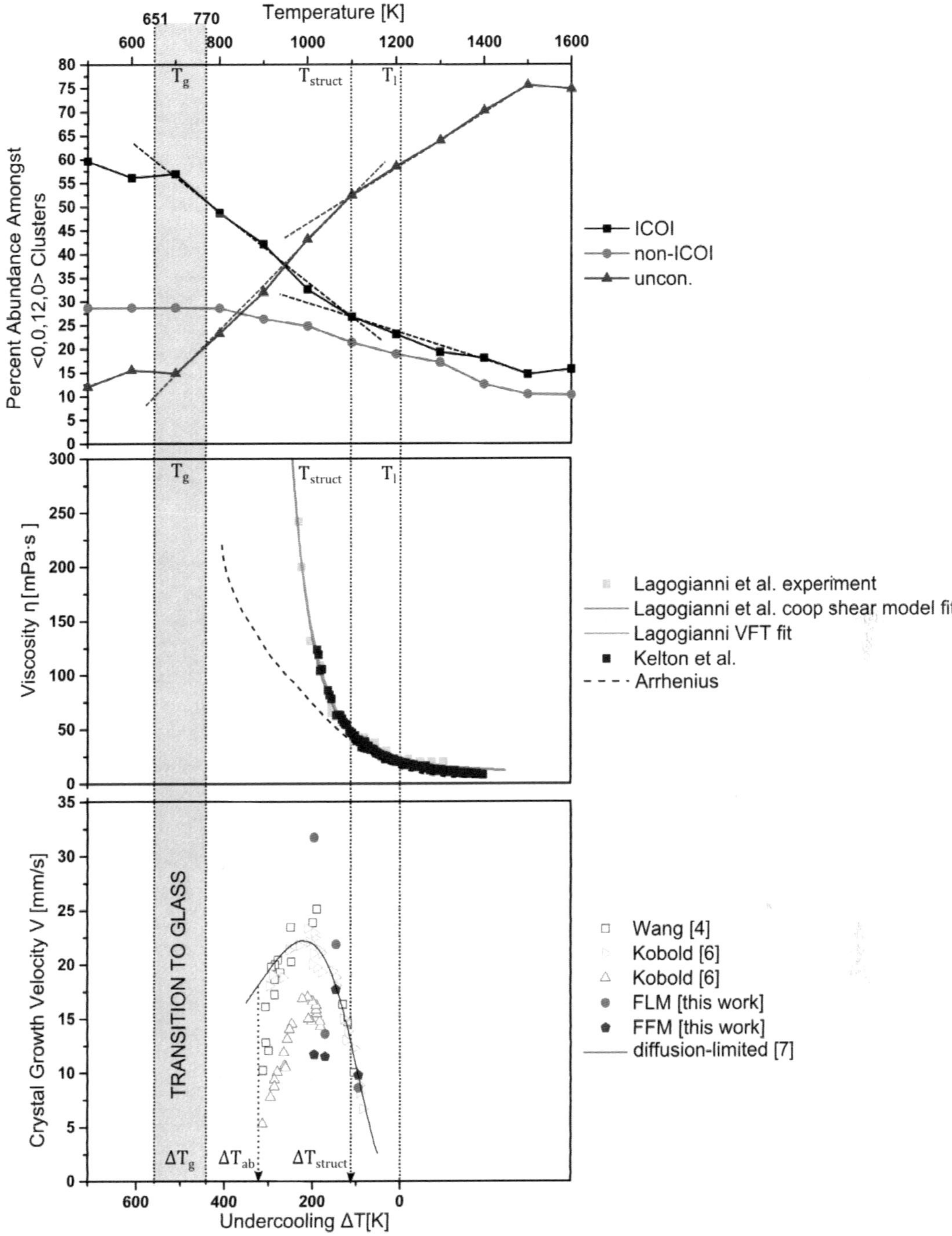

Figure 5. Top: structural ordering in the $Cu_{50}Zr_{50}$ melt obtained via molecular dynamics simulations and Voronoi tesselation analysis [12]. Uncon.- IC, non-IC, pinning and/or bridging connected structures are shown in Fig. 4. Middle: viscosity data by Kelton and Laggagio, fitting curves calculated using the cooperative shear model and Vogel-Fulcher-Tamman equation (VFT) and sketch of Arrhenius behavior (dashed line). Bottom: dendrite growth velocity as a function of undercooling. Groups of points (obtained in EML under different conditions) clearly show upper and lower branches in solidification kinetics.

Overheated and slightly undercooled liquid, i.e. in the temperature range $T_{struct}<T<T_l$, is characterized by a gradual decrease of the number of unconnected clusters and single atoms. The number of interconnected clusters, IC, gradually increases in this region. At $T=T_{struct}$, the intensity of cluster formation drastically changes, exhibiting a breakpoint from which IC forms more readily. Interconnected clusters (IC) rapidly form a cluster net, which results in a structural transition from a strong liquid (atomically disordered liquid) to a fragile liquid (atomically ordered liquid). This fragility transition ends at the glass temperature T_g at which a transition from the liquid state to the amorphous state occurs.

These structural changes strongly influence the viscosity as a function of temperature. Indeed, Kelton et al. [14] and Lagogianni et al. [15] studied the viscosity behavior in superheated and supercooled CuZr melts. In a modeled fit to their data sets [14,14], Lagogianni et al. applied the viscosity data most accurately in the whole temperature range of the viscosity change. With a cooperative shear model for the glass transition the derive a double-exponential dependence of viscosity from temperature (non-Arrhenius behavior) according to

$$\frac{\eta(T)}{\eta_0} = \exp\left\{\frac{V_C C_G}{kT} \exp\left[(2+\lambda)\alpha_T T_g \left(1 - \frac{T}{T_g}\right)\right]\right\} \tag{3}$$

with η_0 being a normalisation constant set by the high temperature limit of η, V_C being the characteristic atomic volume, C_G the shear modulus value at the glass transition temperature T_g, λ a variable expressing all salient features of interatomic interactions and containing the effect of repulsion steepness, and α_T the anharmonicity. A plot of Eq. (3) is shown in Fig. 5(middle), fitting the data sets of the groups of both Lagogianni and Kelton.

Taking into account Fig. 5(top) and evaluating the changes in behavior for unconnected / connected clusters / IC, the temperature for the beginning of the structural changes can be estimated as $T_{struct}\approx 1100$ K. At $T=T_{struct}$, a change from Arrhenius behavior to super-Arrhenius behavior of viscosity may occur as suggested by Eq. (4) and as drawn in Fig. 5 (middle). The change in behavior can be displayed more clearly when plotting $\ln\frac{\eta}{\eta_0}$ over 1000/T as an Arrhenius plot; the temperature below which a deviation from Arrhenius behavior occurs can then clearly be seen[1]. Such structural changes may drastically influence the solidification kinetics of the undercooled $Cu_{50}Zr_{50}$ alloy melts.

The transition to the glassy state in Fig. 5 is shown as a shaded region of temperatures ranging from 651 K to 770 K. The existence of a temperature range for the transition is not surprising, because the temperature T_g characterizes a freezing of the non-equilibrium state which may vary due to different solidification conditions into the non-ergodic glassy state. Therefore, we have just marked the temperature interval in which various authors found the amorphous phase in solidification of $Cu_{50}Zr_{50}$ melts [4,15,17,18,19].

5.2. Solidification kinetics

The described changes in the liquid cluster structure and the viscosity behavior of liquids drastically influence the kinetics of dendritic growth in the undercooled $Cu_{50}Zr_{50}$ melts. This follows straightforwardly from the solidification growth velocity V described in Ref. [20]

$$V = \beta_K\left[1 - \exp\left(-\frac{\Delta G}{RT}\right)\right], \tag{4}$$

[1] Kelton et al. [16] have estimated this significant temperature as $T_{coop}\approx 1283$ K, 183,72 K higher than T_{struct} and indicating that the change in the cluster behavior proceeds at higher temperatures.

where the Gibbs free energy change upon solidification, ΔG, has been estimated by Thompson and Spaepen [21] as

$$\Delta G(T) = \frac{\Delta H_m \Delta T}{T_l}\left(\frac{2T}{T_l+T}\right) \quad (5)$$

with the undercooling $\Delta T = T_l - T$. As stated by Jäntsch [21], the kinetic growth coefficient β_k is inversely proportional to the dynamic viscosity η, i.e. $\beta_k \propto \eta^{-1}$. For the diffusion-limited process of atom attachment to the solid through the solid-liquid interface, Orava and Greer suggest the following expression [20]:

$$\beta_k = \frac{kT}{3\pi a^2 \eta}, \quad (6)$$

where a is an effective atom diameter. As a result, cluster evaluation and viscous behavior give a qualitative answer on the non-linear relationship velocity vs. undercooling for the dendrite growth kinetics in undercooled melts of $Cu_{50}Zr_{50}$.

Figure 5 (bottom) further demonstrates dendrite growth velocities as a function of undercooling in consistence with the transitive temperature intervals of the cluster changes and viscosity behavior. The undercooling ΔT_{struct} is marked at which the deviation from the linear behavior $V = \beta_k \Delta T$ begins. Comparing with the marked temperature $T=T_{struct}$, one should state that the beginning of the sharp increase of connected clusters, Fig. 5 (top), and the beginning of super-Arrhenius law in viscosity, Fig. 5 (middle), are directly related to the beginning of non-linearity in the solidification velocity by the full equation (4).

At undercoolings below the one that correlates with the velocity maximum (see short explanations for Fig. 3), Fig. 5 (bottom) also exhibits the second critical undercooling ΔT_{ab} which shows an abrupt decrease of the solidification velocity (also marked in the diagram "velocity-undercooling" of Fig. 3). This abrupt drop of the growth velocity at the largest undercooling has also been observed in $Zr_{50}Cu_{30}Ni_{20}$ [11] and in $Ni_{40}Al_{60}$ [23]. Considering that this drop occurs at a temperature close to the glass temperature ($\approx$ 60 K + T_g), it can be attributed to the fact that in this temperature region the single atom movement changes to a cooperative movement of atoms according to the mode coupling theory [24]. This is consistent with the fact that also the temperature dependence of the self-diffusion coefficient shows deviations from an Arrhenius-like behavior in this temperature range [25]. To describe this steep decrease at ΔT_{ab}, the model of diffusion-limited growth [7] should be expanded.

6. Conclusions

In the present work, a systematic review of experimental data on solidification of $Cu_{50}Zr_{50}$ alloy melts is presented. The data are collected from results of droplet solidification in different levitators (EML and ESL) used on Ground-based research and during parabolic flights in the TEMPUS facility. Structural changes of clusters in the melt drastically change dynamic viscosity and dendrite growth velocity in the undercooled melt. Theoretical models of solidification and amorphization can be developed using the present analysis of clustering and structural changes in undercooled glass forming alloy melts. In this perspective, the model of the diffusion-limited growth [7] should be expanded concerning two issues: first, the model should incorporate metastable phase formation, taking into account phase selection; second, the model should include the specific processes at T_{struct} and T_{ab} to describe the abrupt decrease of the growth velocity at high undercoolings.

7. References

[1] Herlach D M, Galenko P K and Holland-Moritz D 2007 *Metastable Solids From Undercooled Melts* (Elsevier: Amsterdam)
[2] Binder S, Galenko P K and Herlach D M 2014 The effect of fluid flow on the solidification of Ni_2B from the undercooled melt *Journal of Applied Physics* **115** 053511-1-11
[3] Turnbull D 1950 Formation of crystal nuclei in liquid metals *J. Appl. Phys.* **21** 1022–1028
[4] Wang Q et al. 2011 Diffusion-controlled crystal growth in deeply undercooled melt on

approaching the glass transition *Physical Review B* **83** 014202-1-014202-5

[5] Blodgett M E, Egami T, Nussinov Z. and Kelton K F 2015 Proposal for universality in the viscosity of metallic liquids *Sci. Rep.* **5** 13837

[6] Kobold R 2016 Crystal growth in undercooled melts of glass forming Zr-based alloys *PhD-Dissertation* (*Ruhr-University: Bochum)*

[7] Wang H, Herlach D M, Liu R P 2014 Dendrite growth in Cu50Zr50 glass-forming melts, thermodynamics vs. kinetics *EPL (Europhysics Letters)* **105** 36001

[8] Herlach D M 1991 Containerless Undercooling and Solidification of Pure Metals *Annual Review of Materials Science* **21** 23-44

[9] Herlach D M, Willnecker R and Lohöfer G 1989 *German Patent* DE 3639973

[10] Volkmann T 2012 *In Solidification of Containerless Undercooled Melts (*Wiley-VCH)

[11] Galenko P K, Hanke R, Paul P, Kobold R, Koch S and Rettenmayr M 2016 Unpublished results from parabolic flights

[12] Soklaski R, Nussinov Z, Markow Z, Kelton K F and Yang L 2013 Connectivity of the Icosahedral Network and a Dramatically Growing Static Length Scale in Cu-Zr Binary Metallic Glasses *Phys. Rev. B* **87** 184203

[13] Holland-Moritz D, Yang F, Kordel T, Klein S, Kargl F, Gegner J, Hansen T, Bednarcik J, Kaban I, Shuleshova O, Mattern N and Meyer A 2012 Does an icosahedral short-range order prevail in glass-forming Zr-Cu melts? *EPL* **100** 56002

[14] Mauro N A, Blodgett M, Johnson M L, Vogt A J and Kelton K F 2014 A structural signature of liquid fragility *Nature Communications* **5** 4616

[15] Lagogianni A E, Krausser J, Evenson Z, Samwer K and Zaccone A 2016 Unifying interatomic potential, g(r), elasticity, viscosity, and fragility of metallic glasses: analytical model, simulations, and experiments *Journal of Statistical Mechanics: Theory and Experiment* 084001

[16] Blodgett M, Egami T, Nussinov Z and Kelton K F 2014 Unexpected Universality in the Viscosity of Metallic Liquids arXiv:1407.7558

[17] Lee D, Zhao B, Perim E, Zhang H, Gong P, Gao Y, Liu Y, Toher C, Curtarolo S, Schroers J and Vlassak J J 2016 Crystallization behavior upon heating and cooling in Cu50Zr50 metallic glass thin films *Acta Materialia* **121** 68-77

[18] Wang W H, Lewandowski J J and Greer A L 2005 Understanding the glass-forming ability of Cu50Zr50 alloys in terms of a metastable eutectic *J. Mater. Res.* **20** 2307–2313

[19] Yu P, Bai H Y and Wang W H 2006 Superior glass-forming ability of CuZr alloys from minor additions *Journal of Materials Research* **21** 1674–1679

[20] Orava J, Hewak D W and Greer A L 2015 Fragile-to-Strong Crossover in Supercooled Liquid Ag-In-Sb-Te Studied by Ultrafast Calorimetry *Advanced Functional Materials* **25** 4851–4858

[21] Thompson C V ,Spaepen F 1979 On the approximation of the free energy change on crystallization *Acta Metallurgica* **27** 1855-1859

[22] Jäntsch O 1956 *Zur Theorie der linearen Kristallisationsgeschwindigkeit unterkühlter Schmelzen - first part of dissertation* (Humboldt-University: Berlin)

[23] Lengsdorf R, Holland-Moritz D and Herlach D M 2009 Anomalous dendrite growth in undercooled melts of Al–Ni alloys in relation to results obtained in reduced gravity *Scripta Materialia* **62** 365–367

[24] Das S P 2004 Mode-coupling theory and the glass transition in supercooled liquids *Rev. Mod. Phys.* **76** 785-845

[25] Yang F, Holland-Moritz D, Gegner J, Heintzmann P, Kargl F, Yuan C C, Simeoni G G and Meyer A 2014 Atomic dynamics in binary Zr-Cu liquids *EPL* **107** 46001

IOP Conf. Series: Materials Science and Engineering **192** (2017) 012028 doi:10.1088/1757-899X/192/1/012028

Acknowledgements
P.K.G. R.H., Ph.P., S.K., M.R., D.M.H. and W.D. are grateful to the German Space Agency for funding contract number 50WM1541, Project MULTIPHAS. P.K.G. also acknowledges the support from the RSF (project no. 16-11-10095). E.V.H. acknowledges financial support from the Russian Space Agency and TSNIIMash under project "Kinetika" (contract No. 0521).

IOP Conf. Series: Materials Science and Engineering **192** (2017) 012029 doi:10.1088/1757-899X/192/1/012029

Temperature and concentration dependences of viscosity of $Fe_{85-x}Cr_{15}C_x$ (x=10-17) melts

I V Sterkhova, L V Kamaeva, V I Lad`yanov and A Yu Korepanov

Physical-Technical Institute of Ural Branch of Russian Academy of Sciences, 132 Kirov Street, Izhevsk, Russia

E-mail: irina.sterkhova@mail.ru

Abstract. The investigations of temperature and concentration dependences of viscosity of $Fe_{85-x}Cr_{15}C_x$ (x=10-17) melts have been performed. It has been shown that the viscosity polytherms of the studied melts do not have any peculiarities in the entire studied temperature range and are well described by Arrhenius-type monotonic dependences. It has also been shown that there is no strong dependence between the viscosity of $Fe_{85-x}Cr_{15}C_x$ melts at x=10-17 at.%C and carbon concentration, and only close to eutectic concentration a sharp viscosity increase is observed.

1. Introduction

Among many physicochemical characteristics of metal melts viscosity is especially important, since being highly sensitive to the structural changes in a system. It attracts much attention as a basic technological parameter of a metallurgical process.

A Fe-C system is a basic system for the majority of steels and alloys that are widely applied as functional and structural materials. The addition of Cr into the system results in carbide phase formation, which increases strength and wear properties of these materials. Owing to these improved mechanical characteristics, including high resistance to various types of wear (abrasion, erosion, corrosion, etc.) the alloys Fe-Cr-C are widely used in modern industry as hardfacing materials [1, 2]. Depending on a practical purpose composition of these alloys can be considerably different. Of special importance in estimating abrasive wear at low stress is carbon content [1, 2]. In alloys with high content of carbon a carbide M_7C_3 is formed surrounded by softer austenitic matrix. This results in cracking at the interfaces and hinders the application of such materials without additional treatment. The optimum relationship of components in a ternary Cr-Fe-C system depends on the region of existence of $M_{23}C_6$ and eutectic [1, 2]. Apart from the choice of the chemical composition of an alloy its structure and properties depend on the liquid phase state prior to hardening as well as cooling conditions [3]. Direct structural research devoted to high-temperature melts is hindered by the complexity of experiment and data interpretation, and hence, is not practically performed. Abrupt changes of atomic short-range order in a melt with temperature or concentration change are manifested on temperature and concentration dependences of structural-sensitive properties of a liquid phase. Therefore, the basic information about structural changes in liquid alloys is derived from the data about temperature and concentration dependences of their structural-sensitive properties. In this connection in this paper the investigation of kinematic viscosity of $Fe_{85-x}Cr_{15}C_x$ (x=10-17 at.%) melts has been performed. The concentration area studied in the Fe-Cr-C system includes near-eutectic compositions with carbide phases $(Fe,Cr)_{23}C_6$ or $(Fe,Cr)_7C_3$.

2. Experimental procedure

The alloys were prepared by alloying the corresponding proportions of $Fe_{75}Cr_{15}C_{10}$ and $Fe_{68}Cr_{15}C_{17}$ ligatures in a furnace of the viscometer at 1680°C for 30 min. The synthesis of ligatures was carried out according to the traditional metallurgical technology by alloying carbonyl iron powder (Ultra High Purity 13-2), chromium and graphite (HP) powders in a vacuum furnace at 1700°C for 30 min.

Kinematic viscosity, (ν), was measured by the method of damping torsional vibrations of the crucible filled with the melt. The detailed measurement and calculation techniques are described in [4]. The experiments were carried out in a crucible made of Al_2O_3 in a purified helium atmosphere. To prevent the uncontrolled effect of the oxide film a Al_2O_3 cover was placed into a crucible on top of the sample as a second face surface [5]. The viscosity polytherms were obtained in heating conditions starting from the liquidus temperature to 1680°C with subsequent cooling. The temperature is changed gradually with the step equal to 10°C. The specimens were held under isothermal conditions for 20 min at every temperature before measurement. Then the samples were cooled up to room temperature. Replicates of the experiments on the viscosity measurements (during heating and cooling) were made using the method described above. Maximum error of the viscosity measurement does not exceed 4%.

According to the results of a previous investigation on the viscosity of carbon-containing systems including a certain fraction of a carbide phase [6], thermodynamic equilibrium of the system after melting took place for a long time. For this reason prior to viscosity measurements the ingots were remelted in vacuum at 1600°C for 30 min.

Differential thermal analysis (DTA) carried out in a high-temperature thermal analyzer [7] was used to plot the phase diagram of the Fe-Cr-C system in a concentration area of interest. DTA of alloys was performed under heating in the range from 300°C to 1680°C at a heating rate of 20°C/min. The measurements were performed in Al_2O_3 crucibles in helium atmosphere. The resultant thermograms were used to determine the temperatures of phase transitions in the alloys in the temperature range under study, and these temperatures were afterwards used to plot the phase diagram.

3. Results and discussion

Figure 1 shows equilibrium temperature dependences of viscosity obtained at heating and further cooling of $Fe_{85-x}Cr_{15}C_x$ melts, where x=10-17 at.%. It is seen that with allowance for the measurement error, the heating and cooling polytherms coincide. They do not have any peculiarities (Figure 1a) and are well described by the Arrhenius equation (Figure 1b):

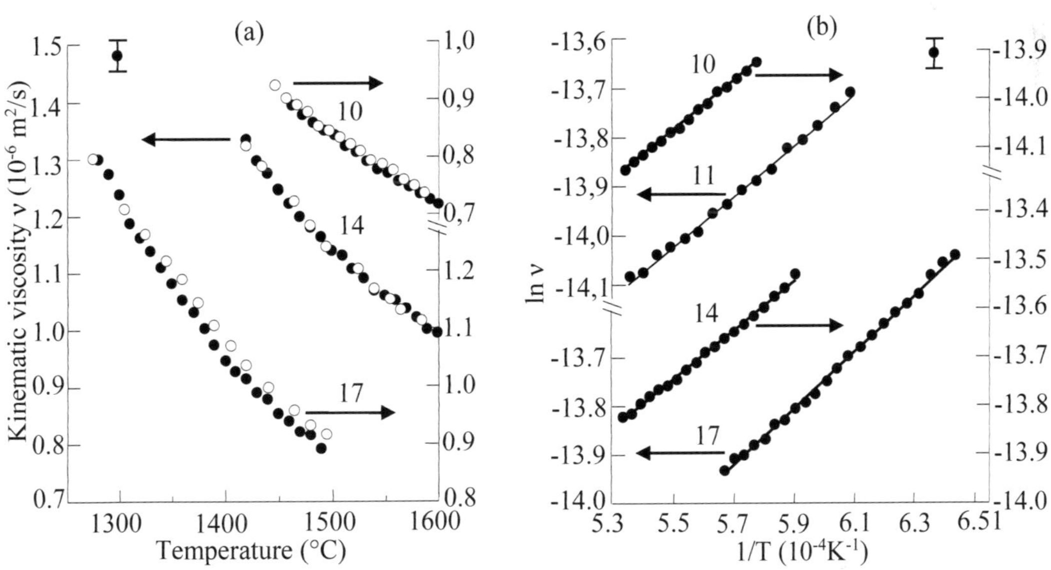

Figure 1. The temperature dependences of viscosity of $Fe_{85-x}Cr_{15}C_x$ melts (x=10-17 at.%C - the numeral near dependence), • – in heating regime, ○ – in cooling regime.

$$\nu = A_\nu \exp\left\{\frac{E_\nu}{RT}\right\}, \tag{1}$$

where Av - constant, Ev - activation energy of viscous flow, R – universal gas constant, T - absolute temperature. The repeated heating and cooling polytherms are highly reproducible, with allowance for an error.

The calculation of the concentration dependences of viscosity was based on two series of measurements of temperature dependences of ν (at the first heating – cooling regime and the repeated heating – cooling regime after the melt crystallization). The viscosity isotherms obtained are shown in Figure 2. Also, Figure 2 presents a pseudobinary section of the equilibrium phase diagram of the Fe-Cr-C system at 15 at.%Cr built using the DTA data. According to the results obtained, the viscosity values of the majority of the melts under study close to the melting point make up $9.5\cdot10^{-7}m^2/s$, which is in good agreement with the literature data. However, a sharp increase of ν values is observed close to 14 at.%C, which is appeared as the maximum on concentration dependences of viscosity in this concentration area. According to the phase diagram built, 14 at.%C corresponds to the eutectic point. A maximum on the viscosity isotherms close to eutectic is not typical of viscosity behavior for the systems metal – metalloid, where eutectic components in solid state are a metal-based solid solution and a chemical compound. The concentration dependence of viscosity in the eutectic area for such systems is usually characterized by a minimum [8, 9]. For better understanding of the concentration-related behavior of viscosity of Fe-Cr-C melts it is possible to use the analysis of the

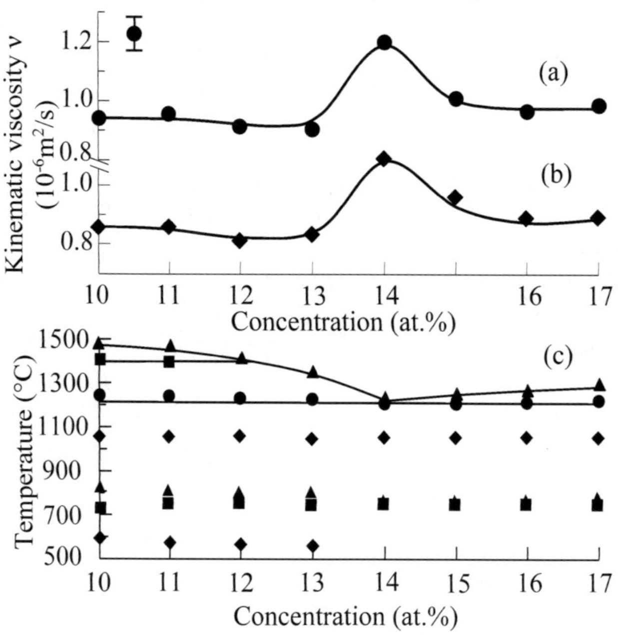

Figure 2. (a) and (b) The concentration dependences of viscosity of $Fe_{85-x}Cr_{15}C_x$ (x=10-17 at.%C) melts at different temperatures ((a) - at 1450°C, (b) - 1500°C). (c) The phase diagram of Fe-Cr-C system in investigated area.

concentration dependences of viscosity of binary Fe-Cr [10], Fe-C [11], and Cr-C [12] systems.

Although concentration dependences of viscosity of Fe-Cr melts are non-monotonic, the viscosity changes observed are far less than in systems Fe-C and Cr-C, and hence, can be neglected. Concentration dependences of viscosity of Fe-C melts at low content of C are also non-monotonic. However, at carbon content more than 10 at.% a sharp decrease of ν and no peculiarities in the viscosity polytherms are observed. For Cr-C melts the increase of carbon content from 10 to 17 at.% results in a sharp increase of viscosity, with viscosity maximum at 14 at.%C (which corresponds to an eutectic point (L→Cr(C)+$Cr_{23}C_6$) for a binary system Cr-C) [12].

It thus follows that the observed concentration dependence of viscosity of ternary liquid Fe-Cr-C alloys is a result of the combined influence of the concentration-related behavior of viscosity of its binary subsystems Fe-C and Cr-C.

4. Conclusion

The investigations of temperature and concentration dependences of viscosity of $Fe_{85-x}Cr_{15}C_x$ (x=10-17) melts have been performed by the method of damping torsional vibrations. It has been shown that the viscosity polytherms of the studied melts do not have any peculiarities in the entire studied temperature range and are well described by Arrhenius-type monotonic dependences. It has also been shown that there is no strong dependence between the viscosity of $Fe_{85-x}Cr_{15}C_x$ melts at x=10-17 at. %C and carbon content, and only close to eutectic concentration a sharp viscosity increase is observed. The concentration dependences of viscosity obtained for the ternary Fe-Cr-C system in the concentration area studied have been stated to be the result of concentration-related changes of viscosity in the binary systems Fe-C and Cr-C.

5. References

[1] Fan C, Chen M C and Wu W 2006 *Surface and Coating Technology* **201** 908
[2] Lin C M, Chang C M, Chen J H, Hsieh C C and Wu W 2009 *Metall. Mater. Trans. A* **40** 1031
[3] Lad`yanov V I, Sterkhova I V, Kamaeva L V, Chueva T R and Molokanov V V 2010 *J. Non-Cryst. Sol.* **356** 65
[4] Bel`tyukov A L and Lad`yanov V I 2008 *Instrum. Exp. Tech.* **2** 304
[5] Olyanina N V, Bel'tyukov A L and Lad'yanov V I 2015 *AIP Conference Proceedings* **1673** 020015
[6] Sterkhova I V, Kamaeva L V and Lad`yanov V I 2013 *Rasplavy* **3** 3
[7] Sterkhova I V and Kamaeva L V 2014 *J. Non-Cryst. Sol.* **401** 250
[8] Lad'yanov V I, Volkov V A, Kamaeva L V and Bel'tyukov A L 2003 *Russ. Metall.* **6** 38
[9] Lad'yanov V I, Kamaeva L V and Volkov V A 2006 *Russ. Metall.* **1** 8
[10] Kamaeva L V, Sterkhova I V and Lad`yanov V I 2012 *Inorganic Materials* **48** 318
[11] Elanskii G N and Kudrin V A 2015 *Bulletin of the South Ural State University. Serias Metallurgy* **15(3)** 11
[12] Sterkhova I V and Kamaeva L V 2014 *J. Non-Cryst. Sol.* **401** 241

Acknowledgments

This work has been carried out within the research project № AAAA-A17-117022250039-4.

IOP Conf. Series: Materials Science and Engineering **192** (2017) 012030 doi:10.1088/1757-899X/192/1/012030

Influence of tiny amounts of impurity on dendritic growth in undercooled melts

O V Kazak[1] P K Galenko[1,2] D V Alexandrov[1]

[1] Ural Federal University, Laboratory of Multi-Scale Mathematical Modeling, Department of Theoretical and Mathematical Physics, 620000 Ekaterinburg, Russian Federation
[2] Friedrich-Schiller-Universität Jena, Physikalisch-Astronomische Fakultät, D-07743 Jena, Germany

E-mail: olegkazak@yandex.ru · peter.galenko@uni-jena.de · dmitri.alexandrov@urfu.ru

Abstract. An influence of small amounts of impurity on the dendrite growth is investigated analytically and numerically. For investigations, a model for anisotropic dendrites rapidly growing in pure and binary melts is used. The dependence of the dendrite growth velocity and of the dendrite tip radius from undercooled melt from amounts of impurity is analyzed. Calculations are carried out for the pure nickel and binary nickel-zircon diluted melts.

1. Introduction

Crystallisation process is one of the most important phase transformations in industrial production. It plays a crucial role in casting and foundry industry. The present work is devoted to the kinetics of dendritic growth from an undercooled pure nickel melt in comparison with the dendrite growth kinetics in highly diluted and diluted nickel-zircon melts.

Solidification of undercooled Ni-Zr alloys has been investigated in [1,2] at small concentrations of Zr using dendritic growth velocities V measured in Ref. [1] as a function of undercooling ΔT in levitated droplets of $Ni_{99}Zr_1$ alloys using electromagnetic levitation technique [3]. These results have been described within the LKT/BCT model of dendrite growth [4, 5]. As it has been shown in [1,2], the increase of undercooling leads to solute trapping with the deviation from local chemical equilibrium at the solid-liquid interface in rapid solidification. In the LKT/BCT model of dendrite growth [4, 5], this phenomenon is introduced by the solute partitioning function $k(V)$ which adopts interfacial diffusion speed V_{DI} as a kinetic parameter of rapid solidification [6]. Experimentally, the diffusion speed V_{DI} has been independently determined by pulsed laser experiments [1] on thin specimens for the $Ni_{99}Zr_1$ alloy. Together with preliminary used parameters of Ni-Zr alloys, the measured results for V_{DI} encouraged the prospects of a parameter-free test of the LKT/BCT model. In addition, the dendritic structure of solidified droplets from $Ni_{99.5}Zr_{0.5}$ and $Ni_{99}Zr_1$ alloys has been evaluated [2] to find morphological transitions in dendritic pattern. The results of metallographic analysis were compared with the predictions of Karmas model [7, 8] on grain refinement *via* dendritic fragmentation.

Selected and novel results of computational experiments on rapid dendritic growth in Ni and Ni-Zr alloys are summarized in the present paper. Predictions of a model for rapid dendritic growth, which assumes deviation from local equilibrium at the interface and in the solute diffusion field, are presented. An influence of the small quantity of Zr on growth of nickel dendrites are shown.

2. Governing equations

The mathematical model for the dendrite growth is described and tested in a growth of TiAl crystals in [9]. Particularly, the growth of a parabolic needle-like dendrite with the four-fold symmetry of anisotropic properties has been analyzed with taking into account of local non-equilibrium conditions in the solute diffusion field and at the solid-liquid iterface. Neglecting the diffusion in the solid phase, the concentration field in the liquid is governed by the following mass transport equation

$$\tau_D \frac{\partial^2 C_l}{\partial t^2} + \frac{\partial C_l}{\partial t} + (\vec{w} \cdot \vec{\nabla}) C_l = D_C \nabla^2 C_l, \tag{1}$$

where C_l is the solute concentration in the liquid, $\vec{w}$ is the flow velocity, D_C is the diffusion coefficient of the solute, t is the time and τ_D is the relaxation time of the diffusion flux to its steady state.

Experimental measurements of fast dendritic growth cover the range of crystal growth velocities which can be smaller, comparable or even greater than the characteristic solute diffusion speed in bulk phases but they are still much smaller than the characteristic speed of heat propagation [10]. Therefore, while the mass transport can be described by the hyperbolic equation 1 which includes the finite value of the diffusion speed $V_D = (D_C/\tau_D)^{1/2}$, the evolution of the temperature field is described by a parabolic equation which assumes an infinite speed for the heat propagation. Designating T_l and T_s as temperatures in the liquid and solid phases, one can write

$$\frac{\partial T_l}{\partial t} + (\vec{w} \cdot \vec{\nabla}) C_l = D_T \nabla^2 T_l, \quad \frac{\partial T_s}{\partial t} = D_T \nabla^2 T_s, \tag{2}$$

where D_T stands for the thermal diffusivity.

At the dendritic interface, conservation of mass and energy yields boundary conditions of the form

$$\tau_D \frac{\partial}{\partial t} \left[(C_l - C_s) \vec{v} \cdot \vec{n} \right] + (C_l - C_s) \vec{v} \cdot \vec{n} + D_C \vec{\nabla} C_l \cdot \vec{n} = 0, \tag{3}$$

$$T_Q \vec{v} \cdot \vec{n} = D_T \left(\vec{\nabla} T_s - \vec{\nabla} T_l \right) \cdot \vec{n}, \tag{4}$$

where $\vec{n}$ is the unit vector normal to the dendrite interface, $\vec{v}$ is the interface velocity, $C_s = k_v C_l$ is the concentration in the solid phase at the dendritic interface, k_v is the velocity dependent solute distribution coefficient, $T_Q = Q/c_p$, Q is the latent heat released per unit volume of solidified matter and c_p is the heat capacity.

The temperature at the dendrite interface $T_i = T_l = T_s$ is connected to the crystallization temperature T_0 of the pure liquid, the local curvature $2/R$ of the front, the liquid concentration C_l, and the intensity of atomic flux providing the normal growth velocity $(\vec{v} \cdot \vec{n}) = v_n$ by

$$T_i = T_0 - m_v C_l - Q c_p^{-1} 2 R^{-1} d(\theta) - \tilde{\beta}(\theta) v_n. \tag{5}$$

Here θ is the angle between the normal to the dendrite interface and its preferred growth direction (e.g., $< 001 >$ directions for fcc crystals), $\tilde{\beta}$ is the anisotropic coefficient of growth, $d(\theta)$ is the anisotropic capillary length, and m_v is the velocity-dependent liquidus line slope in the kinetic phase diagram of a binary system.

Anisotropic properties of the interface, such as surface energy, depend on the spherical angles which define the orientation of the normal to the dendrite interface to its growth direction. Considering a case of axisymmetric needle-like crystal one can apply averaging over one spherical angle and essentially simplify analytical form of the anisotropic properties of the crystal-liquid interface [11]. As a result of such averaging for the cubic crystal symmetry, the capillary length $d(\theta)$ and kinetic coefficient $\beta(\theta)$ are represented as

$$d(\theta) = d_0 \left\{1 - \alpha_d \cos\left[4\left(\theta - \theta_d\right)\right]\right\}, \tag{6}$$

$$\tilde{\beta}(\theta) = \beta_0 Q c_p^{-1} \left\{1 - \alpha_\beta \cos\left[4\left(\theta - \theta_\beta\right)\right]\right\}, \tag{7}$$

where d_0 and β_0 are the capillary and kinetic constants, $\alpha_d \ll 1$ and $\alpha_\beta \ll 1$ are the anisotropy parameters whereas θ_d and θ_β represent the angles between the direction of growth and the preferred growth directions, which correspond to the minima of $d(\theta)$ and $\beta(\theta)$.

Far from the dendritic surface the solute concentration, the temperature and the flow velocity in the liquid are fixed, i.e.

$$C_l = C_\infty, \quad T_l = T_\infty, \quad |\vec{w}| = U, \tag{8}$$

U being the characteristic flow velocity far from the dendrite. A family of exact solutions of the hydrodynamic problem can be found in two cases when the fluid flow is potential or of Oseen type. Here, the more realistic case of viscous flow described by the so-called Oseen and mass conservation equations is considered:

$$U\frac{\partial \vec{w}}{\partial z} = -\frac{1}{\rho_1}\vec{\nabla} p + \nu \nabla^2 \vec{w}, \quad \vec{\nabla} \cdot \vec{w} = 0, \tag{9}$$

where z is the direction of dendritic growth, ρ_1 is the liquid density and ν is the kinematic viscosity. Equations 6 should be solved with no-slip boundary conditions for the fluid velocity.

The system of equations (1)-(6) couples the concentration and the thermal problem and describes the growth of a parabolic dendrite under local non-equilibrium conditions in the diffusion field and at the solid-liquid interface. It extends the previous problem considered by Ben Amar and Pelcé [12] for anisotropic dendrite growth under local equilibrium conditions, generalizes the study of Ref. [13] for sufficiently large growth rates in the local equilibrium limit, and, finally, advances the isotropic description [14, 15] of the local non-equilibrium dendrite growth.

In the modeling we consider the steady state tip of a needle-like dendrite with radius R propagating with constant velocity $V = v_n$. The selection criterion for the steady state dendritic growth can be found by the combination of previously obtained solutions, which describe the anisotropic dendrite growing in the presence of convective flow [16] and the local non-equilibrium dendritic growth in rapid solidification processes without convection [17]. On the basis of these advancements, a sharp interface model for steady-state dendrite has been formulated [9] which is used in the present work to evaluate growth kinetics of a pure nickel dendrites with a small amount of zircon.

3. Results and discussion

The sharp interface model [9] based on equations (1)-(6) has been used without convective flow, i.e. at $U = 0$. To study the influence of a change of growth velocity by small amounts of strongly partitioning impurities a pure Ni-dendrites with small amounts of zircon (0.1 at.%Zr, 0.5 at.%Zr and 1 at.%Zr) have been taken for the calculations. For evaluation of the dendrite growth kinetics the relationship "dendrite tip radius - undercooling" and "dendrite growth velocity - undercooling" have been computed using the materials parameters from Table 1. The results of computations are shown in Figs. 1 and 2.

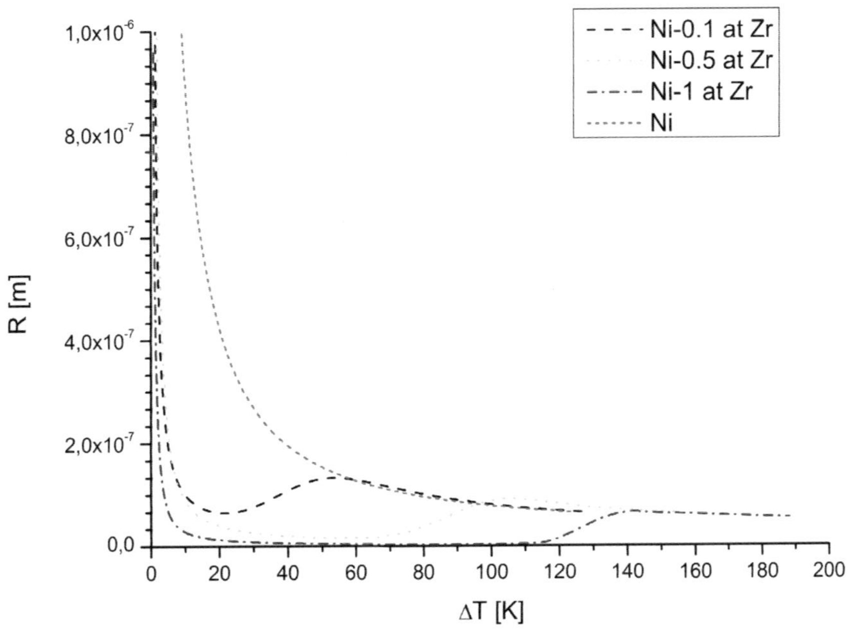

Figure 1. Computed dendrite tip radius for pure Ni, and different dilute Ni-based alloys $Ni_{99.9}Zr_{0.1}$, $Ni_{99.5}Zr_{0.5}$ and $Ni_{99}Zr_{1}$.

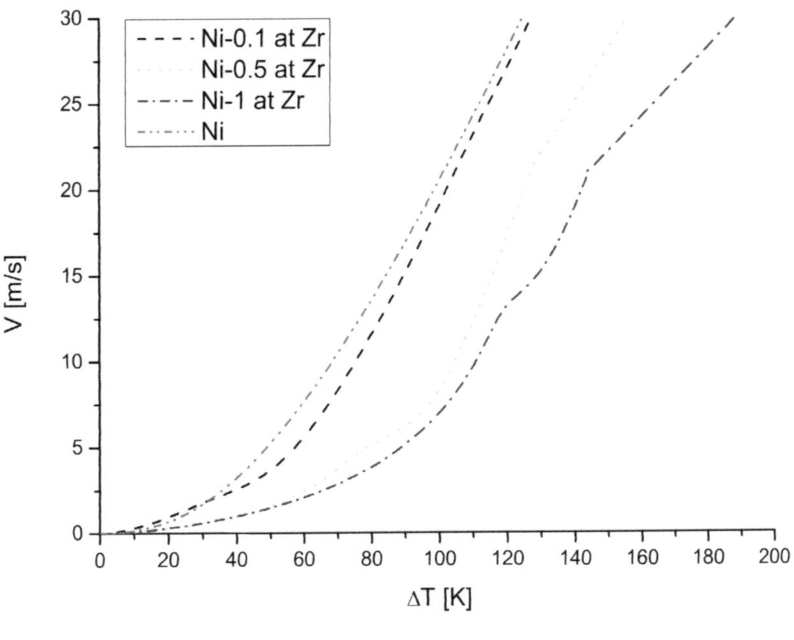

Figure 2. Comparison of dendrite growth velocities as a function of undercooling for pure Ni, and different dilute Ni-based alloys: $Ni_{99.9}Zr_{0.1}$, $Ni_{99.5}Zr_{0.5}$ and $Ni_{99}Zr_{1}$, as calculated within the sharp interface model [9].

Table 1. Material parameters for Ni and NiZr used in calculations as is given in Refs. [18, 19].

Material parameter	Dimension	Pure Ni	Ni–0.1 at.%Zr	Ni–0.5 at.%Zr	Ni–1 at.%Zr
Nominal concentration, C_∞	at.%	0	0.1	0.5	1
Melting temperature, T_m	K	1728	1728	1728	1728
Thermal diffusivity, D_T	m^2/s	$1.2 \cdot 10^{-5}$	$1.2 \cdot 10^{-5}$	$1.2 \cdot 10^{-5}$	$4.2 \cdot 10^{-5}$
Diffusion coefficient, D_C	m^2/s	–	$2.5 \cdot 10^{-9}$	$2.5 \cdot 10^{-9}$	$2.5 \cdot 10^{-9}$
Equilibrium partition coefficient, k_e		–	0.02	0.03	0.04
Liquidus slope, m_e	K/at.%	–	−9	−10	−11
Capillary constant, d_0	m	$7.89 \cdot 10^{-10}$	$7.89 \cdot 10^{-10}$	$7.89 \cdot 10^{-10}$	$2.02 \cdot 10^{-10}$
Kinetic growth coefficient, μ_k	$m/s/K$	0.59	0.59	0.42	0.22
Strength of the surface energy anisotropy, α_d		0.66	0.66	0.675	0.75
Bulk diffusion speed, V_D	m/s	–	23.5	21.5	21
Interface diffusion speed, V_{DI}	m/s	–	22.5	20	20.5
Latent heat of crystallization, Q	J/mol	17150	17150	16700	16230
Heat capacity, c_p	J/mol/K	41	41	41	41

The relation "tip radius – undercooling" clearly demonstrates the influence of doping a pure metal with very small amounts . Figure 1 shows the dendrite tip radius of pure Ni and various dilute Ni-based alloys as a function of undercooling. While in the case of pure Ni the tip radius is purely thermally controlled and decreases monotonically with the increase of undercooling, the addition only of 0.1 at.%Zr (as a strongly partitioning element in Ni-based alloys) leads to an essential change in the dendrite tip radius. The tip radius decreases essentially in comparison with the thermally controlled growth of dendrite tip of the pure Ni. Dendrite growth in the Ni-0.1 at.%Zr alloy melt becomes purely solute diffusion limited with the characteristic lengths much smaller than the thermal length which is dictated the tip radius of a pure nickel dendrite. This is typical example of the growth of the so-called "chemical dendrite" (which is also entitled as "solutal dendrite"). As it can be seen in example of the Ni-0.1 at.%Zr alloy's dendrite, the gradual decreasing of the tip radius occurs up to approximately $\Delta T = 22$K (see Fig. 1). As the sharp interface model [9] predicts, around this undercooling, the characteristic length for diffusion Zr atoms becomes comparable with the capillary length of the solid-liquid interface and the transition from the curved paraboloidal dendrite to the planar interface begins. Therefore, the dendrite tip radius growing in the Ni-0.1 at.%Zr alloy begins to increase at $\Delta T > 22$K. However, transformation of the chemical dendrite to the planar interface can not be proceed to its end due to solidification of the non-isothermal melt that exhibit thermal lengths scale to define a growth of the thermal dendrite. The Ni-0.1 at.%Zr alloy's dendrite reaches thermal length scale approximately at $\Delta T = 60$K (see Fig. 1). In this particular case, the transition from chemical dendrite to thermal dendrite occurs in the range of undercooling $22 \leq \Delta T\text{K} \leq 60$ and the tip radius of the Ni-0.1 at.%Zr alloy's dendrite almost confluence with the behavior of the tip radius of the pure nickel dendrite at $\Delta T\text{K} > 60$ (see Fig. 1). The transition is completely finished at the dendrite tip velocity V equals the diffusion speed V_D in the bulk melt, $V = V_D$ [14, 15]. For the Ni-0.1 at.%Zr alloy's dendrite, this occurs with the undercooling $\Delta T[V = V_D = 23.5(\text{m/s})] = 110$K (see Table 1 and Fig. 1). With $\Delta T \geq \Delta T[V = V_D]$, the solidification proceeds by the diffusionless mechanism and the main stem of the alloy's dendrite growth as the dendritic stem in a pure melt. The diffusionless mechanism of solidification means chemically partitionless solidification without solute redistribution ahead and at the solid-liquid interface. Such mechanism is the result of the complete solute trapping by the dendritic tips. In this case, solid phase forms with the initial (nominal) chemical compositions. Because we consider steady state growth of the dendrite tip, the dendrite stem behind the tip forms without chemical inhomogeneity at $C_s(V \geq V_D) = C_\infty$.

The same tendency have other diluted alloys, Ni-0.5 at.%Zr and Ni-1 at.%Zr (see Fig. 1). However, the minimum and maximum in the function $R(\Delta T)$ of these alloying dendrites are shifted to the larger undercooling as the concentration of Zr increases. This indicates more pronounced constitutional effect with the increase of Zr in the diluted nickel based alloys (for the present case, in the diluted nickel based alloys).

Table 2. Material parameters for Ni and NiZr used in calculations as is given in Refs. [18, 19].

Material parameter	Dimension	Pure Ni	Ni–0.01 at.%Zr	Ni–0.1 at.%Zr
Nominal concentration, C_∞	at.%	0	0.01	0.1
Melting temperature, T_m	K	1728	1728	1728
Thermal diffusivity, D_T	m^2/s	$1.2 \cdot 10^{-5}$	$1.2 \cdot 10^{-5}$	$1.2 \cdot 10^{-5}$
Diffusion coefficient, D_C	m^2/s	–	$2.5 \cdot 10^{-9}$	$2.5 \cdot 10^{-9}$
Equilibrium partition coefficient, k_e		–	0.02	0.02
Liquidus slope, m_e	K/at.%	–	-9	-9
Capillary constant, d_0	m	$7.89 \cdot 10^{-10}$	$7.89 \cdot 10^{-10}$	$7.89 \cdot 10^{-10}$
Kinetic growth coefficient, μ_k	$m/s/K$	0.59	0.59	0.59
Strength of the surface energy anisotropy, α_d		0.66	0.66	0.66
Bulk diffusion speed, V_D	m/s	–	23.5	23.5
Interface diffusion speed, V_{DI}	m/s	–	22.5	22.5
Latent heat of crystallization, Q	J/mol	17150	17150	17150
Heat capacity, c_p	J/mol/K	41	41	41

Figure 2 shows the results of the growth velocity as a function of undercooling in the pure Ni-melt and in the dilute Ni-Zr alloy melts. In the very dilute alloy $Ni_{99.9}Zr_{0.1}$ an enhancement of the velocity is observed compared with the velocity of pure Ni that occurs for the uncerooling $\Delta T < 35$K. The other alloys with higher concentrations show a decrease of the velocity with the undercooling in comparison with the velocity of the pure nickel dendrite. Furthermore the decrease in velocity of the $Ni_{99}Zr_1$-dendrite being more pronounced than the velocity of the $Ni_{99.5}Zr_{0.5}$-dendrite. Apparently, there are two counteracting effects of the addition of impurities, one leading to an increase of the growth dynamics in particular at the very dilute limit and the other one decreasing the growth velocity the latter one dominating at higher concentrations. Indeed, since the chemical diffusion coefficient is about three to four orders of magnitude smaller than the thermal diffusivity the chemical dendrites become narrow because of the strong concentration gradient and they are broaden when reaching the limit of absolute chemical stability. This effect is responsible for the enhancement of the dendrite growth velocity of very dilute alloys at small undercooling since a thin dendrite propagates faster than a thick dendrite. In addition to this, constitutional effect in the stability of a growing dendrite the solute causes a constitutional undercooling that requires solute redistribution and therefore, makes the dendrite growth more sluggish. The effect of velocity enhancement dominates at small concentration whereas the effect of constitutional undercooling predominates at larger concentrations.

To confirm the enhancement of velocity due to the tiny amount of impurity in comparison with the velocity of the pure dendrite, we specially computed growth velocity for dendrites growing from undercooled pure Ni, binary $Ni_{99.99}Zr_{0.01}$ and $Ni_{99.9}Zr_{0.1}$ alloys. The material parameters for pure Ni, binary $Ni_{99.99}Zr_{0.01}$ and $Ni_{99.9}Zr_{0.1}$ alloys are presented in the Rable 2. Figure 3 shows behavior of the tip velocity as the logarithmic function of small undercoolings. The velocity of the $Ni_{99.99}Zr_{0.01}$-dendrite also larger then the velocity of the pure Ni-dendrite in the range $\Delta T < 35$K. This, indeed, confirms the fastest growth of the narrow chemical dendrites at small quantity of impurity in comparison with the thicker thermal dendrites. However, Fig.

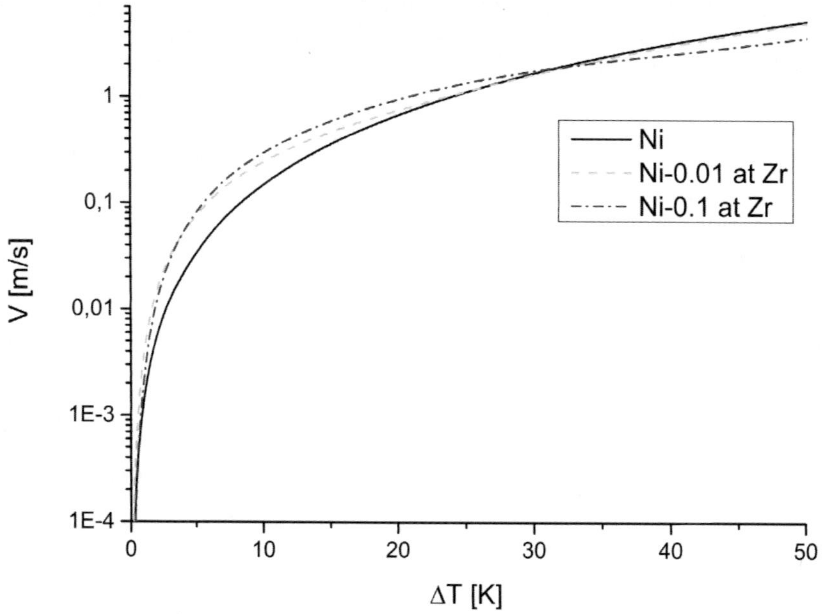

Figure 3. Effect of dendrite growth velocity enhancement in very diluted alloys. It is shown that the tips of alloying dendrites, $Ni_{99.99}Zr_{0.01}$ and $Ni_{99.9}Zr_{0.1}$, are growing faster than the tip of the pure nickel dendrite in the range of undercooling $\Delta T < 35$K.

3 also demonstrates that the $Ni_{99.99}Zr_{0.01}$-dendrite grows slower than the $Ni_{99.9}Zr_{0.1}$-dendrite. As a results of such non-linear dependence at $\Delta T < 35$K, we have obtained the following sequence of the growth velocities:

$$V_{Ni_{99.5}Zr_{0.5}} < V_{Ni} < V_{Ni_{99.99}Zr_{0.01}} < V_{Ni_{99.9}Zr_{0.1}}.$$

This means, between concentrations 0.1%Zr and 0.5%Zr, there is a Ni-Zr content which leads to the biggest enhancement of the dendrite velocity (in comparison with the velocity of the pure Ni-dendrite). This effect of non-linear influence of the impurity on the growth velocity can be considered in our future work.

4. Conclusions

Computational results on primary dendrite solidification have been obtained using the sharp interface model [9]. The kinetics has been eveluated by the dendrite tip radius and growth velocity of dendrites growing in undercooled melts of nickel and dilute Ni-Zr alloys. A transition from chemical dendrite to the thermal dendrite has been demonstrated in a whole range of undercooling. It has been demonstrated that small amounts of Zr lead to the increase of the growth velocity in the range of small undercoolings. This effect of the dendrite velocity enhancement by tiny amonts of impurity should be investigated more detailed due to its non-linearity: there is a critical concentration of Zr at which one may find the biggest enhancement of the dendrite velocity (in comparison with the velocity of the pure dendrite). These outcomes might be investigated experimentally and computationally using a phase field computational technique.

5. References

[1] Arnold C B, Aziz M, Schwarz M, Herlach D M 1999 *Phys. Rev. B* **59** 334-343
[2] Schwarz M 1998 Kornfeinung durch Fragmentierung von Dendriten *Ph.D. Thesis (Ruhr-Universitt Bochum*
[3] Herlach D M, Annu. 1991 *Rev. Mater. Sci.* **21** 23-44
[4] Lipton J, Kurz W, Trivedi R 1987 *Acta Metall.* **35** 957-964
[5] Boettinger W J, Coriell S R, Trivedi R 1988 Rapid Solidification Processing: Principles and Technologies IV *Editors: R. Mehrabian and P. Parrish (Claitors, Baton Rouge, Louisiana* 13
[6] Aziz M, Kaplan T 1988 *Acta Metall.* **36** 2335-2347
[7] Schwarz M, Karma A, Eckler K, Herlach D M 1994 *Phys. Rev. Lett.* **73** 1380-1383
[8] Karma A 1998 Int. J. Non-Equilibrium Processing **11** 201-233
[9] Galenko P K, Danilov D A, Reuther K, Alexandrov D V, Rettenmayr M, Herlach D M 2017 Effect of convective flow on stable dendritic growth in rapid solidification of a binary alloy *Journal of Crystal Growth* **457** 349355
[10] Herlach D, Galenko P, and Holland-Moritz D 2007 Metastable Solids from Undercooled Melts *Elsevier, Amsterdam*
[11] Barbieri A, and Langer J S 1989 Predictions of dendritic growth rates in the linearized solvability theory *Phys. Rev. A* **39** 5314-5325
[12] Ben Amar M, Pelcé P 1989 Impurity effect on dendritic growth *Phys. Rev. A* **39** 4263-4269
[13] Alexandrov D V, Galenko P K 2013 Selection criterion of stable dendritic growth at arbitrary Peclétnumbers with convection *Phys. Rev. E* **87** 062403
[14] Galenko P K, Danilov D A 1997 Local nonequilibrium effect on rapid dendritic growth in a binary alloy melt *Physics Letters A* **235** 271-280
[15] Galenko P K, Danilov D A 1999 Model for free dendritic alloy growth under interfacial and bulk phase nonequilibrium conditions *Journal of Crystal Growth* **197** 992-1002
[16] Alexandrov D V, Galenko P K 2015 Thermo-solutal and kinetic regimes of an anisotropic dendrite growing under forced convective flow *Phys. Chem. Chem. Phys.* **17** 19149-19161
[17] Alexandrov D V, Danilov D A, and Galenko P K 2016 Selection criterion of a stable dendrite growth in rapid solidification *Int. J. Heat Mass Transfer* **101** 789799
[18] Galenko P K, Phanikumar G, Funke O, Chernova L, Reutzel S, Kolbe M, Herlach D M 2007 Dendritic solidification and fragmentation in undercooled NiZr alloys *Materials Science and Engineering A* **449451** 649653
[19] Herlach D M, Galenko P K 25 March 2007 Rapid solidification: in situ diagnostics and theoretical modelling *Materials Science and Engineering: A* **449-451** 34-41

Acknowledgments

O. V. K. acknowledges support of RFBR by the project No. 16-38-60172 mol_a_dk and a program "Postdocs of UrFU". P. K. G. and D. V. A. acknowledge support from the RSF, grant number 16-11-10095. P. K. G. also acknowledges support under the project "MULTIPHAS", contract Nr. 50WM1541 by the European Space Agency, German Aerospace Center and Friedrich-Schiller-Universität Jena.

IOP Conf. Series: Materials Science and Engineering **192** (2017) 012031 doi:10.1088/1757-899X/192/1/012031

Prediction of steel corrosion resistance based on EBSD-data analysis

D S Chezganov, M A Borovykh and O A Chikova

Ural Federal University, Ekaterinburg, 620000, Russian Federation

E-mail: chezganov.dmitry@urfu.ru, chik63@mail.ru

Abstract. We experimentally compared the samples of chromium-manganese steel prepared with various heat-treatment modes to predict the corrosion resistance of the steel relative to hydrocarbon influence. The study was done using scanning electron microscopy, X-ray energy dispersive spectroscopy and electron backscattered diffraction. The surface visualization and elemental analysis allowed to reveal the existence of inclusions and discontinuities and their elemental composition. The peculiarities of crystal structure, grain sizes and orientations were revealed by the analysis of misorientation distribution histograms, and orientation, band contrast and Schmidt factor maps. The analysis allowed to conclude which of the heat-treatment modes provides the increase of corrosion resistance relative to the hydrocarbons influence.

1. Introduction

The corrosion resistance is the important characteristic of the modern functional materials. The preliminary estimation of the material corrosion resistance is based on the study of its microstructure, phase composition and crystal structure. Detection and analysis of the material corrosion signs are usually realized by optical and electronic microscopy, X-ray phase analysis and measurement of micro hardness [1-3].

It is known that the low-energy and special boundaries of metallic grains have a high resistance to destruction [4-5]. The free energy of grain boundaries plays central role in boundary diffusion and segregation of impurity atoms [6]. The increase of a number of the low-angle special boundaries leads to raise of the metal corrosion resistance [7-8]. The analysis of diffraction patterns of backscattered electrons makes it possible to estimate the fraction of the low-angle and special (low-energy) grain boundaries of a metallic material, to construct a Schmidt factor (or Taylor factor) maps and a grain misorientation histograms (correlated distribution). The correlated misorientations display data between neighboring points. The analysis of grain misorientation histograms determines the fraction of the low-angle boundaries. The effects of the inhomogeneous stress-strain state, the criterions of the intergranular destruction of the metal materials based on analysis of Schmidt and Taylor factor maps are discussed in works [9-11].

In this work, we presented an analysis of capabilities of electron backscattered diffraction (EBSD) method for the prediction of the steel corrosion resistance relative to the hydrocarbon influence. The study of microstructure and crystal structure of samples was made to optimize the heat treatment mode.

IOP Conf. Series: Materials Science and Engineering **192** (2017) 012031 doi:10.1088/1757-899X/192/1/012031

2. Materials and experiment

The comparative analysis of EBSD data obtained for different samples cut from metal pipes made of chromium-manganese steel and heat-treated by various methods was made. The studied chromium-manganese steel contained components in following ratio [11-12] (Table 1):

Table 1. Components of studied chromium-manganese steel.

Component	Content, weight%
Carbon	0.31-0.38
Silicon	0.17-0.37
Manganese	0.95-1.25
Chromium	1.0-1.30
Vanadium	< 0.06-0.12
Sulfur	< 0.035
Phosphorus	< 0.035

The samples were selected from ready-made pipes at the "Sinarskiy trubnyy zavod" (Sinarskiy pipe plant) and differed from each other by heat treatment mode: Sample 1 - normalization at 910 °C and high-speed tempering at 700 °C in the furnace; Sample 2 - high-speed tempering at 700 °C in the furnace; Sample 3 - normalization at 910 °C; Sample 4 - long tempering at 700 °C.

Before study the sample surfaces were grinded and polished by diamond abrasive down to 1 μm and colloidal silica during 20 min at the final stage.

The sample surfaces were studied by the three methods: (1) scanning electron microscopy (SEM), (2) X-ray energy dispersive spectroscopy (EDS) and (3) EBSD. The study was made by scanning electron microscope AURIGA CrossBeam (Carl Zeiss, Germany) equipped with the EBSD analysis system HKL Channel 5(Oxford Instruments, UK) with spatial resolution down to 20 nm and EDS system IncaEnergy (Oxford Instruments, UK) with 350X-MAX X-ray detector with spatial resolution about 1 μm and spectral resolution of 125 eV at Kα line of Mn.

The SEM was used for visualization of surface morphology and structural defects in secondary electron mode with the resolution down to 2 nm. The EDS was applied for revealing of elemental composition of defects and inclusions. The EDS data collection and processing were made by means of IncaEnergy software.

The EBSD was used for detailed study of structural state of the sample surface. The EBSD data acquisition was made by Flamenco Acquisition software (Oxford Instruments, UK). The accelerating voltage of 20 kV and electron beam current of 8 nA were used during scanning. The areas with a size of 20×20 μm^2 were scanned with a step size of 80 nm. The obtained data was processed and analyzed by Tango software (Oxford Instruments, UK). The band contrast (BC) map with coincident site lattice boundaries (CSL), Schmidt factors maps and histograms of misorientation angle distribution were constructed. The histograms consist of 2000 columns with width of 0.3°. The angle of tolerance was chosen to be 5°.

3. Results and discussion

The reason of the intergranular corrosion for ferrites steels, such as chromium-manganese steel, is an emergence of local distortion of crystal lattice and presence of large-angle non-special grain boundaries. The appearance of local or pitting corrosion relates to localization of the non-metal inclusions such as the oxides of iron, silicon and aluminum. This was a reason for selection of regions for EBSD-study associated with metal defects of technological origin, e.g. the cracks filled with gas or dross.

The study of the surface by SEM allowed to reveal existence of inclusions and discontinuities with sizes about 1-10 μm in subsurface layer of the pipes. The EDS analysis showed oxygen content in the inclusions. Moreover, the increased of carbon content in surface layer was found. The elemental analysis of the inclusions revealed a presence of Fe, Mn, Si, Cr, Al, O and C. The systematic segregation of atoms wasn't found. The extended non-metal inclusions with the size about 50 μm were observed in the sample 2, which was oriented into the pipe depth. In sample 4 the non-metal inclusions demonstrated spherical shape and size about 1 μm. It was supposed that these inclusions consist of aluminum oxide Al_2O_3.

The figures 1(a) – 1(d) represents the BC maps with superimposition of colored CSL boundaries. The colors are coded by CSL type as shown in the histogram of CSL boundaries distributions (Fig. 1(e) – (1(h)). The band contrast is an EBSD pattern (EBSP) quality factor derived from the Hough transform that describes the average intensity of the Kikuchi bands with respect to the overall intensity within the EBSP[13]. The values are scaled to a byte range from 0 to 255 (i.e., low to high contrast). With this scale mapped to a grayscale from black to white, image-like maps can be plotted. These "images" show the microstructure in a qualitative fashion that were used to seeing either in the SEM or light microscope. Because EBSPs along grain boundaries tend to show poor BC they appear dark in a map. Conversely, EBSPs in undeformed regions of a grain appear light.

It was observed that the sample 3 have the smallest grains which might indicate to greater plasticity and corrosion resistance of this sample (Fig. 1(c)). The revealed orientation distribution demonstrated the wide range of colors close to discontinuities which indicated a high degree of grain misorientations and weakly pronounced metallographic texture.

The revealed CSL boundaries, close to Brandon's criterion for special boundaries, marked by light colored lines while other grain boundaries were kept in the form of dark lines. CSL boundaries are the special boundaries which fulfil the coincident site lattice criterias whereby the lattices are sharing some lattice sites. CSLs are characterized by Σ where Σ is the ratio of the CSL unit cell compared to the standard unit cell. CSL boundaries typically have a significant impact on the material properties, which means that from a materials engineering viewpoint it is important to determine both the ratio of CSL boundaries and their distribution within the material.

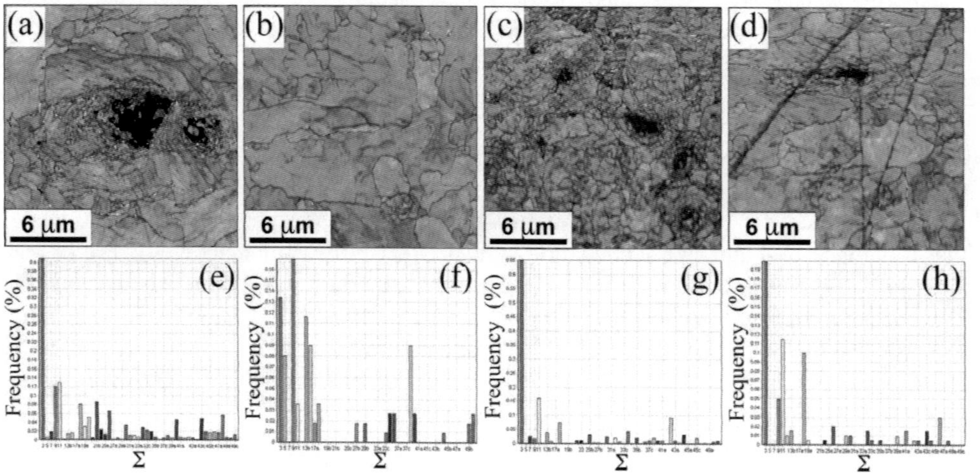

Figure 1. (a) – (d) Band contrast maps with colored CSL boundaries obtained at different samples: (a) 1, (b) 2, (c) 3, (d) 4. (e) – (h) CSL boundaries distributions and color code for sample: (e) 1, (f) 2, (g) 3, (h) 4.

The "dark" grain boundaries form a continuous network while twin boundaries do not form one. With all the differences in the location of twin grain boundaries their relative proportion are about the same for all samples. All the twins are inside the grains and do not cross the grain boundaries. It means that they arose after a common grain structure was formed during rolling process. The revealed twins are the annealing twins. The part of special boundaries is small (~ 1.5-2.0%) and the largest number of them in the microstructure of the sample 4 (Fig. 1(d)), which indicates its higher corrosion resistance [14-15].

The Figure 2 shows the histograms of misorientation angle distribution. The uncorrelated misorientations (light gray lines) indicate misorientations between randomly selected points in the data set. Correlated misorientations (dark gray lines) display data between neighboring points, in the other words, the angle distribution of the grain boundaries. A theoretical curve (solid black curve) shows what one would expect from a random set of orientations. The difference between uncorrelated misorientation and the theoretical curve arises, as a rule, due to the strong texture, which revealed itself in all samples (Fig. 2). According to this fact, the highest texturing was revealed in sample 2 (Fig. 2b). The presence of a larger specific fraction of low-angle boundaries is the most pronounced for sample 3, which indicates its high plasticity and corrosion resistance (Fig. 2(c)).

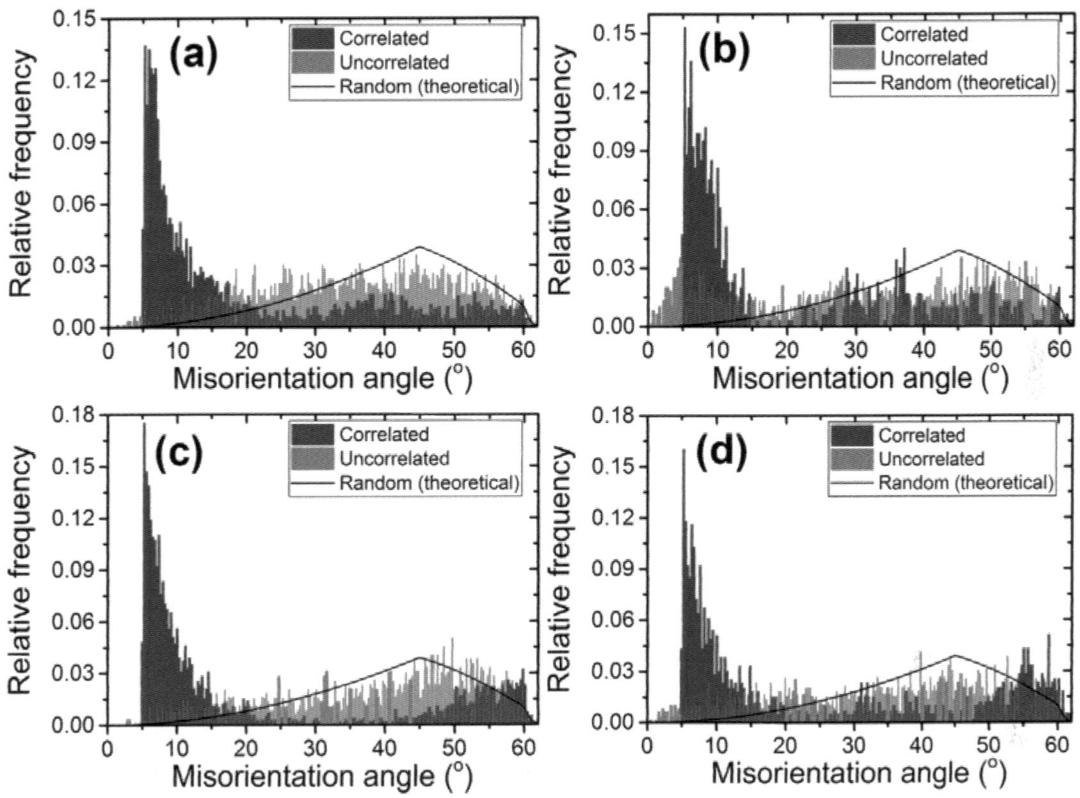

Figure 2. Misorientation angle distribution shows the frequency of grain boundaries in the sample: (a) 1, (b) 2, (c) 3, (d) 4.

The Schmidt factor maps are used to determine the uniformity degree of possible deformation. In the presence of an external load, for example, during operation, deformations begin to develop in light grains and gradually move to dark ones. In external mechanical loading in cubic crystals with a bcc lattice type, such as α-Fe, plastic deformation by slipping can take place.

The Schmidt factor maps were constructed for a slip system typical for BCC crystals: {110} <111> with loading direction parallel to OZ axis. The Fig. 6 demonstrated that there is a wide range of shades of gray close to the discontinuities due to local distortions of the crystal lattice of the metal, which indicates a low deformation uniformity of the metal and a high probability of the pitting corrosion. The Schmidt factor map for sample 3 is characterized by the greatest deformation uniformity, which is an indirect evidence of its greater resistance to intercrystalline corrosion.

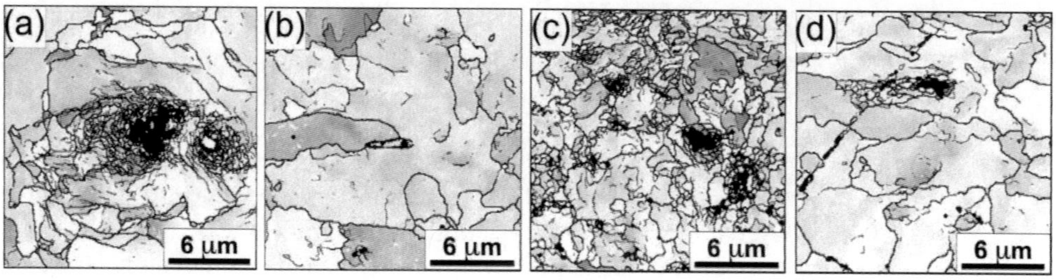

Figure 3. Schmidt factor maps for sample: (a) 1, (b) 2, (c) 3, (d) 4.

4. Conclusion

The comparative study of different samples cut from metal pipes made of the chromium-manganese steel and heat-treated by various methods was made by SEM, X-ray EDS and EBSD to predict the corrosion resistance of steel relative to the hydrocarbon influence. SEM surface visualization revealed the existence of inclusions and discontinuities with sizes about 1 – 50 μm. The EDS analysis showed oxygen content in the inclusions and elevated content of carbon in the surface layer. The EBSD analysis allowed to reveal peculiarities of crystal structure, grain sizes and orientations and obtain the data of local mechanical stresses. We have shown that the heat treatment mode influences on the grain size, misorientation distribution and Schmidt factor map. The analysis of EBSD-data allowed to conclude that heat-treatment mode for sample 3 provides the increase of the corrosion resistance relative to the hydrocarbons influence.

5. References

[1] Latanision R M 1995 *Corrosion* **51** 270
[2] Jargelius-Pettersson R F A 1998 *Corrosion* **54** 162
[3] Bell T 2002 *Surf. Eng.* **18** 453
[4] Watanabe T 1984 *Trans. Jap. Inst. Metals* **27** 73
[5] Watanabe T 1993 *Mater. Sci. Eng.: A* **166** 11
[6] Gupta G, Ampornrat P, Ren X, Sridharan K and Allen T R Was G S 2007 *J. Nucl. Mater.* **361** 160
[7] Lin P, Palumbo G, Erb U and Aust K T 1995 *Scr. Mater.* **33** 1387
[8] Palumbo G, King P J, Aust K T, Erb U and Lichtenberger P C 1991 *Scr. Mater.***25** 1775
[9] Watanabe K, Matsuda K, Miura N, Uetani Y, Ikeno S., Yoshida T and Murakami S 2014 *Keikinzoku/Journal of Japan Institute of Light Metals* **64** 368
[10] Lopatin N V 2012 *Mater. Sci. Eng.: A* **556** 704
[11] Zhou W and Wang Z L 2007 *Scanning microscopy for nanotechnology: techniques and applications* (New York: Springer) 522
[12] Lin P, Palumbo G, Harase J and Aust K T 1996 *Acta Mater.* **44** 4677
[13] Chun H, Na S M, Mudivarthi C and Flatau A B 2010 *J. Appl. Phys.* **107** 09A960

Acknowledgments
The equipment of the Ural Center for Shared Use "Modern nanotechnology" UrFU was used.

IOP Conf. Series: Materials Science and Engineering **192** (2017) 012032 doi:10.1088/1757-899X/192/1/012032

Hydrodynamics and heat and mass transfer during formation of thin metal glass tape

Yu A Bazdyreva, V V Belousov, V V Bodryaga, V I Bondarenko, F V Nedopekin and S F Prohorenko

Donetsk National University, 83001 Donetsk, Ukraine

v.v.bilousov@gmail.com

Abstract. The mathematical model of hydrodynamic and heat transfer processes in a system, which rotates a disk, a metal tape, an environment, is developed. The adequacy of both the mathematical model and the computing algorithm is confirmed by comparison the computing results with the laboratory experiment. It has been shown that increase of the disc's rotational speed reduces the thickness of the amorphous metal tape.

1. Introduction

In production of the amorphous metals, while providing a high melt cooling rate, which results in formation of the glassy state, we cannot completely avoid the influence of the crystallization process on an amorphous material. The glassy state is metastable with respect to the crystalline one and converts into it at various thermal processes. In this regard, one of the central places in the physics of amorphous materials is taken by the study of the melt crystallization processes at high cooling rate. Hydrodynamic processes play a rather significant role in this case. Despite of the short duration of tape production and its small geometric dimensions, hydrodynamic and heat transfer processes significantly affect on the uniformity of the amorphous tape thickness.

2. Problem setting and methods of its implementation

An alloy $Fe_{80}B_{20}$ ($m = 25\ g$) is supplied to a rotating copper disk ($R = 0.3\ m$, $w_0 = 3000$ rpm). The disk surface temperature is maintained constant [1]. The disk-melt system is in the helium atmosphere, and therefore there is no heat removal from the top and side borders.

The mathematical model of the process is described by the equations of heat transfer, motion and continuity and by the boundary conditions.

It is deemed that at these cooling rates a high degree of melt supercooling, about 150°C below the solidus temperature, is observed. Due to this, the position of the solid crust is determined by isotherm [2].

The model has the following assumptions:

• the system is two-dimensional;

• metal properties do not depend on temperature (except for the viscosity);

• melt flow is laminar.

The latter statement is justified by the fact that the Reynolds criterion $\mathrm{Re} = 2100$, which is far below the turbulence threshold [3]

Thermophysical properties of the melt are as follows: $T_l = 1173\ K$ –melting point; $\lambda = 71\ W/(m \cdot K)$ – thermal conductivity coefficient; $T_l + 50 \div 100\ K$ – casting temperature; $a = 0.5 \cdot 10^{-5}\ m^2/s$ – thermal diffusivity coefficient; $\nu = 0.2 \cdot 10^{-5}\ m^2/s$ - kinematic viscosity coefficient.

Thermophysical parameters of the substrate are as follows: $c = 544\ J/(kg \cdot K)$ - specific thermal capacity; $\alpha = 2 \cdot 10^5\ W/(m^2 \cdot K)$ - heat transfer coefficient; $\lambda = 20\ W/(m \cdot K)$ - thermal conductivity coefficient; $\rho = 7500\ kg/m^3$ - density.

The mathematical model for solving this problem can be constructed from the equations where velocity and pressure are the unknowns. However, it is more convenient to introduce the other variables: stream function ψ and vorticity ω (Helmholtz variables) that are associated with the speed by vector relations:

$$V_1 = \frac{\partial \psi}{\partial y}; \quad V_2 = -\frac{\partial \psi}{\partial x}; \quad \vec{\omega} = \nabla \vec{V}. \tag{1}$$

In this case continuity equation is satisfied automatically and certain arbitrary rule is allowed in selecting stream function.

Navier–Stokes equation in Boussinesq approximation, continuity and heat transfer equations will be written as follows [4]:

$$\frac{\partial V_1}{\partial t} + \frac{\partial V_1^2}{\partial x} + \frac{\partial V_1 V_2}{\partial y} = -\frac{1}{\rho}\frac{\partial p}{\partial x} + \frac{\partial \nu}{\partial x}\frac{\partial V_1}{\partial x} + \frac{\partial \nu}{\partial y}\frac{\partial V_1}{\partial y}; \tag{2}$$

$$\frac{\partial V_2}{\partial t} + \frac{\partial V_1 V_2}{\partial x} + \frac{\partial V_2^2}{\partial y} = -\frac{1}{\rho}\frac{\partial p}{\partial y} + \frac{\partial \nu}{\partial x}\frac{\partial V_2}{\partial x} + \frac{\partial \nu}{\partial y}\frac{\partial V_2}{\partial y} + g\beta(T - T_0); \tag{3}$$

$$\frac{\partial V_1}{\partial x} + \frac{\partial V_2}{\partial y} = 0; \tag{4}$$

$$c\rho\left[\frac{\partial T}{\partial t} + \frac{\partial V_1 T}{\partial x} + \frac{\partial V_2 T}{\partial y}\right] = \frac{\partial \lambda}{\partial x}\frac{\partial T}{\partial x} + \frac{\partial \lambda}{\partial y}\frac{\partial T}{\partial y} + \lambda\left(\frac{\partial^2 T}{\partial x^2} + \frac{\partial^2 T}{\partial y^2}\right); \tag{5}$$

where p – pressure, c – melt heat capacity, ρ – melt density, g – gravity acceleration, β – volumetric expansion coefficient, λ – thermal conductivity of the medium, T_0 – initial temperature. Let us evaluate contribution of thermal convection into the process of amorphous tape formation using Richardson criterion [5] ($Ri = Gr/\mathrm{Re}^2$): $Ri = 0.000722$, which indicates a substantial predominance of dynamic forces over the buoyancy forces. Hence, volumetric forces can be neglected in the equation (3).

Equations (2), (3) and (5) may be written in the criterial form in the variables of vortex (ω) – stream function (ψ) [4]:

$$\frac{\partial \omega}{\partial Fo} + \frac{\partial V_1 \omega}{\partial x} + \frac{\partial V_2 \omega}{\partial y} = \frac{\partial \Pr(\nu)}{\partial x}\frac{\partial \omega}{\partial x} + \frac{\partial \Pr(\nu)}{\partial y}\frac{\partial \omega}{\partial y}; \tag{6}$$

$$\frac{\partial^2 \psi}{\partial x^2} + \frac{\partial^2 \psi}{\partial y^2} = -\omega; \tag{7}$$

$$\frac{\partial \theta}{\partial Fo} + \frac{\partial V_1 \theta}{\partial x} + \frac{\partial V_2 \theta}{\partial y} = \frac{\partial a}{\partial x}\frac{\partial \theta}{\partial x} + \frac{\partial a}{\partial y}\frac{\partial \theta}{\partial y} + a\left(\frac{\partial^2 \theta}{\partial x^2} + \frac{\partial^2 \theta}{\partial y^2}\right); \tag{8}$$

where Fo, $\Pr$, Gr – Fourier, Prandtl and Grashof numbers [3], $\theta = T/T_0$, T – current temperature, a – thermal diffusivity coefficient.

Initial conditions:

$$Fo = 0: \ \psi = 0; \ \ \omega = 0; \ \ \theta = 1.$$

Boundary conditions:

$$y = 0: \ \ \frac{\partial \psi}{\partial y} = 0; \ \ \psi = 0; \ \ \frac{\partial \theta}{\partial y} = 0;$$

$$y = L_y: \ \ \frac{\partial \psi}{\partial y} = V_m; \ \ \psi = 0; \ \ \theta = \theta_{disk};$$

$$x = 0: \ \ \psi = 0; \ \ \frac{\partial \psi}{\partial y} = 0; \ \ \frac{\partial \theta}{\partial x} = 0;$$

$$x = L_x: \ \ \frac{\partial^2 \psi}{\partial y^2} = 0; \ \ \frac{\partial \psi}{\partial y} = 0; \ \ \frac{\partial \theta}{\partial x} = 0.$$

The short duration of the process ($0.001\,s$), high temperature (up to $1600^0 C$) and metal aggressiveness do not allow to solve the problem by means of the experiment. The problems of this class can only be solved by numerical methods, in particular using finite differences [4]. The essence of this method lies in the fact that the region of the continuous variation of the argument is replaced by a finite set of points (nodes) forming the spatial-temporal finite-difference grid. In order to implement numerically the mathematical model of the processes, it is necessary to formulate a computational algorithm which is a logical sequence of operations in which the solution of the problem can be found. Proceeding from the requirements of solution accuracy and stability, as well as machine time efficiency, the method of alternating directions was chosen.

The algorithm for the software product is as follows: using the input data (the boundary conditions), vorticity and stream function are defined on the base of the solution of Navier-Stokes and continuity equations. When stream function field is known, it is possible to calculate the distribution of the velocity components.

Temperature distribution on the tape surface is determined using the velocity field. The process is repeated at a new time step. Graphical output of fields of stream functions, velocity and temperature at different times is used for easing the analysis of the obtained data.

3. Research results

In [6] the authors consider the classical representations of the stresses that arisen in the process of tape formation. They suggested that the non- solidified section of the volume will separate from the solidified one at stresses making approximately 10^8 Pa. In this case the horizontal component of velocity and viscosity are determining parameters. Since stress is defined as

$$\tau = \eta \left| \frac{\partial V_1}{\partial y} \right|, \tag{9}$$

an assumption can be made that separation will be observed at the point where product of viscosity by the derivative of horizontal velocity is maximized.

As Figure 1 shows, the tape layers, adjacent to the disc, have the values of the horizontal component of velocity close to the linear velocity of disc rotation. At the point of separation the velocity value sharply decreases, and in some cases it has nay an opposite sign. It happens at the point of a sharp velocity variation that there exists the boundary of tape solidification, as we have assumed.

Varying the disc rotation velocity leads to changes in the tape thickness. To confirm this, several options of linear rotational velocity were calculated on the disk surface:

1) $U = 70.7\ m/s$; 2) $U = 47.1\ m/s$; 3) $U = 23.5\ m/s$; 4) $U = 11.8\ m/s$.

As a consequence, the solidification velocity decreases. This result agrees qualitatively with the experimental studies conducted in the Donetsk Physico-Technical Institute of the National Academy of Sciences of Ukraine. The error amounted to 16%. Figure 1 shows that the change of the tape thickness is linear to the velocity value of 47.1 *m/s;* at higher velocities, the curve monotonically gets broken.

Figure 2 shows the effect of the disk rotation velocity on the distribution of the horizontal component of velocity along the thickness tape.

As can be seen, under the influence of the viscous forces, the velocity decreases along the tape thickness, while at the outermost sections of the substrate the reverse motion is observed. Moreover, the larger the disk rotation velocity becomes, the higher reverse motion speed is. In turn, such a velocity distribution affects on the stresses in the tape (Fig. 3), and therefore, on the possibility for the non-solidified portion of tape volume to be separated. The higher the disk rotation velocity becomes, the greater the likelihood of separation is.

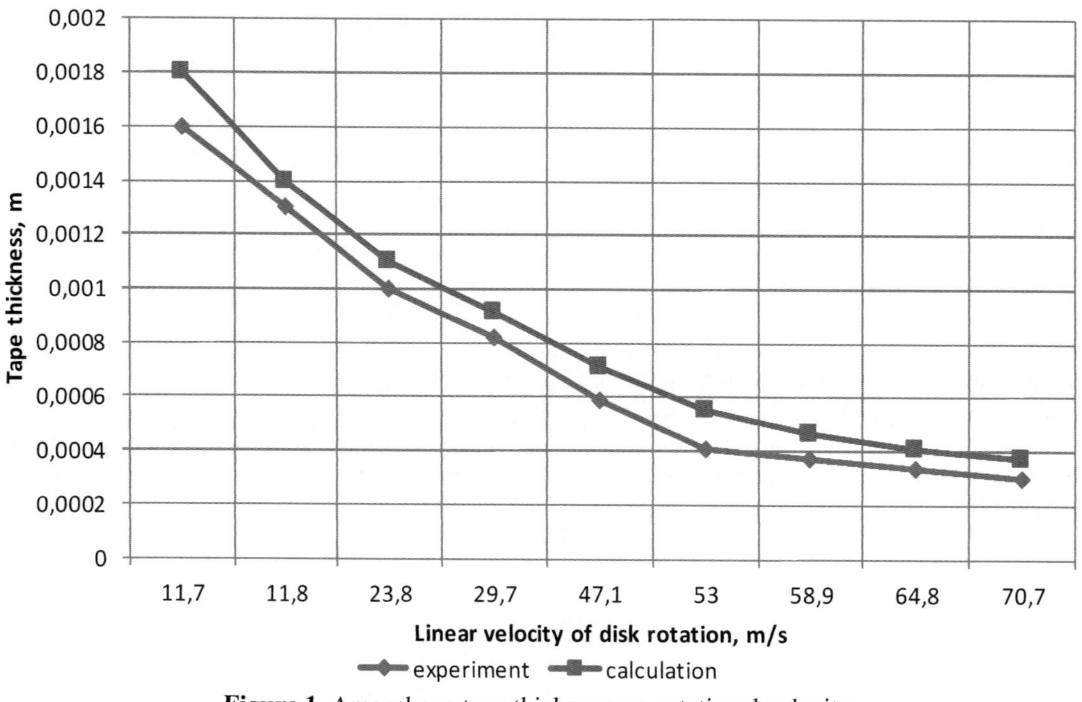

Figure 1. Amorphous tape thickness vs. rotational velocity.

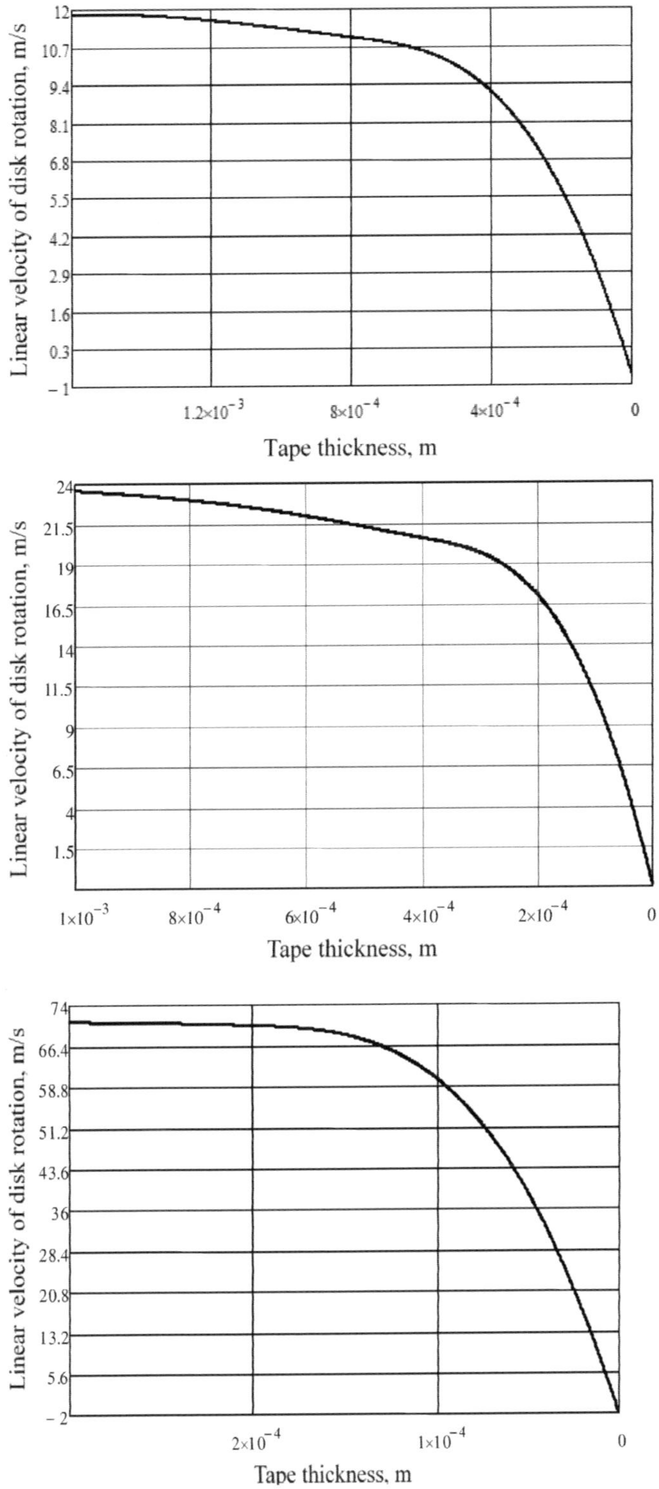

a)

b)

c)

Figure 2. Horizontal component of the velocity distributed along the tape thickness with linear rotational velocity: a) 11.8 m/s; б) 23.5 m/s; в) 70.7 m/s.

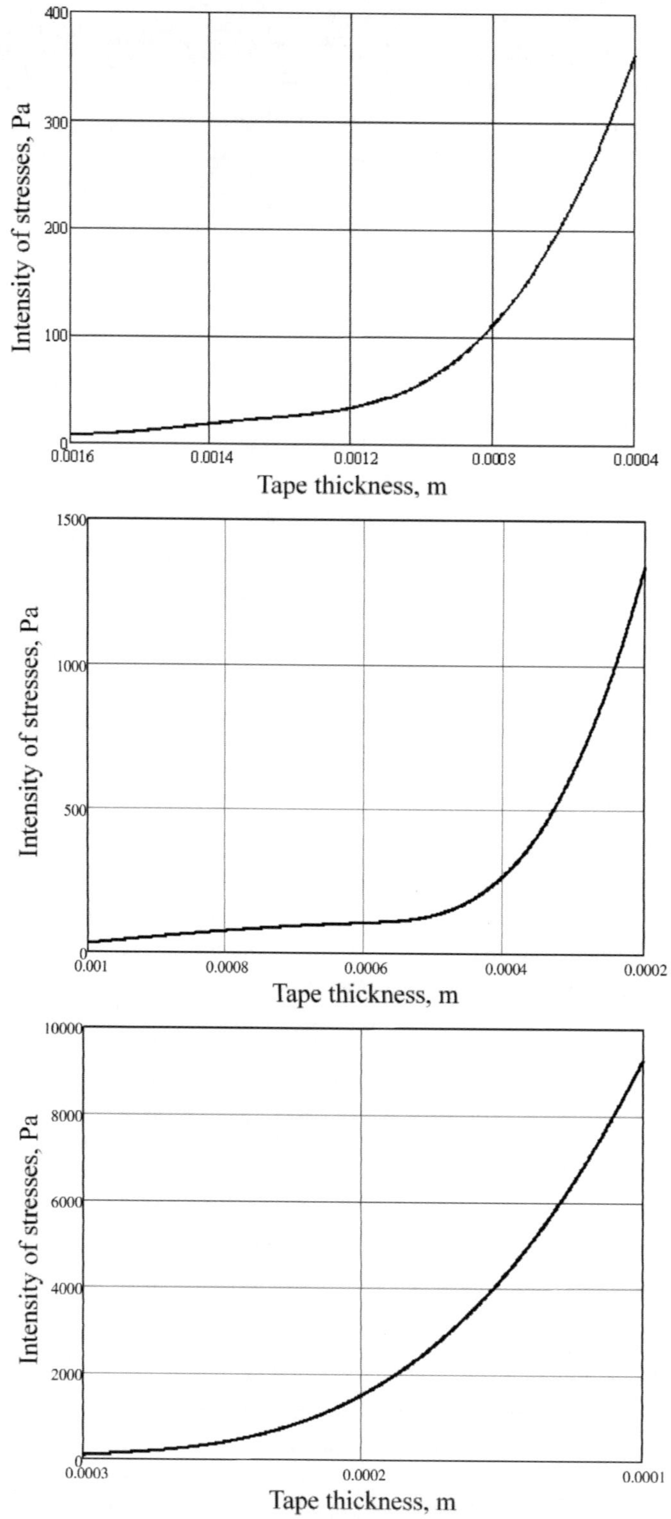

a)

b)

c)

Figure 3. Stresses distributed along the tape thickness with linear rotational velocity: a) 11.8 m/s; б) 23.5 m/s; в) 70.7 m/s.

IOP Conf. Series: Materials Science and Engineering **192** (2017) 012032 doi:10.1088/1757-899X/192/1/012032

4. Conclusions

1. The mathematical model and computational algorithm for calculating the hydrodynamic and heat transfer processes have been developed for the amorphous tape production.

2. The software has been developed which is capable to calculate the fields of temperature, velocity and stresses, as well as to visualize the distribution of velocity, stream function and voltage components on the display at any given time.

3. Increase in the disc rotation velocity results in decreased thickness of the solidified tape.

4. Comparison of the results of numerical experiments with the laboratory research described in [6] confirms the adequacy both of the developed mathematical model and of the computational algorithm.

5. References

[1] Naberezhnye V P, Tkach V I et al. 1987 *The analysis of metallic glasses of the crystallization process and* $Fe_{80}B_{20}$ $Fe_{40}Ni_{40}F_{14}B_6$ *in non-isothermal conditions* (Donetsk: Preprint)

[2] Nemoshkalenko V V, Romanov A V, et al. 1987 *Amorphous metal alloys* (Kiev: Naukova Dumka)

[3] Bilousov V V and Bologna N I 2003 *Fundamentals of heat and mass transfer and thermal physics of closed volumes* (Donetsk: Southeast)

[4] Bilousov V V, Nedopekin F V, Hrychikov V E, Leibenzon V A, et al. 2009 *Solidification of metals and metal compositions* (Kiev: Naukova Dumka)

[5] Bilousov V V, Babanin A Y, Belov B F and Pavlov Y V 2015 *Processing of iron-carbon melts highly active agents* (Donetsk: East Publishing House)

[6] Kester W and Gerold W 1983 *Crystallization of metallic glasses Met. Art.* (Moscow: Mir) 325-367

[7] Yavari A R, Hicter P and Desre P 1981 The effect of quenching stresses on the kinetics of crystallization and structure of metallic glasses *Proc. 4th Conf. on Rapidly Quenched Metals* 713-716

IOP Conf. Series: Materials Science and Engineering **192** (2017) 012033 doi:10.1088/1757-899X/192/1/012033

Application of the Saint-Venant model and the modified Stefan model for modeling the formation of the ice cover at the thermal growth stage

Ilya Starodumov[1], **Anna Derevianko**[2] **and Dmitri Alexandrov**[1]
[1] Laboratory of Multi-Scale Mathematical Modelling, Ural Federal University, 620000 Ekaterinburg, Russia
[2] Kiev National University, 01601 Kiev, Ukraine

E-mail: ilya.starodumov@urfu.ru

Abstract. River ice processes are complex phenomena that are affected by many factors, including meteorological conditions, thermal inputs, hydraulic conditions and channel geometry. Because of the complexity of river ice phenomena and the limited current understanding, the development ofa river ice model requires the maximum use of existing theories and mathematical techniques. In the river hydraulics component, the flow condition is often determined by the solution of the Saint-Venant unsteady flow equations. The complex phenomenon of ice formation at the stage of thermal growth in turn is usually described by models based on the Stefan equation. In current paper we consider the opportunity to use the modified Stefan model, which includes the model of mushy layer in the task of ice thermal grows together with the solution of the problem of Saint-Venant.

1. Introduction

The process of river ice formation is a complex and incompletely studied natural phenomenon. It includes the formation, evolution, transport, dissipation and deterioration of various forms of ice. This process depends not only on the geometry of the riverbed, the atmospheric conditions, but also by parameters of the water flow in the river. Thus, the ice formation mechanisms are complex interconnected system, parts of which are not completely understood. However, by using existing theories, an in-depth mathematical models can be proposed [1–6]. Such models allow us to predict a continuous process of ice formation using a limited amount of data. The result of predictions can be used in design and operation of river control works. A mathematical model can also be used by researchers to identify crucial gaps and weaknesses in the current theories of river ice.

The process of ice cover formation is a very complex physical phenomenon that can be divided into a number of essential processes. A detailed schematic description of the process of ice formation is given in [7]. A comprehensive one-dimensional river ice model called RICE is developed in conformity with that scheme and makes maximum use of the existing information, with improvements on analytical and mathematical formulations. Because of using a separate subroutine for the modeling of each ice process, the general model allows modifications of every submodel. In this paper we propose to consider possible the modification for the problem of the ice cover at the thermal growth stage.

2. Modeling of the ice cover at the thermal growth stage

2.1. Unsteady flow modelling

The hydraulics of one-dimensional river flow can be described by Saint Venant equations. These equations represent the conservation ofmass and momentum in the river:

$$\frac{\partial A}{\partial t}+\frac{\partial Q}{\partial x}=q,\ \frac{\partial Q}{\partial t}+\frac{\partial}{\partial x}\left(\frac{Q^2}{A}\right)+\rho gA\left(\frac{\partial h}{\partial x}+S_f\right)=0, \qquad (1)$$

where :

x – the length of watercourse (m),
t – time (s),
$A(x,t)$ – net flow cross-sectional area (m^2/s),
$Q(x,t)$ – water flow (m^3/s),
$q(x,t)$ – external power per unit length of the watercourse (m^2/s),
g – the gravitational acceleration (m/s^2),
$h(x,t)$ – depth (m),
$S_f(x,t)$ – the friction slope flow.

In case of ice cover problem $(S_f(x,t)=p_b\tau_b+p_i\tau_i)/g\rho\overline{A}$, where:

p_b – wetted perimeter formed by the channel bed,
p_i – wetted perimeters formed by the ice cover,
τ_b – shear stresses at the channel bottom,
τ_i – shear stresses at the ice/water interface,
$\overline{A}$ – mean net flow cross-sectional area.

For such a problem, implicit finite-difference or finite-element numerical methods are usually used. That methods requires the river to be divided into a sufficient number of reaches. Moreover, stability and accuracy are achieved under certain conditions of discretization of the system [7]. We believe that for the calm flow of water in a channel of a rather large width of a river bed, it is possible to use more simple approximations of the Saint-Venant model and, as a consequence, less demanding methods of solution. Such an approach can be the kinematic approximation and its solution proposed in [8]. In this paper, an explicit numerical scheme is proposed that uses analytical solutions within computational cells. This makes it possible to solve the problem of water flow using large cells where possible, but retaining an acceptable level of accuracy due to approximation by analytical methods.

2.2. Thermal growth of ice covers

The thickness of an ice cover can be changed by thermal growth as a result of heat exchange with the environment and the water. The ice cover thickness can influent on the flow cross-sectional area and by this way on the velocity of the flow. It's important in determining the stability of an ice cover and its break-up. Many methods have been used to model the thickness of ice covers. Many of them are basically based on the degree-day method or the Stefan model [1, 9, 10]. But as rule the model of planar front is used under consideration. We propose to consider a modification of the Stefan model which includes the model of mushy layer – regions of mixed phase (solid and liquid) in which the solid forms a rigid matrix and the liquid fills its interstices. In classical degree-day method the ice thickness h is given as

$$h=\alpha_h\sqrt{S}, \qquad (2)$$

$S = \int_{t_0}^{t} (T_m - T_a)\mathrm{dt}$ – cumulative freezing degree-days of air temperature since the formation of the cover at time t_0.

t – time from the formation of the ice cover

T_m – freezing temperature of water

T_a – air temperature

α_h – empirical degree-day constant. In case of planar front this constant can be expressed as $\sqrt{2k_i/L_v}$, where L_v is the latent heat released as the ice fraction increases and k_i is the thermal conductivity.

In [11] the modified equation of the Stefan model is presented taking into account the mushy layer approach. The scheme of the difference between classical model of planar front and mushy-layer model is presented in Figure 1. This non-linear problem under consideration is solved

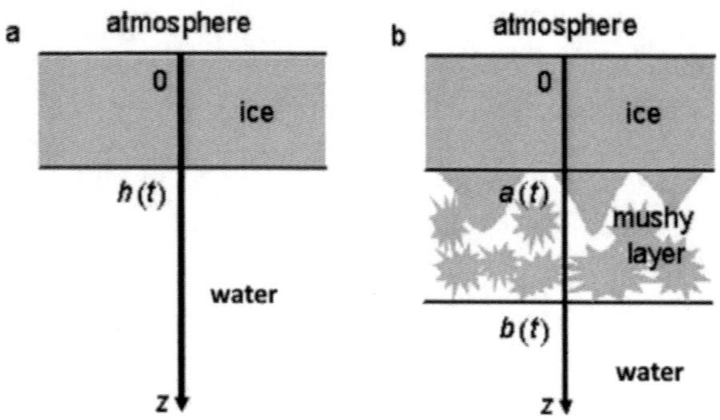

Figure 1. A scheme illustrating the models which are taking into consideretion: (a) planar front and (b) mushy layer.

analytically in [11]. Solutions represented by the following equations:

$$a(t) = -\frac{D_w L_v \varphi_b}{T_o \Phi} b(t), \tag{3}$$

$$b(t) = \alpha_h \sqrt{S}. \tag{4}$$

In a result the degree-day constant is modified and expressed by $\sqrt{2/I}$, where:

$$I = \frac{L_v \varphi_b}{\Phi}[1 - \frac{D_w L_v \varphi_b}{T_o \Phi}(\frac{L_v D_w}{k_i T_o} + 1 - K)(\varphi_b - 1)]. \tag{5}$$

Here where φ_b is the solid fraction and T_o is the constant temperature of the water determined on the interface mushy-layer/water. $\Phi = k_i\varphi_b + k_w(1-\varphi_b)$. k_w and D_w are the thermal conductivity and diffusion coefficient, respectively, of the water. $K = \frac{k_w}{k_i}$. [11] shows that analytical solutions for the mushy layer transform to their analogues for the planar front. But in contrast to it, the modified model allows for taking into account the mushy layer, which can have a significant effect on the hydraulic effects in water flow under certain conditions.

3. Conclusions

The paper considers the complex phenomenon of ice formation on the rivers. Mathematical modeling of such a phenomenon is an extremely difficult task. One of the successful projects to create a comprehensive mathematical model of ice formation is the RICE system. This system includes sub-models of individual physical processes. This makes it possible to modify the general model by changing of components. The authors consider a submodel of the thermal growth, which can be described by various techniques. However, all these techniques are based on the degree-day Stefan model, taking into account only planar front of ice formation. The authors propose to consider the possibility of using a modified model that takes into account the mushy layer of ice formation. This model allows to evaluate not only the ice-water interface, but also the mushy zone in which the ice formation is not yet complete, but the hydraulic processes of the water flow can have significant features. In addition, the authors considered the main hydraulic model of Saint-Venant. For it, in appropriate cases, it is suggested to use a simplified model for which there is a numerical solution that uses the analytical approximation in the computational cells. Such an approach can simplify the formulation of the problem for data deficit cases.

References

[1] Shen H and Yapa P 1985 *Canadian Journal of Civil Engineers* **12** 54–62

[2] Simonsen C and Carson R 1977 *Proceedings of the Third National Hydrotechnical Conference* 19

[3] Petryk S and Boisvert R 1978 *Proceedings of the Speciality Conference on Computer Applications in Hydrotechnical and Municipal Engineering* 239–259

[4] Michel B and Drouin M 1982 *Canadian Journal of Civil Engineering* **8** 351–363

[5] Petryk S, Panu U, Cartha V and Clement F 1981 *Proceedings of IAHR International Symposium on Ice* 426–436

[6] Calkins D 1984 *Proceedings of Workshop on the Hydraulics of River Ice* 247–277

[7] Lai A W and Shen H T 1993 *A Mathematical Model for River Ice Processes* (US Army Corps of Engineers)

[8] Starodumov I 2014 *The Bulletin of Adyghe State University* **2** 37–43

[9] Stefan J 1891 *Wien Sitzunsber.Akad. Wiss.Series A* **42** 965–983

[10] Billleo M 1980 *Maximum thickness and subsequent decay oflakes, rivers and fast sea ice in Canada and Alaska, CRREL Report* (US Army Corps of Engineers)

[11] Alexandrov D, Malygin A and Alexandrova I 2006 *Annals of Glaciology* **44** 118–122

4. Acknowledgments

Authors achnowledge support from the Russian Science Foundation under project number 16-11-10095.

IOP Conf. Series: Materials Science and Engineering **192** (2017) 012034 doi:10.1088/1757-899X/192/1/012034

Mathematical modeling of the stationary nucleation and crystallization process in supersaturated systems with a crystallizer

E V Makoveeva[1], A P Malygin[1] and D V Alexandrov[1]

[1] Ural Federal University, Department of Theoretical and Mathematical Physics, Laboratory of Multi-Scale Mathematical Modeling, 620000 Ekaterinburg, Russian Federation

E-mail: eugenya1m@gmail.com

Abstract. Motivated by important industrial applications we consider the growth prosess of solid particles in a supersaturated (supercooled) system with allowance for a crystallizer. The particle-radius distribution function satisfies the second order kinetic equation supplimented by different boundary conditions. An exact steady-state analytical solution is found. We show that two different types of analytical solutions for the kinetic equation exist.

1. Introduction

The process of growth of a new phase in metastable systems can be divided into three stages: nucleation, growth of critical nuclei and relaxation of a new phase to equilibrium. The division is conditional, because all these processes occur simultaneously. Modeling the processes of formation of a new phase is important both for fundamental research in the field of phase transitions and for industrial applications. The particle size distribution can have a decisive influence on the properties of the systems in the production of powder materials from melts, precipitation from solutions, synthesis of colloidal solutions and so on.

In this study, the process of formation and growth of particles of a new phase under stationary conditions is considered. Such conditions can be ensured when the particles of the solid phase are continuously removed from the liquid and the supersaturation (supercooling) is artificially maintained. The removal of particles can be realized both as a result of processes of precipitation of large particles due to sedimentation, fragmentation and sedimentation of large volumes of the new phase, and also due to artificial deposition, for example, during centrifugation.

The problem of stationary distribution was previously considered by Buevichi and Mansurov [1]. However, they considered the determenistic growth of solid crystals without their fluctuations. The present paper takes into account random processes in the Fokker-Planck equation for the particle-radius distribution function.

2. Model

It is assumed that the physical properties of both the solution and the crystals are independent of supersaturation (hereinafter, supersaturation will also be understood as supercooling) and time. Supersaturation is homogeneous throughout the reservoir volume. The nucleation kinetics is stationary. This is permissible as the time of nucleation process, as a rule, is several orders of magnitude smaller than the time of continuous crystallization for any real supersaturation. The size of spherical crystals is determined by using one parameter, their radius.

The crystal radius is a random variable satisfying the stochastic equation [2]

$$dr = g(t,r)dt + \sqrt{2D}W, \tag{1}$$

where t is the growth time, $g(t,r)$ is the growth rate of a particle that has no oscillations, D is a function that expresses the speed of free oscillations (the coefficient of mutual Brownian diffusion of a particle) and W is the Wiener process. The distribution function f of particles of radius r for the stochastic equation (1) considering the crystallizer (crystal derivation) is described by a kinetic equation of the Fokker-Planck type [3].

$$\frac{\partial f}{\partial t} + \frac{\partial}{\partial r}(gf) + h(r)f = \frac{\partial}{\partial r}\left(D\frac{\partial f}{\partial r}\right), \; r > r_*. \tag{2}$$

The function $h(r)$ describes the rate of crystal extraction, r_* is a critical radius. The exact definition of D is a difficult problem in statistical physics. However, in the first approximation one can assume that D is proportional to the nucleation rate of growth [3, 4]

$$D = d_1 g(t,r), \tag{3}$$

where d_1 is a coefficient.

The rate of particle growth g can be found from a problem of the Stefan type in a metastable system [4, 5]. The result can be expressed in the form

$$g(t,r) = \frac{\beta_* \triangle \theta}{1+\beta_* q r}, q = \frac{L_V}{\lambda_l} \tag{4}$$

if the system is a supercooled melt and in the form

$$g(t,r) = \frac{\beta_* \triangle C}{1+\beta_* q r}, q = \frac{C_p}{D_l} \tag{5}$$

if the system is a supersaturated solution. Here, β_* is the kinetic coefficient, λ_l and D_l are the temperature conductivity and diffusion coefficients, $\triangle C$ is the supersaturation, $\triangle\theta$ is the supercooling. The heat (mass) balance equation for the dimensionless supercooling $W = \triangle\theta / \triangle \theta_0$ (supersaturation $W = \triangle C / \triangle C_0$) can be written as [4].

$$w(\tau) = 1 - b\int_{r_*}^{\infty} r^3 f(\tau,t)dr, \tau > 0, \tag{6}$$

where $b = 4\pi L_V/(3\rho_m C_m \triangle \theta_0)$ if the system under consideration is a supercooled melt, and $b = 4\pi C_p/(3 \triangle C_0)$ if the system is represented by a supersaturated solution. Here L_V is the latent heat parameter, C_p is the concentration at saturation, ρ_m and C_m are the density and specific heat of the mixture and subscript 0 designates the initial state.

For a complete investigation of the problem, it is necessary to set the initial and boundary conditions. As the initial condition for the distribution function, it is usually assumed that there are not nuclei in the system:

$$f = 0, \; \tau = 0. \tag{7}$$

The initial condition for w becomes

$$w = 1, \; \tau = 0. \tag{8}$$

The first boundary condition for the density distribution function takes the form

$$gf - D\frac{\partial f}{\partial r} = I(w),\ r = r_*,\ \tau > 0. \tag{9}$$

The empirical expressions for the nucleation rate which are frequently used in analyzing many industrial processes can be expressed as

$$I = I_*(\triangle\theta)^p = (I_*(w \triangle \theta_0)^p \tag{10}$$

for supercooled melts and

$$I = I_*(\triangle C)^p = (I_*(w \triangle C_0)^p \tag{11}$$

for supersaturated solutions. Expressions (10) and (11) are frequently called the Meirs nucleation kinetics. It follows from the normalization condition that the distribution function vanishes for sufficiently large r.

Generally speaking, two boundary conditions are possible. The first of them reflecting the distribution function decaying is

$$f = 0,\ r \to \infty. \tag{12}$$

The second one describes the case when the density distribution function vanishes for a finite r

$$f = 0,\ r = r_0. \tag{13}$$

3. Steady-state regime

We are dealing with two processes: the formation and growth of crystals and their removal. These processes can balance each other. Thus, if the supersaturation is artificially maintained, it is possible to achieve the existence of a solution in the form of a stationary (time-independent) distribution-density function. The equation becomes an ordinary differential equation, and the initial conditions are no longer required. I this paper, we assume that the rate of the crystal removal is independent of their radii. The experiment shows that the crystal growth rate g is frequently constant. In this case, the kinetic equation takes the form

$$D\frac{d}{dr}\left(\frac{df}{dr}\right) - g\frac{d}{dr}(f) - hf = 0, \tag{14}$$

$$\frac{d}{dr}\left(\frac{df}{dr}\right) - A\frac{d}{dr}(f) - Bf = 0,\ A = \frac{g}{D},\ B = \frac{h}{D}. \tag{15}$$

The solution of this equation has the form

$$f = C_1 e^{\lambda_1 r} + C_2 e^{\lambda_2 r}, \tag{16}$$

$$\lambda_1 = \frac{A + \sqrt{(A^2 + 4B)}}{2},\ \lambda_2 = \frac{A - \sqrt{(A^2 + 4B)}}{2}, \tag{17}$$

where C_1 and C_2 are determined from the boundary conditions, consequently two cases are possible.

In the first case, (12), we have

$$C_1 = 0, \tag{18}$$

$$C_2 = \frac{I}{g(1 - d_1\lambda_2)e^{\lambda_2 r_*}}. \tag{19}$$

IOP Conf. Series: Materials Science and Engineering **192** (2017) 012034 doi:10.1088/1757-899X/192/1/012034

In the second case, (13), we get

$$C_1 = -\frac{Ie^{\lambda_2 r_0}}{g((d_1\lambda_1 - 1)e^{\lambda_2 r_0}e^{\lambda_1 r_*} - (d_1\lambda_2 - 1)e^{\lambda_1 r_0}e^{\lambda_2 r_*})}, \tag{20}$$

$$C_2 = \frac{Ie^{\lambda_1 r_0}}{g((d_1\lambda_1 - 1)e^{\lambda_2 r_0}e^{\lambda_1 r_*} - (d_1\lambda_2 - 1)e^{\lambda_1 r_0}e^{\lambda_2 r_*})}. \tag{21}$$

The analytical solutions (16)-(21) can be rewritten in the dimensionless form as

$$F(x) = \frac{f}{C_2} = Ce^{\mu_1 x} + e^{\mu_2 x}, \tag{22}$$

$$C = \frac{C_1}{C_2},\ x = \frac{r}{d_1},\ \Lambda = \frac{4B}{A^2}, \tag{23}$$

$$\mu_1 = \frac{1}{2}\left(1 + \sqrt{1 + \frac{4B}{A^2}}\right),\ \mu_2 = \frac{1}{2}\left(1 - \sqrt{1 + \frac{4B}{A^2}}\right). \tag{24}$$

Here the reduced constant C is introduced as

$$C = 0 \tag{25}$$

in the case of boundary conditions (18), (19) and

$$C = -e^{(\mu_2 - \mu_1)x_0} \tag{26}$$

in the case of boundary conditions (20), (21).

4. Conclusions

Figures 1 and 2 show graphs of the distribution function for different boundary conditions. Of special interest is the case when the distribution function vanishes for finite $r = x_0 d_1$. As we can see, the function F in Figures 1 and 2 does not have extremal points and is a monotonically decreasing function. Thus, we come to the conclusion that under stationary conditions with increasing radius of the nuclei the probability of finding such nuclei in the crystallizer decreases. The difference between two boundary conditions (12) and (13) shown in Figures 1 and 2 consists in the fact that $F = 0$ at $x = x_0$ in Figure 2 while $F \to 0$ at $x \to \infty$ in Figure 1. This is connected with different boundary conditions (12) and (13) initially used in our model. Generally speaking, both of them are possible if the system approaches to its steady-state. This, in turn, depends, first of all, on system equilibria and how it tends to them. It depends on the level of system supercooling (supersaturation) as well. In addition, Figure 2 demonstrates the presence of inflection points appearing as a result of the boundary condition (13).

Note that increasing Λ (withdrawal rate h) decreases the distribution function and the number of crystals as is seen from comparison of the left and right panels in Figures 1 and 2.

The distribution funstion behavior illustrated in Figures 1 and 2 is a direct consequence of the second spatial order of the kinetic equation (2) and corresponding boundary conditions. Indeed, if we are dealing with the first order kinetic equation the distribution function breaks down at a finite crystal radius [6, 7]. An important point is that the distribution function (22) is always decreasing because its second derivative is negative at any x.

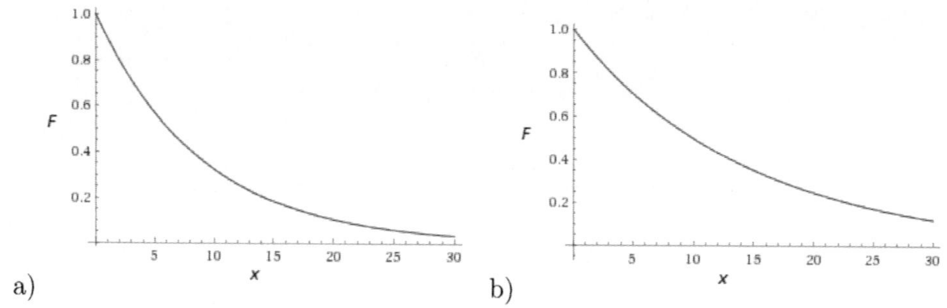

Figure 1. Dimensionless distribution function versus dimensionless cristal radius at a) $\Lambda = 0.5$ and b) $\Lambda = 0.3$ plotted for the case of boundary condition (12).

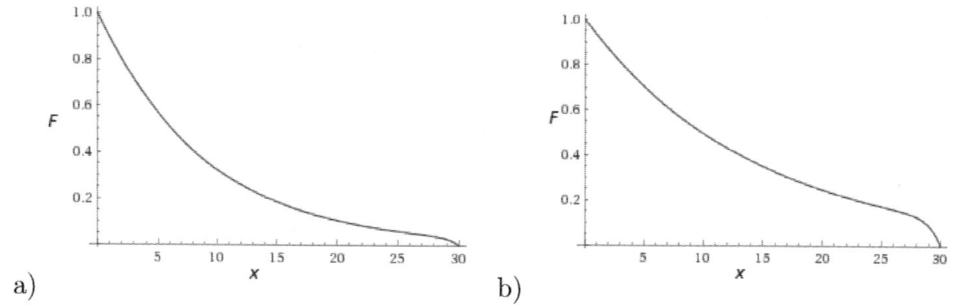

Figure 2. Dimensionless distribution function versus dimensionless cristal radius at a) $\Lambda = 0.5$ and b) $\Lambda = 0.3$ plotted for the case of boundary condition (13), $x_0 = 30$.

5. References

[1] Buyevich Yu A and Mansurov V V 1990 Kinetics of the intermediate stage of phase transition in batch crystallization *J. Cryst. Growth* **104** 861-867
[2] Gardiner C W 1983 *Handbook on stochastic methods: for physics, chemistry and the natural sciences* (Berlin, Germany: Springer)
[3] Lifshitz E M and Pitaevskii L P 1981 *Physical kinetics*(New York, NY: Pergamon Press)
[4] Buyevich Yu A, Alexandrov D V and Mansurov V V 2001 *Macrokinetics of crystallization* (NewYork, NY: Begell House)
[5] Kirkaldy J S and Young D J 1987 *Diffusion in the condensed state* (London, UK: Institute of Metals)
[6] Alexandrov D V and Malygin A P 2013 Transient nucleation kinetics of crystal growth at the intermediate stade of bulk phase transitions *J. Phys. A: Math. Theor.* **46** 455101
[7] Alexandrov D V 2014 Nucleation and crystal growth kinetics during solidification: Tye role of crystallite withdrawal rate external heat and mass sources *Chem. Eng. Sci.* **117** 156-160

Acknowledgments

This work was supported by the Russian Foundation for Basic Research (grant no. 16-08-00932) and the Ministry of Education and Science of the Russian Federation (project no. 1.9527.2017).

Institute of Physics
Dirac House, Temple Back
Bristol BS1 6BE UK

ISSN: 1757-8981
ISBN 978-1-5108-3961-8

9 781510 839618

CURRAN ASSOCIATES INC.
proceedings.com

International Conference on Forest Products (ICFP 2018)

Efficiency and Quality Improvement of Forest Products Towards Technology 4.0

IOP Conference Series: Earth and Environmental Science
Volume 359

Bogor, Indonesia
2 November 2018

IOP
Institute of Physics